ENVIRONMENTAL
PRAGMATISM

Environmental pragmatism is a new strategy in environmental thought: it argues that theoretical debates are hindering the ability of the environmental movement to forge agreement on basic policy imperatives. This new direction in environmental philosophy moves beyond theory, advocating a serious inquiry into the practical merits of moral pluralism. Environmental pragmatism, as a coherent philosophical position, connects the methodology of classical American pragmatist thought to the explanation, solution and discussion of real issues.

This concise, well-focused collection is the first comprehensive presentation of environmental pragmatism as a new philosophical approach to environmental thought and policy.

Contributors: Larry Hickman, Bryan G. Norton, Paul Thompson, Anthony Weston, Kelly Parker, Sandra Rosenthal and Rogene Bucholz, Andrew Light, Eric Katz, David Rothenberg, Emery Castle, Ari Santas, Edward Schiappa, and Gary Varner, Susan Gilbertz and Tarla Peterson.

Andrew Light is a research fellow in the Environmental Health Program and adjunct professor of philosophy at the University of Alberta, Canada. **Eric Katz** is Director of the Science, Technology and Society Program at the New Jersey Institute of Technology.

ENVIRONMENTAL PHILOSOPHIES SERIES
Edited by Andrew Brennan

Philosophy, in its broadest sense, is an effort to get clear on the problems which puzzle us. Our responsibility for and attitude to the environment is one such problem which is now the subject of intense debate. Theorists and policy analysts often discuss environmental issues in the context of a more general understanding of what human beings are and how they are related to each other and to the rest of the world. So economists may argue that humans are basically consumers sending signals to each other by means of the market, while deep ecologists maintain that humans and other animals are knots in a larger web of biospheric relations.

This series examines the theories that lie behind different accounts of our environmental problems and their solution. It includes accounts of holism, feminism, green political themes, and the other structures of ideas in terms of which people have tried to make sense of our environmental predicaments. The emphasis is on clarity, combined with a critical approach to the material under study.

Most of the authors are professional philosophers, and each has written a jargon-free, non-technical account of their topic. The books will interest readers from a variety of backgrounds, including philosophers, geographers, policy makers, and all who care for our planet.

ENVIRONMENTAL PRAGMATISM

Edited by Andrew Light and Eric Katz

London and New York

First published 1996
by Routledge
11 New Fetter Lane, London EC4P 4EE

Simultaneously published in the USA and Canada
by Routledge
29 West 35th Street, New York, NY 10001

Selection and editorial matter © 1996 Andrew Light and Eric Katz;
individual chapters © the contributors

Typeset in Adobe Garamond by
Keystroke, Jacaranda Lodge, Wolverhampton

Printed and bound in Great Britain by
Clays Ltd, St Ives PLC

British Library Cataloguing in Publication Data
A catalogue record for this book is available from the British Library

Library of Congress Cataloguing in Publication Data
A catalogue record for this book has been requested

ISBN 0–415–12236–8
0–415–12237–6 (pbk)

This book is dedicated to my parents
A. Ronald Light
and
Elizabeth Pelligrino Light
– AL

This book is dedicated to my mother
Irene Davis Katz
and to the memory of my father
Mortimer Katz
– EK

CONTENTS

Part 2 Pragmatist theory and environmental philosophy

Part 3 Pragmatist approaches to environmental problems

Part 4 Environmental pragmatism: an exchange

CONTENTS

FIGURES

NOTES ON CONTRIBUTORS

Rogene A. Buchholz is the Legendre-Soule Professor of Business Ethics at Loyola University of New Orleans. Articles by Professor Buchholz have appeared in, among other places, the *Journal of Management Studies, Industrial and Labor Relations Review, Harvard Business Review* and the *Journal of Business Ethics*. He is the author of nine books in the areas of business and public policy, business ethics and the environment. He is past Chair of the Social Issues in Management Division of the Academy of Management.

Emery N. Castle is Professor in the University Graduate Faculty of Economics at Oregon State University. He has authored numerous articles and edited books on resource and environmental economics and policy. From 1975 to 1985 he served as Vice-president and then President of Resources for the Future. Since 1987 he has served as the Chair of the National Rural Studies Committee, a multi-disciplinary group of scholars, financed by the W. K. Kellogg Foundation, engaged in a study of the problems of rural America.

Susan J. Gilbertz is a senior lecturer in the Department of Speech Communication at Texas A&M University. She specializes in conflict negotiations and persuasion theory. She is currently pursuing a PhD in geography with emphasis on cultural interpretations of landscapes.

Larry A. Hickman is Director of the Center for Dewey Studies and Professor of Philosophy at Southern Illinois University, Carbondale. In addition to numerous articles on American pragmatism and the philosophy of technology, he is the author of *Modern Theories of Higher Level Predicates* and *John Dewey's Pragmatic Technology*. He is the co-editor of *Technology and Human Affairs* and the editor of *Technology as a Human Affair*.

Eric Katz is Director of the Science, Technology and Society Program at the New Jersey Institute of Technology, where he teaches environmental philosophy, engineering ethics and the philosophy of technology. Katz is the author of two comprehensive annotated bibliographies of the field of environmental ethics, as well as over two dozen articles. He has recently completed a book on deep ecology for Routledge, *Deep Green: A Critical Introduction to the Philosophy of Deep Ecology*.

Andrew Light is a research fellow in the Environmental Health Program and Adjunct Professor of Philosophy at the University of Alberta. In addition to a number of articles on philosophy of technology, environmental philosophy and Marxist political theory, he is the editor of the forthcoming volumes, *The Environmental Materialism Reader* and *Anarchism, Nature, and Society: Critical Perspectives on Social Ecology*. He is the co-editor of the annual *Philosophy and Geography*.

Bryan G. Norton is Professor of Philosophy in the School of Public Policy, Georgia Institute of Technology. He writes on intergenerational equity, sustainability theory, and biodiversity policy. He is the author of *Why Preserve Natural Variety? Toward Unity Among Environmentalists*, editor of *The Preservation of Species* and co-editor of *Ecosystem Health: New Goals for Environmental Management* and *Ethics on the Ark*. Norton has served on numerous panels, including the Ecosystem Valuation Forum and The Risk Assessment Forum (US EPA).

Kelly A. Parker teaches in the Philosophy and Liberal Studies Programs at Grand Valley State University in Allendale, Michigan. His main research interests are in American philosophy. Besides his work on environmental philosophy, he has written extensively on the philosophy of Charles S. Peirce.

Tarla Rai Peterson has an MA in speech communication and an interdisciplinary PhD. She is an Associate Professor in the Department of Speech Communication at Texas A&M University. She performs rhetorical criticism of environmental conflicts, focusing on implications for public participation in decision-making.

Sandra B. Rosenthal is Professor of Philosophy at Loyola University of New Orleans. In addition to over a hundred articles relating to American pragmatism, she has authored or co-authored seven books in the area. She has served as president of several philosophical societies including the Charles Peirce Society and the Society for the Advancement of American Philosophy. Rosenthal was also a member of the executive committee of the American Philosophical Association, Eastern Division and is currently President of the Metaphysical Society of America.

David Rothenberg is Assistant Professor of Philosophy at the New Jersey Institute of Technology and acting director of the Program in Science, Technology and Society. He is the editor of the journal *Terra Nova: Nature and Culture* and author of *Hand's End: Technology and the Limits of Nature*. He edited the recently published collection of essays critiquing the meaning of wilderness, *Wild Ideas*.

Ari Santas is an Assistant Professor of Philosophy at Valdosta State University, Georgia, where he teaches Applied Ethics, American Pragmatism, and History of Philosophy. He is the Coordinator of The Center for Professional and Applied Ethics (CPAE) and List Manager of CPAE, an electronic discussion group on the Internet.

Edward Schiappa is Associate Professor of Communications at the University of Minnesota. His work on classical and contemporary philosophy of language appears in such journals as *Philosophy and Rhetoric, Ancient Philosophy, American Journal of Philology* and *Argumentation*. He has published *Protagoras and Logos: A Study in Greek Philosophy and Rhetoric* and edited *Warranting Assent: Case Studies in Argument Evaluation*.

Paul B. Thompson is Professor of Philosophy and Agricultural Economics at Texas A&M University. A prolific author, his most recent book is *The Spirit of the Soil: Agriculture and Environmental Ethics* (Routledge, 1995). Work on the paper for this volume was graciously supported by the Program on Agrarian Studies, Yale University, during Thompson's year in residence as a program fellow, 1994–1995.

Gary E. Varner is an Assistant Professor of Philosophy at Texas A&M. He is the author of over twenty papers on conceptual and normative questions in environmental ethics, animal rights and the philosophical underpinnings of environmental law. He is currently finishing a book manuscript titled *In Nature's Interests? Interests, Animal Rights, and Environmental Ethics*. He has spoken on related topics in a variety of forums, ranging from the American Bar Association's National Judicial College to the Columbus Zoo.

Anthony Weston teaches philosophy and interdisciplinary studies at Elon College, NC. He has previously taught at SUNY-Stony Brook, the Adult Education Division of the New School for Social Research, and his daughter's preschool in Raleigh, NC. He is the author of *A Rulebook for Arguments, Toward Better Problems* and most recently *Back to Earth*. He has also published a number of articles in philosophy of education, ethics and value theory, and philosophy of technology, as well as environmental ethics.

ACKNOWLEDGMENTS

This book was conceived when the editors met at a meeting of the Society for Philosophy and Technology in Peniscola, Spain, in the spring of 1993. After much discussion, we felt that there was a need for a comprehensive appraisal of the position we called environmental pragmatism. This collection is the result of those initial discussions.

Thus we would first like to thank the Society for Philosophy and Technology for organizing a conference on the theme of *Technology and Ecology*. The conference brought us together, with time for planning this project, and it also enabled us to discuss the project with Larry Hickman, a contributor to this volume and then Vice-president of the Society. Andrew Light received partial funding for the conference through a travel grant from the Graduate Student Council at the University of California, Riverside, and full administrative and secretarial support to work on the topic of environmental pragmatism that year from the Center for Ideas and Society at UC Riverside, then under the direction of Bernd Magnus. Eric Katz was able to attend the conference because of funds provided by a Separately Budgeted Research Grant from the New Jersey Institute of Technology.

The idea for this collection was field tested at a meeting of the International Society for Environmental Ethics (ISEE) in Atlanta in December 1993, where papers by Kelly Parker and Andrew Light were read. Other contributions came from a meeting of the ISEE in Kansas City in May 1994.

In the planning stages of this project, Bryan Norton was an immense help – the volume would not have proceeded without his enthusiastic support. We would also like to thank Scott Christensen, Wes Dean, Meredith Garmon, Eric Higgs, Bernd Magnus, John

McDermott, Dorit Naaman, Herman Saatkamp, Paul Thompson, Gary Varner and Anthony Weston for their advice, inspiration and helpful comments on some of the manuscripts. The suggestions of Andrew Brennan, the editor of this series in Environmental Philosophies, were indispensable. We would like to thank the Senior Editor at Routledge, Richard Stoneman, our extremely capable editorial assistant, Vicky Peters and desk editor, Joanne Snooks. We are also grateful to Constance Macintosh for her fine work on the index.

Finally, Andrew Light wishes to thank Steve Hrudey, the Eco-Research Chair in Environmental Risk Management at the University of Alberta, for providing funding, direction and a home in which to work. The Eco-Research Chair in Environmental Risk Management is supported by the Tri-Council Secretariat, representing the three Canadian federal research granting councils, and twenty-four public and private sponsors. Without Professor Hrudey's uncompromising support and enthusiasm for the project, as well as the stimulating atmosphere provided by the faculty, staff and graduate students supported by his Chair, and colleagues in the Philosophy Department at the University of Alberta, this project would not have been completed.

INTRODUCTION

Environmental pragmatism and environmental ethics as contested terrain

Andrew Light and Eric Katz

As environmental ethics approaches its third decade it is faced with a curious problem. On the one hand, the discipline has made significant progress in the analysis of the moral relationship between humanity and the non-human natural world. The field has produced a wide variety of positions and theories[1] in an attempt to derive morally justifiable and adequate environmental policies. On the other hand, it is difficult to see what practical effect the field of environmental ethics has had on the formation of environmental policy. The intramural debates of environmental philosophers, although interesting, provocative and complex, seem to have no real impact on the deliberations of environmental scientists, activists and policy-makers. The ideas within environmental ethics are, apparently, inert – like Hume's *Treatise*, they fall deadborn from the press.

The problematic situation of environmental ethics greatly troubles us, both as philosophers and as citizens. We are deeply concerned about the precarious state of the natural world, the environmental hazards that threaten humans, and the maintenance of long-term sustainable life on this planet. The environmental crisis that surrounds us is a fact of experience. It is thus imperative that environmental philosophy, as a discipline, address this crisis – its meaning, its causes and its possible resolution.

Can philosophers contribute *anything* to an investigation of environmental problems? Do the traditions, history and skills of philosophical thought have any relevance to the development of environmental policy? We believe that the answer is yes. Despite the problematic (and, heretofore, ineffectual) status of environmental ethics as a practical discipline, the field has much to offer. But the fruits of this philosophical enterprise must be directed towards the practical resolution of environmental problems – environmental

1

ethics cannot remain mired in long-running theoretic debates in an attempt to achieve philosophical certainty. As Mark Sagoff has written:

> [W]e have to get along without certainty; we have to solve practical, not theoretical, problems; and we must adjust the ends we pursue to the means available to accomplish them. Otherwise, method becomes an obstacle to morality, dogma the foe of deliberation, and the ideal society we aspire to in theory will become a formidable enemy of the good society we can achieve in fact.[2]

In short, environmental ethics must develop for itself a methodology of *environmental pragmatism* – fueled by a recognition that theoretical debates are problematic for the development of environmental policy.

This collection is an attempt to bring together in one place the broad range of positions encompassed by calls for an environmental pragmatism. For us, environmental pragmatism is the open-ended inquiry into the specific real-life problems of humanity's relationship with the environment. The new position ranges from arguments for an environmental philosophy informed by the legacy of classical American pragmatist philosophy, to the formulation of a new basis for the reassessment of our practice through a more general pragmatist methodology.

From the perspective of environmental pragmatism, we can return to our question: Why has environmental ethics failed to develop its practical task? Perhaps one reason is methodological and theoretical dogmatism. Mainstream environmental ethics has developed under a narrow predisposition that only a small set of approaches in the field is worthwhile – that only some ways of developing an environmental philosophy will yield a morally justifiable environmental policy. Although a wide variety of positions is discussed in the literature, the consensus it seems, is that an adequate and workable environmental ethics must embrace non-anthropocentrism, holism, moral monism, and, perhaps, a commitment to some form of intrinsic value.[3] Those who wish to defend or develop different positions are rarely heard or taken seriously, and are always assumed to have the burden of proving just cause for deviating from the norms of current theory. It seems that anyone who is still questioning which is the correct side in the debates over individualism/holism, anthropocentrism/non-anthropocentrism, instrumental/intrinsic value and pluralism/

monism is seen as being unnecessarily obfuscatory. According to the consensus view, it is time to move on to other projects – namely the unification of theories on the right side of these divides.[4]

Given the relative youth of environmental philosophy as a recognizable discipline in its own right, it is surely odd that the community of scholars has agreed, nearly completely, on the right direction for the further development of the field. The failure of this unified vision to effect practical policy should give us further pause. Viewing this problematic situation, it is the conclusion of environmental pragmatists that it is time for environmental ethics to consider some new positions in the field, and more importantly, to reassess its direction. The small set of acceptable approaches to environmental ethics may be inapplicable to the development of an acceptable environmental policy – it may be necessary to explore other possible sources and foundations for a truly moral environmentalism. Thus methodological dogmatism may account for the failure of environmental ethics in the realm of practical affairs.

This volume is part of a network of attempts to reassess the field of environmental philosophy as a reaction to these sorts of problems. One of the primary arguments supporting this reassessment – a claim that in part inspired this collection of essays – is presented in this volume by Anthony Weston in his 1992 paper, "Before Environmental Ethics." In a presumption creating argument, Weston claims that given the early, or "originary" stages of our field of inquiry, should we not assume that the right direction for our theories is less settled and sure, rather than more settled?

> At the originary stage we should . . . expect a variety of fairly incompatible outlines coupled with a wide range of proto-practices, even social experiments of various sorts, all contributing to a kind of cultural working-through of a new set of possibilities. . . . [O]riginary stages are the worst possible times at which to demand that we all speak with one voice. Once a set of values is culturally consolidated, it may well be possible, perhaps even necessary, to reduce them to some kind of consistency. But environmental values are unlikely to be in such a position for a very long time. The necessary period of ferment, cultural experimentation, and thus *multi*-vocality is only *beginning*.[5]

Thus, the burden of proof – given the moment we are in – is on those who would restrict discussion of environmental values to the prevailing theories. Following Weston, we resist the dominant trend to homogenize environmental philosophy.

But one can imagine a response that those same demands which make re-evaluation of the discipline important, also make stabilization on a narrow path of work equally important. Given the crisis we face, how could we afford the sorts of delays seemingly implicit in talk of "social experiments"? The sort of moral pluralism suggested by Weston at times sounds dangerously close to the abject relativism of deconstructive postmodernism. And this relativism is something that we may agree ought to be avoided during periods of crisis.[6]

But before answering this question, let us get clear on the kind of pluralism called for by environmental pragmatists in the continuing development of environmental ethics. On our interpretation environmental pragmatism recognizes two distinct types of pluralism: theoretical and metatheoretical. Theoretical pluralism is the acknowledgment of distinct, theoretically incommensurable bases for direct moral consideration.[7] One example of this theoretical distinction would be a position which holds a concern for the moral consideration of different individual animals, based on both Peter Singer's criterion of sentience, and Paul Taylor's criterion of respect for all teleological centers of life. Metatheoretical pluralism involves an openness to the plausibility of divergent ethical theories working together in a single moral enterprise – as both ecofeminists and ecological holists can work towards the preservation of the same natural habitats, based on different foundational claims supporting their actions.

Now the question becomes whether a metatheoretical pluralism, or a theoretical pluralism, must necessarily feed some sort of postmodern relativism. The various accounts of environmental pragmatism presented in this volume argue that they do not: pluralism of either type is at least not incommensurable with a workable, robust and critical environmental philosophy, and at best may provide the foundations or guidelines for the types of theory development needed at this stage in the growth of environmental philosophy. The *pragmatist* claims of all the papers here, as we hope is clear, is towards finding workable solutions to environmental problems now. Pragmatists cannot tolerate theoretical delays to the contribution that philosophy may make to environmental questions. If some framework is provided to prevent pluralism from lapsing into an indecisive form of relativism, and if pluralism of either type can be argued to be important for the health of environmental ethics, then this collection can be read as a plea for pragmatism as the framework of choice for a pluralist environmental philosophy.

The call for moral pluralism, the decreasing importance of theoretic debates and the placing of practical issues of policy consensus in the foreground of concern, are central aspects of our conception of environmental pragmatism. But over the years others have attempted to look at pragmatism's relevance to environmental philosophy in a more limited framework. Many others have focused on the contribution to environmental ethics that could be made by a re-reading of the fathers of pragmatism in light of the emergence of environmental philosophy.[8] Some have attempted to draw a more general approach to environmental philosophy that is informed by pragmatism, but not necessarily wedded to any particular authors in its rich heritage.[9] Still others have tried to extend the context of contemporary neo-pragmatism (found in the work of Richard Rorty, Richard Bernstein, Stanley Fish et al.), to the field of environmental ethics.[10]

With the preceding ideas in mind, we believe that environmental pragmatism, as it has emerged in environmental philosophy today, can take *at least* four forms:

1 Examinations into the connection between classical American philosophical pragmatism and environmental issues;
2 The articulation of practical strategies for bridging gaps between environmental theorists, policy analysts, activists, and the public;
3 Theoretical investigations into the overlapping normative bases of specific environmental organizations and movements, for the purposes of providing grounds for the convergence of activists on policy choices; and among these theoretical debates,
4 General arguments for theoretical and meta-theoretical moral pluralism in environmental normative theory.

While these topics are diverse they are none the less related and arguments for one form of pragmatism can draw strength from the positions of the others – as the essays in this collection will demonstrate. Although its net is so widely cast, environmental pragmatism is clearly a distinct and identifiable perspective in and for environmental philosophy. We believe that environmental pragmatism as such is primarily a new *strategy* for approaching environmental philosophy and environmental issues – it more accurately refers to a cluster of related and overlapping concepts, rather than to a single view.

None of the four forms of environmental pragmatism listed above is hegemonic as *the* characterization of the role of pragmatism in environmental philosophy. Neither are any of the different facets of the practice of environmental pragmatism *necessarily* mutually exclusive. As the subfield and literature on environmental pragmatism develops, distinctions between types of environmental pragmatism may become necessary.[11] But given the appreciation that all environmental pragmatists have for either theoretical or meta-theoretical pluralism, such distinctions are not cause for alarm when considering the coherence of this type of environmental philosophy.

We have divided this collection of essays into four main parts, in order to exhibit the diversity of pragmatic approaches to environmental philosophy and environmental problems. Although the sections do not correspond exactly with the four forms of environmental pragmatism noted above, all of the issues raised within the various forms are represented.[12] Part 1 contains essays that investigate the connections between historical American pragmatism and environmental philosophy. Part 2 focuses on the use of environmental pragmatism in the theoretical debates of environmental philosophy, including the examination of moral pluralism and anthropocentrism as alternative bases of environmental ethics. Part 3 looks at the application of pragmatic thought to a variety of practical environmental issues. And Part 4 reprints and updates a theoretical debate over the merits of pragmatism as an environmental philosophy. As a whole, this collection thus provides a comprehensive introduction to the meaning and use of pragmatic thought in environmental philosophy and policy. Moreover, it should be noted that most of the essays are original contributions written especially for this volume.

The essays in Part 1 focus primarily on the philosophical tradition of classical American pragmatism – the philosophy of Peirce, Royce, James, Mead, and Dewey. At issue here are two main questions: first, what elements of traditional pragmatic thought can be used to develop an appropriate environmental philosophy – an ethic for dealing with environmental issues? and second, what elements of mainstream environmentalism are fundamentally pragmatic, even if not articulated as an overt expression of classical American pragmatism? Several essays reveal the practical upshot to these questions: an understanding of the basic ideas of pragmatism and the pragmatic method can resolve many of the controversies in contemporary environmental philosophy – as well as resolving actual environmental problems. The basic ideas and methodology of pragmatism are necessary for a proper understanding of the human relationship with the natural world.

In the first essay, "Pragmatism and Environmental Thought," Kelly Parker presents a brief overview of the fundamental ideas of American pragmatism and their relationship to the central concerns of environmental philosophy. This essay is essentially critical in that it emphasizes the pragmatic challenge to both the Western philosophical tradition and recent doctrines in ecophilosophy. For Parker, pragmatism revises traditional ideas in epistemology, metaphysics and value theory by stressing that humans – as well as other organisms – are embedded in a particular environment, so that knowledge and value are the result of transactions or interactions with the world. Pragmatic epistemology is a radical form of empiricism, highly critical of any notion of absolutes in either knowledge or metaphysics. Pragmatic metaphysics emphasizes those qualities of the world that are grasped by active experience with the world: pluralism, indeterminacy, change and the primacy of relations. And pragmatic value theory focuses on what is good for a particular organism in its environment, thus acknowledging a plurality of values. Pragmatic ethics is thus the systematic understanding of these values in all their multiplicity, complexity and indeterminacy. Parker then uses these basic pragmatic concepts to analyze major issues in current environmental philosophy. He is particularly critical of the notion of an "environment" distinct from human activity and experience and of the idea that non-anthropocentric and intrinsic values are the basis of environmental ethics. And he cites with approval the development of a moderate moral pluralism in environmental thought and the commitment to make philosophy practical by an active engagement with social issues. All of these themes – both critical and positive – reverberate throughout the essays in this volume.

The second essay, "How Pragmatism *is* an Environmental Ethic," by Sandra B. Rosenthal and Rogene A. Buchholz, continues the explication of pragmatism as particularly appropriate for the resolution of problems in environmental philosophy. Rosenthal and Buchholz begin with an analysis of pragmatism as a critique of the modern scientific worldview – a view that objectifies Nature and posits a clear separation between humanity and the natural world. Pragmatism offers a "radical correction of modernity" by focusing on the method of science, science as creative human activity. Humans exist in the world as active experimenters, consciously organizing lived experience. It is within this conscious organization that value emerges – for what is valuable connects present experience with future possibilities. Rosenthal and Buchholz also emphasize the relational context of pragmatic thought – the organic unity of the individual organism embedded in its environment. This organic unity makes pragmatism an environmental ethic. The rational organization of experience to produce future value – pragmatism's concept of ethics – only makes sense with a full consideration of the agent or organism as a relational entity within the natural world. Humans are biological creatures, continuous with nature. For Rosenthal and Buchholz, this pragmatic worldview serves to dissolve the facile and problematic dualisms of contemporary environmental philosophy: anthropocentrism/biocentrism, individualism/holism and intrinsic/instrumental values. From a pragmatic perspective the controversy over anthropocentrism, for example, is meaningless, for it is impossible to draw a line between human well-being and the well-being of the environment in which it is situated.

The third and fourth essays focus more exclusively on the thought of individual pragmatists. In "Nature as Culture: John Dewey's Pragmatic Naturalism," Larry A. Hickman presents a comprehensive overview of Dewey's naturalism, and then compares the "guiding stars" of Dewey's thought to the land ethic of Aldo Leopold. Although Hickman surveys many of Dewey's key ideas, the most important for understanding the pragmatic view of nature are his instrumentalism and his constructivism. For Dewey, nature does not exist independently from humanity, either as a self-contained machine or as a self-directed transcendent being, as in the Gaia hypothesis. Nature is a human construct, a cultural artifact – nature-as-culture – constructed out of the immense variety of human tools of inquiry. Nature-as-culture can be contrasted, in retrospect, with nature-as-nature, i.e. nature as raw experience, with immediate value. But it is only nature as a constructed cultural artifact that connects the experiences of nature into a coherent verified whole and makes them valuable as guides to future human experiences. Hickman then compares Dewey's naturalism with Leopold's attempt to develop a postmodern or "arcadian" ecology emphasizing the human connection to the natural environment – and he ends his essay with a practical example of the value of restoration ecology.

In the fourth essay, "The Environmental Value in G. H. Mead's Cosmology," Ari Santas examines two central ideas in the philosophy of nature of the most often-neglected pragmatist, George Herbert

Mead. Mead argued for a continuum of existence that connected all entities and, particularly, all living beings. In addition, Mead stressed that the notion of an individual that is distinct from its environment is physically impossible – the individual is defined by its physical interactions with the environment. For Santas, these two basic ideas in Mead's descriptive philosophy of nature provide a groundwork for developing an environmental ethic. Mead's cosmology offers a coherent account of value as the functional properties of living things within their environment.

The final essay of Part 1 is a reprint of Bryan G. Norton's "The Constancy of Leopold's Land Ethic" from *Conservation Biology* (1988). The essay develops in more detail the connection between the environmental philosophy of Aldo Leopold and the basic elements of pragmatic thought. Norton presents a revision of the traditional understanding of Leopold's "conversion" from intensive game management to ecosystemic preservation. The shift was not due, Norton argues, to a change in metaphysical or moral views; instead, Leopold was guided by a pragmatic recognition that human knowledge of ecosystemic relations was inadequate for the management of natural systems. Norton speculates on the influence of A. T. Hadley, an American pragmatist who was president of Yale, on the formation of Leopold's ideas – especially the notion, borrowed from James, that the long-term survival of an idea, an entity or a system, is a test and warrant of its validity and value. According to Norton, Leopold maintained this long-range, pragmatic, and anthropocentric sense of right action throughout the development and shift in his style of environmental management.

The three essays in Part 2 focus on theoretical debates within environmental philosophy as the proper targets for the use of the pragmatic method. The goal is to move environmental philosophy into the sphere of practical action, to make environmental ethics a discipline useful for the development of a justifiable environmental policy. Within these essays the main issue is whether or not ethical theory is relevant to the determination of policy. All three essays question the traditional attempt to "apply" theory to practical decision-making.

In "Integration or Reduction: Two Approaches to Environmental Values," Bryan G. Norton criticizes the attempt to develop a unified theory of value and moral ontology. He defends a pluralistic ethic that is instead unified by a theory of physical scale. Norton's pluralistic theory of integrated values takes into consideration both the location and the magnitude of spatial and temporal changes. The attention to the specific context of action reveals a methodology explicitly pragmatic, in that practice precedes the development of theory. Norton also contrasts the notions of applied and practical philosophy. The former applies a valid theoretical principle to a specific situation; it requires a commitment to a theory prior to its application. Practical philosophy arises within a specific problem situation, deriving theories (if need be) from the problem context itself. The role of the applied philosopher in solving practical problems is thus much different from that of the practical philosopher. Norton illustrates the difference by a detailed criticism of a leading environmental philosopher, J. Baird Callicott. Norton considers Callicott's vision of environmental ethics to be a paradigm case of an applied moral monism. The essay is thus an important contribution to a long-running debate in the field of environmental philosophy between monists such as Callicott, Tom Regan and Paul Taylor and pluralists such as Norton, Peter Wenz and Christopher Stone.

Another version of a pragmatic moral pluralism is presented in Anthony Weston's "Before Environmental Ethics," reprinted from *Environmental Ethics* (1992). As we mentioned earlier, Weston argues that the field of environmental ethics is "at an originary stage," i.e. it is at a moment in its historical development when new values are being first considered, tested and consolidated. The values of environmental philosophy are co-evolving with new social practices regarding the human relationship to the natural world. In this stage of flux and uncertainty, when the goal of the new evolution of value is undetermined, it would be a serious mistake to attempt to force a theoretical consistency in environmental philosophy – to require, as Callicott does, one basic formulation of environmental ethics. Weston suggests that this is rather a time for open-ended inquiry and a kind of Deweyan social reconstruction, a time for experimentation in the expression and language of environmental thought. This practical project Weston calls "enabling environmental practice" – for it will enable the creation and evolution of new environmental values and a new relationship with the natural world.

In the final theoretical essay, "Compatibilism in Political Ecology," Andrew Light develops a strategy for resolving competing claims within environmental political theory. He focuses on the claims of two kinds of theorists: ontologists – such as the deep ecologists – and materialists – such as Murray Bookchin and the social ecologists.

The main point of Light's environmental pragmatism is that the urgency of the ecological crisis requires a form of metatheoretical compatibilism between the opposing theories. Inspired by a selective and *critical* reading of the arguments of neo-pragmatist Richard Rorty concerning the distinction between public and private practice, Light argues for a principle of tolerance among theorists, so that some questions would be left to private dispute. The commitment to solving environmental problems becomes not only the precondition for any workable and democratic political theory, but also a regulative ideal, "emanating from practice" – from environmental activism – and guiding the construction of political and normative theories.

Thus all three essays in Part 2 argue for some version of the primacy of practice over theory: appropriate environmental praxis will determine the limits and the content of environmental philosophy and political theory. The essays following in Part 3 demonstrate the validity of this approach, for all five adopt a pragmatic strategy of problem resolution concerning specific environmental issues. Here we see the pragmatic justification of environmental pragmatism, its cash value as a method for doing environmental philosophy.

The first essay in Part 3, Paul B. Thompson's "Pragmatism and Policy: The Case of Water," serves as a fine transition to the use of pragmatic thought in the analysis of policy debates. Thompson's central concern is the resolution of conflicts in water policy, and his essay is based on two case studies, one a classroom exercise and one a real-life policy dispute from Texas. Thompson's analysis focuses on the attempted use of foundational ethical theories – utilitarianism, libertarianism, and egalitarianism – to derive a correct moral policy. The application of ethical theories fails to resolve the dispute because it serves to freeze the debate: its main result is to provide each side with a robust theoretical argument for the policy they favor. Thompson thus argues against the traditional "application" of theory for the resolution of policy conflicts; he prefers a solution based on James' idea of pragmatic necessity and Dewey's notion of the reconstruction of community.

This essay, placed in the middle of this collection, directs the reader's attention in a number of directions. First, it demonstrates the relevance of classical American pragmatism to environmental policy. Second, it develops the criticism of applied ethics as the mere imposition of theoretical viewpoints on a practical situation, and thus continues the use of a distinction introduced by Norton – the difference between applied and practical ethics. Third, it lays the groundwork for a pragmatic methodology in the formulation of policy, as we will see in the other essays in this section. Finally, Thompson's essay begins a debate on the utility of theoretical discussions in philosophy – in the last essay in this section, Gary E. Varner, Susan J. Gilbertz and Tarla Rai Peterson will argue that a discussion of theory in some cases can facilitate the resolution of environmental conflicts.

The second essay in this section, "Toward a Pragmatic Approach to Definition: 'Wetlands' and the Politics of Meaning" by Edward Schiappa, investigates the controversy surrounding the attempt by the Bush administration to redefine the concept of "wetlands." Schiappa defends a pragmatic and political perspective on the function of definition – definitions serve particular social interests. The attempted redefinition of "wetlands" would have drastically reduced the amount of land protected by environmental regulation; the failure to codify the new definition demonstrated the power of environmental interest groups. Environmental activists, scientists and regulators were able to offer a pragmatic definition of wetlands; rather than describing essential qualities, they identified the important functions that wetlands serve. Definitions are central to a policy debate, then, because they help to determine the reality described by policy-makers. But definitions, like the policies they serve, are open to pragmatic revision. Schiappa's essay is thus a warning about the political dimensions of the definitions and concepts we choose.

The third essay, Emery N. Castle's "A Pluralistic, Pragmatic and Evolutionary Approach to Natural Resource Management," is reprinted from *Forest Ecology and Management* (1993). Castle, an economist, further develops the notion of a pragmatic and pluralistic methodology for environmental policy by stressing the need for interdisciplinary cooperation and discussion. Using forest resource management as an example, he criticizes the imposition of any one particular strategy for formulating environmental policy, such as cost-benefit analysis. His argument is based on two conditions for the development of a successful management strategy: (1) the recognition of the dynamism of both social systems and the natural environment, and (2) the variety of human preferences within a given social system. These conditions suggest the need for a resource management approach that is "pluralistic in philosophy and pragmatic in application." The fluidity and variety of human relationships with the natural world preclude any monistic environmental philosophy. Instead, Castle offers an approach that is explicitly pragmatic – like Dewey's notion of truth, adequate environmental policies would be the ones best able to "take new information into account and thereby provide for adaptation and change."

In the fourth essay, "Laws of Nature vs. Laws of Respect: Non-violence in Practice in Norway", David Rothenberg demonstrates another use of environmental pragmatism to resolve policy debates. Rather than mediating between differing theoretical positions, pragmatic principles can evaluate intramural debates within a particular tradition – in this case, the deep ecology position of Norwegian environmentalism. Rothenberg focuses on the current whaling dispute in Norway and the extra-legal and confrontational protest strategy of Paul Watson and the Sea Shepherds. Although the Sea Shepherds claim to be followers of the tradition of deep ecology, Rothenberg contrasts their practical activities with Arne Naess' belief in the non-violent tradition of Gandhi. Pragmatism can help solve the difficulty in determining the correct interpretation of deep ecology, and in doing so, help formulate the best strategy for environmental activism in this case. From the pragmatic perspective, the proper strategy will be the one most effective in the specific situation – and Rothenberg presents a convincing argument that activism based on non-violence, mutual respect, and education is most appropriate in the Norwegian whaling controversy.

The final essay in Part 3, "Teaching Environmental Ethics as a Method of Conflict Management" by Gary E. Varner, Susan J. Gilbertz and Tarla Rai Peterson, redirects the discussion back to the relationship of theory and practice in environmental policy. The argument of Varner, Gilbertz and Peterson can be seen as a rebuttal – or at least an empirical modification – of Paul Thompson's criticism of the use of ethical theory in the discussion of environmental policy disputes. This essay thus also serves to provoke further debate for those authors in Part 2 of this collection, who claim that the formulation of appropriate environmental policies will be more likely if theoretical debates are pushed to the background. Varner, Gilbertz and Peterson report on their use of workshops in environmental ethics as a teaching device to manage regional environmental conflicts. The workshops did not change the positions of the participants in the disputes – but after learning the basic theoretical issues in environmental ethics, the participants in the workshops did modify their views concerning their opponents in the conflicts. They came to see their opponents as reasonable individuals arguing on the basis of principle and not as mere extremists. This empirical study is thus an argument for a pragmatic use of ethical theory, not as an attempt to provide a unified vision of values and policy, but as a tool for effective discussion, mutual respect and the building of community.

We conclude this collection, in Part 4, by presenting an historically important exchange of essays in the field of environmental pragmatism and a contemporary analysis of that debate. The original exchange – between Anthony Weston and Eric Katz – marked the first detailed discussion of environmental pragmatism in the pages of the leading journal in the field of environmental philosophy, *Environmental Ethics*. In many ways, this exchange of arguments concerning the merits of environmental pragmatism served as one of the most important factors behind the creation of this collection.

Weston's essay, "Beyond Intrinsic Value: Pragmatism in Environmental Ethics," from 1985, was the first to raise many of the issues that are developed in more detail in these pages – especially the restrictive nature of theoretical debates in environmental ethics and the subsequent need for a pluralistic approach to the development of environmental values. Weston's primary criticism of environmental ethics was its focus on the development of an independent, non-anthropocentric concept of intrinsic value in nature. He argued instead for an open-ended development of a pluralism of human values – the totality of human experiences of the natural world – as the basis for an environmental ethic. Katz responded to Weston in 1987 in the essay "Searching for Intrinsic Value: Pragmatism and Despair in Environmental Ethics." This essay is especially important in this collection because it is the only contribution critical of a pragmatist approach. Katz argued that Weston's criticisms of environmental ethics were mistaken, for a proper environmental ethic is not based on a search for intrinsic value – it is based on a conception of the holistic, functional value of natural systems. In addition, Katz questioned the subjective relativism of a normative pluralism based on human experiences of nature. Weston responded to Katz's criticisms in 1988, and we here reprint both Weston's response and Katz's brief rebuttal.

Finally, in "Environmental Pragmatism as Philosophy or Metaphilosophy? On the Weston–Katz Debate," Andrew Light reviews this historical exchange from the perspective of recent work in environmental pragmatism. Light expands on one of his distinctions from his earlier contribution to the volume – that between metaphilosophical environmental pragmatism and philosophical environmental pragmatism. He argues that this distinction is necessary to make sense of the competing uses of pragmatism in environmental philosophy. From this claim, he can argue that both Weston and Katz are environmental pragmatists of sorts, and that the coarse identification of pragmatism as only one type of theory prevented them from seeing the points on which their work converged. Light considers both Weston and Katz as advocates for metatheoretical pragmatism. This is in sharp contrast to the moral monism and cultural pluralism of Callicott – who plays a very important role in their debate. Callicott's view is not representative of either form of environmental pragmatism.

To conclude this introduction, we would like to preempt a possible misreading of this collection and suggest a more plausible alternative. Some readers may view these essays as a prolegomena to the full development of a coherent normative position in the field of environmental philosophy – indeed they may argue that if environmental pragmatism is to be a coherent position, the essays must all fit together in a rather tight fashion. From such a perspective, the essays would represent a complete rereading of the origins and direction of environmental philosophy. Thus the essays in Part 1 show that the original environmental philosophy was pragmatism itself. Aldo Leopold, the contemporary patron saint of environmental ethics, becomes the direct heir to this legacy through his pragmatism engendered by A. T. Hadley, president of Yale during his student days there (Norton paper in Part 1). Evolving out of this rich legacy, and helped along occasionally by contemporary neo-pragmatists, are strong arguments for new approaches to environmental theory (Part 2) and environmental practices and policies (Part 3).

This reading of the book should be resisted. Not all the essays can trace their pragmatic contents in so direct a lineage. We offer an alternative interpretation that respects the clustering quality of many philosophical schools of thought. The essays here *as a collection* do not represent an attempt to create a new side to the environmental ethics debates, complete with a historical ground from more established sources. Rather, here is the first concentration of a cluster of relevantly similar attempts to raise important questions in environmental philosophy, all inspired by the philosophical legacy of pragmatism. While some individual authors may wish to argue for a direct philosophical extension of pragmatism to environmental ethics today, others do not. The contributors to this volume maintain a variety of commitments to American pragmatists, neo-pragmatists, moral pluralists and fellow travelers.

15

Is it problematic that not all of the authors here embrace classical American pragmatism as the wellspring of contemporary environmental thought? No. As pragmatists, the last thing we would want to do is inhibit the development of a healthy metatheoretical pluralism. Metatheoretical pluralism argues that environmental philosophy should be treated, to borrow the Gramscian phrase, as a contested terrain.[13] This terrain is up for grabs – thereby producing a variety of theories to account for, encourage and justify values in nature, and to explain and validate environmental policies. This ongoing contest is, given Weston's suggestion, quite appropriate for this early moment in environmental philosophy. It is similarly appropriate that environmental pragmatism should be a contested terrain as well. Tolerance among the different sorts of environmental pragmatists may serve as a model for the metatheoretical pluralism that is required to continue, for decades to come, a robust and realistic environmental ethics.

NOTES

1 J. Baird Callicott, for example, marks three types of non-anthropocentric theories: neo-Kantain (Paul Taylor, Robin Attfield, Holmes Rolston), "Leopoldian" (Callicott, William Godfrey-Smith, Richard Sylvan, Val Plumwood) and Self-Realized (deep ecologists). See Callicott's "The Case against Moral Pluralism," *Environmental Ethics* 12:2, Summer 1990, pp. 101–102. In this paper he gives an excellent genealogy of the development of these areas. One may also include individualist non-anthropocentric theories like those of animal liberationists Peter Singer and Tom Regan, as well as anthropocentric holists such as Bryan Norton, and Gary Varner's biocentric individualism, to name just a few representative theorists.

2 Mark Sagoff, *The Economy of the Earth* (Cambridge: Cambridge University Press, 1988), p. 14.

3 For example, Callicott during a brief refutation of the suggestion that Rolston is not a monist remarks: "Given that even Rolston is not really a pluralist after all, one begins to wonder why *our best, most systematic, and thoroughgoing* environmental philosophers cling to moral monism." See "The Case against Moral Pluralism," op. cit., p. 109. Our emphasis. Obviously, many very well-respected theorists in this field inside and outside of this volume may have some disagreement with such a claim. There are many other examples, some of which are discussed in the papers and commentary in Part 4 of this volume. Other areas of environmental philosophy face similar dilemmas. Environmental political theory, for example, though the benefit of a wealth of different work, is not without its claimants to theoretical hegemony. Says John Clark of social ecology in an introductory textbook: "Social ecology, a form of

16

dialectical naturalism, is the most extensively developed ecophilosophy yet to appear." Though this may not be as strong a claim as Callicott's on moral monism, since it is not clearly evaluative, the potential for hubris in such a claim is obvious. See Clark's "Introduction" to the section on Social Ecology in *Environmental Philosophy: From Animal Rights to Radical Ecology*, ed. Michael E. Zimmerman et al. (Englewood Cliffs, NJ: Prentice Hall, 1993), p. 345.

4 For example, Callicott claims that until Christopher Stone's book on pluralism in environmental ethics came along in 1988 (*Earth and Other Ethics*), he was prepared for the discipline (meaning right thinking non-anthropocentrists) to "begin to work toward the creation of an intellectual federation and try to put an end to the Balkanization of nonanthropocentric moral philosophy." Callicott presumed, in other words, that it was settled that monistic non-anthropocentrism was the agreed upon direction for environmental ethics, and he was thus prepared to move ahead unencumbered. Op. cit., p. 102.

5 Weston, "Before Environmental Ethics," reprinted in this volume, p. 151.

6 Callicott makes this sort of argument against Christopher Stone's pluralism in "The Case against Moral Pluralism," op. cit., pp. 116–120.

7 This is Gary Varner's definition of "robust theoretical pluralism" in his "No Holism without Pluralism," *Environmental Ethics* 13:2, Summer 1991, p. 177. He distinguishes this type of pluralism from what he calls pragmatic pluralism, which is akin to our definition of metatheoretical pluralism.

8 Among those excellent essays which for reasons of space allocation we could not include in this volume (and because of our focus on producing a collection of mostly original essays), are: Robert Fuller's "American Pragmatism Reconsidered: William James' Ecological Ethic," *Environmental Ethics* 14, Summer 1992; William Chaloupka's "John Dewey's Social Aesthetics as a Precedent for Environmental Thought," *Environmental Ethics* 9, Fall 1987; Bob Pepperman Taylor's "John Dewey and Environmental Thought," *Environmental Ethics* 12, Summer 1990; and Susan Armstrong-Buck's "Whitehead's Metaphysical System as a Foundation for Environmental Ethics," *Environmental Ethics* 8, Fall 1986.

9 See, for example, Kelly Parker's "The Values of a Habitat," *Environmental Ethics* 12:4, Winter 1990; Bryan Norton's *Toward Unity Among Environmentalists* (Oxford: Oxford University Press, 1991); Anthony Weston's *Toward Better Problems: New Perspectives on Abortion, Animal Rights, the Environment, and Justice* (Philadelphia: Temple University Press, 1992); Peter Wenz's "Alternative Foundations for the Land Ethic: Biologism, Cognitivism, and Pragmatism," *Topoi* 12, March 1993; John J. McDermott's "To Be Human is to Humanize: A Radically Empirical Aesthetic," in *The Culture of Experience* (New York: New York University Press, 1976); John E. Smith's "Nature as Object and as Environment: The Pragmatic Outlook," in *Man and Nature*, ed. George F. MacLean (New York:

Oxford University Press, 1979); and John C. Thomas' "Values, the Environment, and the Creative Act," *Journal of Speculative Philosophy* 4, 1990.

10 Such applications of Rorty's work, for example, can be found intermittently in Max Oelschlaeger's *The Idea of Wilderness* (New Haven: Yale University Press, 1991), and much more directly in Oelschlaeger and Michael Runer's paper, "Rhetoric, Environmentalism, and Environmental Ethics," *Environmental Ethics* 16:4, Winter 1994. If anti-foundationalism is to be applied to environmental ethics, it makes sense to do it through pragmatism and neo-pragmatism. Particularly with respect to the importance of thinking seriously about the relevance of anti-foundationalism to any form of social philosophy, American pragmatism seems to be a better starting point than other versions of anti-foundationalism which are currently more popular. As Rorty suggests, if you're going to think seriously about anti-foundationalism, it's better to work through James and Dewey than Nietzsche and Heidegger for certain projects, because the former at least have a vision of social hope that is served by their epistemology. See Rorty's "Pragmatism, Relativism and Irrationalism," in *Consequences of Pragmatism* (Minneapolis: Univ. of Minnesota Press, 1982), p. 161.

11 One of us already believes that this time has arrived. See both contributions by Andrew Light in this volume for arguments for the distinction between philosophical and metaphilosophical environmental pragmatism.

12 The biggest gap in this collection is a section devoted to the rich literature on moral pluralism. We had planned such a section in our original proposal for this collection, but had to drop it due to space limitations. It must be recognized though that one need not be a pragmatist of any type in order to embrace moral pluralism, although the two very often go together. It is certainly the case, however, that moral pluralists in environmental ethics influenced many of the contributors in this volume. As is briefly mentioned earlier, an argument could be made that environmental pragmatism is the framework within which moral pluralism should be pursued within environmental ethics. Had we been able to include a section on moral pluralism in this volume, we would certainly have drawn from the work of Christopher Stone, Mary Anne Warren, Peter Wenz, Andrew Brennan, Eugene Hargrove and others.

13 For a nice account of the connection between pragmatism and the Western Marxism of Antonio Gramsci, see Cornel West's *The American Evasion of Philosophy: A Genealogy of Pragmatism* (Madison: University of Wisconsin Press, 1989). West states in his last chapter that his own "prophetic pragmatism" is "inspired by the example of Antonio Gramsci" and that this form of pragmatism "closely resembles and, in some ways, converges with the metaphilosophical perspectives of . . . Gramsci" (pp. 231, 230).

Part 1

ENVIRONMENTAL THOUGHT AND CLASSICAL AMERICAN PHILOSOPHY

1

PRAGMATISM AND ENVIRONMENTAL THOUGHT

Kelly A. Parker

"Pragmatism" here refers to a school of philosophical thought – American pragmatism – and not to that shortsighted, allegedly "practical-minded" attitude towards the world that is a major obstacle to environmentally responsible behavior in our time.[1] The insight behind "environmental pragmatism" is that American pragmatism is a philosophy of environments. Although the founders of pragmatism rarely had occasion to write explicitly on what we would today call environmental concerns, the fundamental insights of environmental philosophy are implicit in their work. The observations that the human sphere is embedded at every point in the broader natural sphere, that each inevitably affects the other in ways that are often impossible to predict, and that values emerge in the ongoing transactions between humans and environments, for example, are all central concepts for the pragmatists – as for many contemporary philosophers of environment.

Part 1 of this essay outlines the main features of American pragmatism. So that readers new to pragmatism may more readily situate its main tenets with respect to other philosophical approaches, the major points are here presented as critical responses to familiar positions in epistemology, metaphysics, and value theory. Part 2 situates pragmatism with respect to some of the major issues in current ecophilosophy. Here, too, the presentation is largely a critical response to prevailing views. It must be stressed, however, that pragmatism is a *constructive* philosophical approach: the purpose of criticism, after all, is to open the way for new insight. Part 2 particularly stresses the question of a metaphysical grounding for environmental ethics, an area of environmental philosophy where pragmatism may have the most to offer.

1 PRAGMATISM

Pragmatism emerged as a school of thought around the beginning of this century.[2] The major early pragmatists were Charles S. Peirce, William James, Josiah Royce, John Dewey and George Herbert Mead. We might also include Alfred North Whitehead and George Santayana as "honorary" pragmatists who rejected the label, but some of whose views bear close affinity to pragmatism.

Although the pragmatists' views are certainly diverse when it comes to particulars, some characteristic themes appear throughout their writings. First, all agree in their rejection of foundationalist epistemology. There are no innate beliefs, intuitions or other indubitable "givens" upon which our knowledge is built, or in terms of which the truth or meaning of concepts can be analyzed. To say that a belief is true, according to James, is to say that the belief succeeds in making sense of the world and is not contradicted in experience.[3] Peirce's version of pragmatism asserts that the meaning of an idea consists entirely in the effects that the idea could in principle have in subsequent thought and experience.[4] We have no absolutely *indubitable* beliefs; only a stock of importantly *undoubted* ones. We have no absolutely clear, immutable concepts; we do have many concepts that are sufficiently clear and stable to let us make pretty good sense of experience. Experience, however, can at any time expose our settled beliefs as false, or reveal an unsatisfactory vagueness or confusion in our concepts. Knowing is thus an open-ended quest for greater certainty in our understanding; if we forget that our understanding is fallible, the philosophical quest for wisdom may devolve into a pathological crusade for absolute certainty.

The most interesting aspect of pragmatist epistemology for ecophilosophers is its rejection of the dualistic "spectator theory" of knowledge and its companion, the simple "correspondence theory" of truth. To object to James' definition (as many have) because it does not make truth consist in the conformity of a belief in the knower's mind to the objective state of things in the external world, is to miss what the pragmatists have to say about the nature of mind, the world and the activity of knowing.

It often comes across, even in the hands of those friendly to pragmatism, that pragmatism is *only* a theory of truth. This is as correct, and as incomplete, as saying that democracy is *only* a theory of political sovereignty. In both cases, the theories have significant practical implications. It is in tracing out these implications that we can begin to see ourselves and our world in a new light.

The founders of pragmatism recognized the philosophical implications of evolutionary theory. The characteristics and activities of any organism are always understood in light of the organism's relations to its environments. The human capacities of thinking and knowing are no exception. Consciousness, reason, imagination, language and sign use (*mind*, in short) are seen as natural adaptations that help the human organism to get along in the world.

The world we inhabit is the world *as known*. Its fabric is woven of a plurality of phenomena which can be functionally distinguished into two general types – though we fall into paradox and confusion if this *functional* distinction is uncritically taken as a *metaphysical* one. The world of experience deals harshly with absolute distinctions, at whatever level they are made. On the one hand, however, we can identify the matrix of conceptual constructs, both tacit and theoretical, that bring order to raw experience. On the other hand, we find the "stuff" of chaotic, unassimilated raw experience. The world we live in is surrounded by a fringe of the unknown, an ineffable but insistent existential reality that is larger than ourselves and our settled knowledge.[5] It is on this fringe, and in those parts of our knowledge that occasionally become unsettled, that the transformative activity of knowing goes on.

Mind is not apart *from* the world; it is a part *of* the world. "Knowing the world" is not a detached activity. It is, rather, a mutual transaction between the organism and its surroundings. In this transaction an uncertain, doubtful, indeterminate situation is reconstructed so as to make more sense, to be more intelligible.[6] The process of reconstruction transforms *both* the knowing subject *and* the known object. T. S. Eliot described his poetry as "a raid on the inarticulate."[7] The phrase aptly characterizes any mode of knowing, and it is crucial to note that, in a raid, *both* sides are liable to be affected in unforeseen ways. In creating a poetic vision, developing a scientific theory, or articulating a conception of ethics, we *literally* transform both ourselves and the world as it previously stood. Subjects and objects are not absolute entities; knower and known are inextricably twined together from the beginning. Subjects and objects are nexus of relations in an ever-shifting universe of complex relationships. The venerable distinction between subject and object is thus a convenience of speech that does not bear up under metaphysical scrutiny. It names an important but objectively vague distinction between two poles in a primordially continuous field of experience.[8] Any reconciliation between self and world in the act of knowing is tentative and fallible. To say that knowledge is *true* means only that the reconciliation is satisfactory. To say that it is *absolutely* true means that it will never stand in need of readjustment – something we can perhaps accomplish, but can never judge with certainty to be the case. Experience may shock us into doubt tomorrow.

Clearly, this epistemology involves a fundamental critique of traditional metaphysics, but the pragmatists' attitude towards metaphysical speculation was ambivalent. Peirce reportedly opened one lecture at the Johns Hopkins University with a wholesale denunciation of metaphysics as mere moonshine unworthy of attention. He ended the same lecture by urging his students to establish a metaphysical club where these crucial issues could be discussed. The story nicely illustrates what I take to be the pragmatists' typical attitude: traditional accounts of reality are so misleading as to be best ignored, but all the same, we *need* a sound metaphysics. As Peirce observed, those who resolve not to engage in metaphysical speculation do not thereby avoid metaphysics – they only condemn themselves to seeing the world through the filter of whatever "crude and uncriticized metaphysics" they have picked up along the way.[9]

Peirce and Royce enthusiastically embraced the project of articulating a metaphysics; James and Dewey were often reluctant to use the word except in a pejorative sense. Whether they called it metaphysics or not, though, the pragmatists were all concerned to develop an analysis of reality that *both* makes sense of experience *and* does not overstep the bounds of knowledge legitimately derived from experience. (Peirce and James frequently cite Hegel as a philosopher whose speculative system was a spectacular failure in both respects.) The value of metaphysical thought depends upon its making only justifiable assumptions and on following a methodology that allows for correction of its assertions.

Immanuel Kant provided the starting point for pragmatic metaphysics. The noumenal world, the world as it is in itself independent of the ordering categories of the mind, is by definition incapable of entering into knowledge or experience. To a pragmatist, the concept of a world, entity or property existing apart from the ordering influence of mind is strictly *meaningless*. To speak of the world at all is thus to speak of what Kant called the phenomenal world. To be real is to be capable of entering into experience; a thing's effects, its relations to other phenomena, are thus *all* there is to be known about the thing. The early pragmatists accordingly dropped talk of forms, essences and substances, and set about developing a

24

new metaphysics born of experience. Their resulting views tend to cut across such standard philosophical dichotomies as "idealism vs. realism."

While it is wrong to suggest that there is a "consensus metaphysics" among the pragmatists (and recognizing that "neo-pragmatists" such as Richard Rorty would maintain that it is a mistake to talk about metaphysics at all), we can identify some characteristic themes in pragmatic thought about the world. There is an irreducible *pluralism* in the world we encounter. There is the idea (supported by contemporary physics) that *indeterminacy* and *chance* are real features of the world. *Change, development,* and *novelty* are everywhere the rule. The pragmatists also attend to certain common – perhaps even universal – *structures* and *relations* that appear throughout our experience. Pragmatism, then, sees *reality* as process and development, and sees *beings* as relationally defined centers of meaning rather than as singular entities that simply stand alongside one another in the world. It emphasizes not substantial beings, but interrelations, connectedness, transactions and entanglements as constitutive of reality. All of this is based on rigorous attention to what is actually there in experience, and not on what this or that philosophy suggests we should find. This commitment to experience itself as the primary authority in speculative matters led James to call his philosophy "radical empiricism."[10]

The pragmatists proposed reforms of epistemology and metaphysics that turn Enlightenment thought inside out. The implications of pragmatic thought about value are no less revolutionary. The central emphases on experience, and on the experimental approach to establishing our knowledge and practices, make for a value theory that highlights the aesthetic dimension, sees ethics as a process of continual mediation of conflict in an ever-changing world and lays the groundwork for a social and political philosophy that places democratic and humanitarian concerns at the center of social arrangements.

All value emerges in experience. The question of ethics – "What is good?" – ultimately brings us back to concrete questions about what is experienced as good in the interaction of the organism with its environment. The inquiry does not *end* with the individual's affective experience, of course, but it recognizes this as the only possible birthplace of value. In determining the aesthetic significance of experience, pragmatists maintain a Jamesian radical empiricism: nothing is introduced that is *not* experienced, but due consideration must be afforded to all that *is* experienced.[11] The first question about value, then, is not "What ought we to desire?" but "What do people in fact desire, and why?" The answers are many and complex, and are not fully reducible, for example, to the categories of a utilitarian pleasure–pain calculus.

In aesthetics, as in metaphysics, the sheer pluralism that appears in lived experience gives us pause. The valued elements are there, and not just in private consciousness. Satisfactions arise in the semi-private, semi-public domain that is the organism-in-environment, and as such they have significance not only for the being that apprehends them but also for the environment itself and for all those other beings that inhabit it. The diversity and tangibility of aesthetic values, though, must give rise to conflict as soon as more than one valuing organism inhabits an environment.[12]

Thus arises the need for ethics, a systematic understanding of the relations that ought to obtain among various values, a theory of what is right. Based as it is on the view that value arises in a dynamic, infinitely complex system of organisms-in-environments, it is a basic tenet of pragmatic ethics that the rightness of an action is largely system-dependent. The Enlightenment dream of a universally valid ethical theory may appear plausible at first glance because many morally problematic situations do resemble one another so closely. The pragmatist, however, attends to difference and change as well as to similarity and constancy. As the world evolves, and as human thought and activities change along with it, new kinds of ethically problematic situations inevitably emerge. To cope, we need to develop new ways of comprehending what is right. No list of virtues, no list of rights and duties, no table of laws, no account of the good should be expected to serve in every possible situation that we confront. Attempts to set down the "final word" on what is right have a disturbing tendency to show up as incomplete, ambiguous or quaintly archaic in the next generation. Pragmatism maintains that no set of ethical concepts can be the absolute foundation for evaluating the rightness of our actions. We know from past experience that some ethical concepts work better than others in given situations, but our past experience is the only thing we have as an ethical "foundation." As Anthony Weston puts it, ethics is an endeavor more like creatively making our way through a swamp than it is like erecting a pyramid on a bedrock foundation.[13] After many trips through the swamp, we arrive at the means that serve best. Tomorrow we may have to readjust, though, because it is the nature both of swamps and of the world of values to shift continually beneath us. The aim of ethics is not perfect rightness, then, since there is no absolute standard for reference, but rather creative mediation of conflicting claims to value, aimed at making life on the planet relatively better than it is.

At the social and political level, this perspective implies that the individual person is of inestimable importance. All individuals are, *prima facie*, worthy of equal consideration. Since it is impossible to comprehend any individual except in a context of relations, however, the individual is always to be seen as an integral part of many communities. Social, political and cultural institutions are there to provide for the needs of individuals. I have elsewhere put this point in terms of providing for the *adequacy* of life and, beyond this, for the *significance* of life.[14] That is, social arrangements need to be constantly re-evaluated and reconstructed to ensure that minimal requirements of the organisms-in-environment are met. Beyond this, *growth* ought to be encouraged. "Growth" here is not reducible to "material growth." To equate the two leads to unfortunate conclusions – for example, that per capita Gross Domestic Product measures well-being, which is ultimately a suicidal concept for a society to embrace. Growth might better be understood in terms of increasing the aesthetic richness of experience, of expanding the available means of finding satisfaction in life.[15] Contrary to what the telecommunications industry tells us, this might well mean *recycling* one's television set rather than upgrading the cable service.

Or it may mean getting involved in the public sphere, which brings us to the other side of pragmatic social and political philosophy. Social institutions constantly need reform. Their direction can legitimately be set only by the people they serve. For the pragmatists, "participatory democracy" is a political expression of the metaphysical idea that reality *is* involvement and transformation. Because the public consists of a vast plurality of people and things valued, and because the world is changing at every moment, the ways and means of best providing for the individual and common good have to be experimentally determined. The experimenters, the political scientists who serve on a vast, ongoing "ways and means committee," should be the people themselves. Innovation is always needed in governance, and innovation typically arises at the level of one or a few people trying to resolve a particular problem, to reconstruct their corner of reality. Pragmatism (especially in Dewey's writings) emphasizes the necessity of these many diverse individuals, actively coming together in the public sphere, to present their demands, offer their insights, and hammer out their differences.[16] *That* is an activity suitable for humans, and it can be an intensely rewarding mode of being. Ultimately, that society works best which makes best use of the diverse intelligence and experience of its citizens.

2 PRAGMATIC ENVIRONMENTAL PHILOSOPHY

The early pragmatists were visionary thinkers, often ahead of their time. They are our contemporaries in many respects. Nonetheless, they naturally wrote for their own time. They addressed the problems and promises the world faced near the beginning of the twentieth century. Even John Dewey, who lived until 1952, was unable fully to envision the environmental crises we face near the beginning of the twenty-first century. The classical American pragmatists provide us with a powerful set of basic philosophical ideas. When it comes to applying these insights to contemporary issues of "the environment," though, to developing the details of legitimate environmental philosophy, we enter new territory. The only thing for us to do is to begin. What follows is only a beginning, a broad attempt to interpret and reconstruct our understanding of some major issues in current environmental philosophy, by showing how they appear in the light of pragmatism. My positions on these issues are here stated and explained only briefly. The work of constructing detailed arguments for these positions and judging their merits for environmental philosophy lies ahead, in the ongoing examination of fundamental concepts, problems, and approaches in our field. The remainder of this essay comprises a brief survey of the matters I take to be most germane to an exploration of "environmental pragmatism." These matters are (1) the concept of *environment*, (2) the place of environmental ethics in philosophical inquiry and (3) the social and political dimensions of environmental ethics. Under (4) I propose pragmatic contributions to the current debates over (a) moral pluralism, (b) anthropocentrism and (c) the intrinsic value of nature.

(1) For the pragmatist, the environment is above all *not* something "out there," somehow separate from us, standing ready to be used up or preserved as we deem necessary. As the French phenomenologist Maurice Merleau-Ponty said, "Our own body is in the world as the

heart is in the organism".[17] We cannot talk about environment without talking about *experience*, the most basic term in pragmatism. All that we or any being can feel, know, value, or believe in, from the most concrete fact ("I am cold") to the most abstract or transcendental idea ("Justice," "God"), has its meaning, first of all, in some aspect of an immediately felt *here and now*. Environment, in the most basic sense, is the field where experience occurs, where my life and the lives of others arise and take place.

Experience, again, is not merely subjective. It has its "subjective" side, but experience as such is just another name for *the manifestation of what is*. What *is* is the ongoing series of transactions between organisms and their environments. The quality of experience – whether life is rich or sterile, chaotic or orderly, harsh or pleasant – is determined at least as much by the quality of the environment involved as by what the organism brings to the encounter. Environment is as much a part of each of us as we are parts of the environment, and moreover, each of us *is* a part of the environment – a part of experience – with which other beings have to contend.

In asserting the fundamental relatedness among organisms and environments, pragmatism commits us to treating all environments with equal seriousness. Urban and rural; wilderness, park and city; ocean and prairie; housing project, hospital and mountain trail – all are places where experience unfolds. The world, in this view, is a continuum of various environments. Endangered environments perhaps rightly occupy our attention first, but environmental philosophy and ecological science are at bottom attempts to understand *all* the environments we inhabit.

Attention to the whole continuum of environments allows us to put into perspective what is truly valuable about each. The environments we inhabit directly affect the kinds of lives that we and others can live. There is an unfortunate tendency to draw crassly instrumentalist conclusions from this line of thought. I want to caution against this tendency. If environment "funds" experience, this reasoning might go, then let us use technology to turn the whole world into an easily manageable, convenient stock of environments that conduce to pleasant human experiences. This *Theme Park: Earth* line of thinking neglects our inherent limitations as finite parts of the world, and sets us up for disaster. Repeated attempts to dominate nature (e.g., our damming the Nile and its damning us right back, or our tragicomic efforts to "tame" the atom) should have begun to teach us something about the limits of human intelligence. Such attempts to dominate nature assume that no part of the environment in question is beyond the field of settled experience. We can indeed exert remarkable control over parts of the experienced world, remaking it to suit our purposes. This may be appropriate, if our purposes make sense in the first place. (I know of no reason to object to the prudent use of natural gas to heat our homes, for example.) But the very idea that the environment funds experience involves the notion that there is an ineffable aspect of the world. It is indeed arrogant to think that we can master nature; it is moreover delusional and ultimately self-negating. If we have our being in the ongoing encounter with environment, then to will that the environment become a fully settled, predictable thing, a mere instrumental resource in which there can *be* no further novelty, is to will that we undergo no further growth in experience. The attempt to dominate nature completely is thus an attempt to annihilate the ultimate source of our growth, and hence to annihilate ourselves.

What we must try to do is not to master the natural world, but to cultivate meaningful lives within various environments. We are exceedingly efficient at altering and destroying parts of the earth, but are for the most part inept at living well on it. To exercise our power wisely would require that we genuinely understand the sources of value in the world and in ourselves. Environmental philosophy must begin with close attention to the quality of experience that arises (or could arise) from inhabiting various environments.[18] We need to ask what is valuable in experiences, what features of environments they are associated with, and what ways of inhabiting environments are most appropriate. All the while we must retain respect for the wild and ineffable aspect of the world. We need to ask once more the aesthetic questions of what is good, and how goodness comes to be in *our* world – a world importantly different from that of Kant, of James or of John Muir – before we can go much further in implementing an ethics of environment.

(2) Pragmatism sees philosophical ethics as an *ongoing* attempt to determine what is good, and what actions are right. The sudden emergence of a new area of ethical inquiry is a signal that something has changed at a very deep level of our collective life. Experience has thrown us a whole new set of problems in recent years, resulting in a batch of new intellectual industries. Environmental ethics is one among several new disciplines that have emerged, first to *extend*, and then to *transform* settled ways of thinking about value.

Medical ethics, business ethics and feminist/feminine ethics are other developments similar in this way to the emergence of environmental ethics. In each of these areas, traditional theories were first applied to new problematics. The new problematics soon outstripped the available conceptual resources, showing the inadequacies of such received theoretical orientations as utilitarianism, contractualism and deontological ethics. Extension of concepts shaded over into the development of new concepts, new theoretical frameworks. The tendency of environmentalists to rely on ecological metaphors in their thinking has led some to embrace an ethic that recognizes the centrality of relations. This ethic, like that of pragmatism, recognizes the intrinsic value, within and for the system, of all the things related. Much work in feminist and feminine ethics also focuses on relations. A number of writers have noted that the notion of an "ethic of care" appears to be a promising direction for environmental ethics to pursue, and the literature on "ecofeminism" is growing steadily.[19] From the pragmatic perspective, this emerging ethic of relationships appears to be ontologically more sound than traditional ethical theories.

(3) Environmental ethics has also been associated with innovative public policy-making procedures, new applications of the legal system, and grass-roots activism. These ways of putting environmental awareness into action have come a long way in the past few decades, but of course there is a long way yet to go. The aim in all these areas, according to the pragmatic view, is to keep experimenting with ways to restructure our social institutions so that the public has a real voice in determining the kind of environments we inhabit. Pragmatism, as noted before, sees individuals as the source of genuine insight into what is needed, and accordingly tries to maximize participation in governing. Pragmatism is, in this respect as in others, closely allied with the ideals of the social ecology movement.[20]

(4) I now want to turn to three debates that currently loom large in environmental ethics. These are the debates over moral pluralism, anthropocentrism and the intrinsic value of the natural world. What pragmatism suggests about each of these debates is perhaps controversial. I hope the controversy will suggest some alternative ways to think about these focal issues.

(a) Moral pluralism can be defined as the view that no single moral principle, or over-arching theory of what is right, can be

31

appropriately applied in all ethically problematic situations. Pragmatism recognizes that there are genuine differences among moral situations, because there are many kinds of entities and possible relations among them. These situations involve a significant variety of values, and hence of kinds of conflict to be resolved. J. Baird Callicott has objected to moral pluralism on the grounds that changing among traditional ethical frameworks involves changing metaphysical assumptions.[21] We cannot in good faith be Kantians in the morning and Leopoldians in the afternoon. Peter Wenz has identified this game, which Callicott calls "metaphysical musical chairs," as an implication of what he calls *extreme moral pluralism*.[22] To shift metaphysical systems at will indeed does suggest shallow commitment to basic beliefs, if not a profound mental instability.

Pragmatism points towards what Wenz calls *moderate moral pluralism*. The movement among moral principles is here grounded in a single metaphysical view that acknowledges irreducible pluralism in the world – some ethically significant situations are simply different from others, since they involve differing goods and kinds of entity. Hence we find ourselves appealing to a variety of principles as we deal with various situations. If we embrace moderate moral pluralism, environmental ethicists inherit the serious task of sorting out what considerations have priority when conflicting principles can be applied in a given situation. Two benchmarks may help in this task: with its emphasis on the quality of the experienced world, pragmatism suggests that the *sustainability* and *diversity* of experiences made possible by a course of action should be promoted wherever possible.[23] Both are crucial not only for the land ethic, as Aldo Leopold noted, but for *any* ethic.

(b) The debate over anthropocentrism is especially tendentious. The question concerns the primary locus of value. Anthropocentrism maintains that value is *of* or *for* human beings. Biocentrism maintains that all forms of life, as such, are valuable. Ecocentrism emphasizes the value of ecological systems as a whole, including natural processes, relationships and non-living parts of the environment. An aspect of this debate concerns whether value attaches to individual entities or whether value must be seen holistically.

The pragmatist would ask why we should be expected to pledge allegiance to any of these flags *a priori*, and exclude the others. Genuine value emerges at all of these focal levels. Indeed there will be conflicts because of this, but the occurrence of such moral conflict is not peculiar to this approach. Antigone found that "family values" can tragically conflict with the values of the state; today's CEO likewise finds that business values conflict with the value of an endangered owl's habitat. *Denying* that one or the other sphere is worthy of consideration may appear to prevent potential moral conflict from arising, but only at the risk of serious moral blindness. Blind anthropocentrism has deplorable consequences for the non-human world, but a blindly misanthropic *eco*centrism is no less deplorable.

Again, pluralism is a fact encountered in experience. Value arises in a variety of relationships among differing parts of the experienced world. Each situation must be appraised on its own distinct terms. As before, the twin values of sustainability and diversity provide reference points. Sometimes we rightly focus on the sustainability of the whole system; sometimes on the unique value of an individual. Sometimes the individual or the system is human and sometimes it is not. From this perspective, environmental ethics can be seen as continuous with other areas of ethics, a distinct but integral part of value inquiry in general.

I have spoken of the experience of organisms-in-environments as centrally important. Pragmatism is "anthropocentric" (or better, "anthropometric")[24] in one respect: the *human* organism is inevitably the one that discusses value. This is so because human experience, the human perspective on value, is the only thing we *know* as humans. Many other entities indeed have experience and do value things. Again, this is not to say that human whim is the measure of all things, only that humans are in fact the measurers. This must be a factor in all our deliberations about environmental issues. We can and should speak on the others' behalf when appropriate, but we cannot speak from their experience. We can in some sense *hear* their voices, but we cannot speak *in* their voices. I see no way out of our own distinctively human bodies. In this sense, the human yardstick of experience becomes, by default, the measure of all things. Although the debate over environmental issues is thus limited to human participants, this is not inappropriate – after all, the debate centers almost exclusively on human threats to the world. Wolves, spotted owls, and old-growth forests are unable to enter the ethics debate except through their human spokespersons, and that is perhaps regrettable. Far better that they should speak for themselves! Lacking this, they do at least *have* spokespersons – and these spokespersons, their advocates, need to communicate their concerns only to other humans. To do this in anthropic value categories is not shameful. It is, after all, the only way to go.

(c) The last point I want to touch upon is one that many take to be *the* most important issue in environmental ethics. It is often repeated that the viability of environmental ethics depends on establishing the intrinsic value of the non-human world. (Perhaps I should use the term "inherent value." I'll deal with that momentarily.) The main concern is that as long as the non-human world is seen as a stock of resources having only instrumental value, there can be no genuine "environmental ethic." To be morally considerable in a strong sense, the non-human world must be more than useful. It must be valuable in its own right.

Pragmatism cuts this Gordian knot by denying that instrumental value and intrinsic value are ever mutually exclusive. The being of any existent thing, human or non-human, is constituted in its relations with other things in a context of meaningful connections. Thus anything that is good is *both* instrumentally valuable (it affects some goods beyond itself) *and* intrinsically valuable (it is good for what it is, a significant entity essential to the constitution of these relations). We can indeed distinguish the two kinds of value, but nothing can ever be instrumentally valuable without at the same time possessing intrinsic value. Thus even the "last man" on earth, in Richard Routley's classic scenario, would be doing something morally wrong in wantonly destroying parts of the natural world.[25] He would be annihilating intrinsically good parts of the field of experience. He would be needlessly damaging not just those supposedly discrete things, but intrinsically good parts of himself and of all other beings potentially or actually in the experiential web.

People may mean something else by "intrinsic value," however. Callicott reserves the term "intrinsic value" for the goodness of something *independent of any consciousness that might value it.*[26] This is sometimes called the "inherent value" or "inherent worth" of natural objects. Now, pragmatism would point out that where there is and could in principle be no valuing agent, there is no conceivable experience – and hence no aesthetic or moral value at all. In a universe of mere objects absent a valuing consciousness, things may have *being* but not *value*. Perhaps intrinsic/inherent value is the contemporary equivalent of the medieval concept of "ontological goodness" – then in so far as it exists, everything is good in God's eyes. Or perhaps whatever is, is good for some non-human consciousness other than God. (These latter two cases conform to what Callicott identifies as *inherent value.*) I respect both of these possibilities, but as a human philosopher I cannot, and need not, comprehend them from the inside. If there were no human agent there would after all be no possibility (and no *need*) for the kind of environmental ethic we seek. I do not know what it is like to be God, nor do I know what it is like to be a bat. The concept of intrinsic/inherent value is thus either meaningless, or else it reduces to the value of something that enters into ecological relations that do not *immediately* affect any human agent. All that *is*, however, does eventually, *mediately*, affect some human agent. Its value can thus be cognized by humans, and its moral considerability can be acknowledged and respected. The lesson here, that we are connected at all points to our environments, and they to us, is the Alpha and the Omega of pragmatic thought about the environment.

ACKNOWLEDGMENTS

An earlier version of this essay was presented at the annual meeting of the International Society of Environmental Ethics in Atlanta, GA, December 1993. I particularly wish to thank Eric Katz, Andrew Light, Stephen Rowe and Beth Singer for their assistance – any flaws that remain are certainly not for want of suggestions on their part.

NOTES

1 This attitude is apparently an outgrowth of another American-bred way of thinking, which John Dewey identified in 1929 as "the business mind." John Dewey, *Individualism, Old and New, The Later Works, 1925–1953*, vol. 5, ed. Jo Ann Boydston (Carbondale, Il: Southern Illinois University Press, 1988), pp. 61, 69.

2 James named the movement "pragmatism" in an 1898 address, and there credited C. S. Peirce with introducing the central insight in an 1878 article. The published version of James' address appears as "The Pragmatic Method," *Essays in Philosophy, The Works of William James*, gen. ed. Frederick H. Burkhardt (Cambridge, MA: Harvard University Press, 1978), pp. 123–139. (References to the Harvard edition of James' writings are subsequently cited by *Volume Title, Works* (publication date).) Peirce's "How to Make Our Ideas Clear," to which James was referring, appears in *Collected Papers of Charles Sanders Peirce*, vol. 5, ed. Charles Hartshorne, Paul Weiss and Arthur Burks (Cambridge: Harvard University Press, 1931–1958), 5.388. (References to Peirce's *Collected Papers* cite volume and paragraph numbers: CP 5.388 indicates volume 5, paragraph 388.)

3 See especially "Lecture VI" in *Pragmatism, Works* (1975) and the sequel to *Pragmatism*, James' *The Meaning of Truth, Works* (1975).

4 See CP 5.1–13 and CP 5.438 for Peirce's differences with James concerning the pragmatic method.

5 William James, "A World of Pure Experience," *Essays in Radical Empiricism, Works* (1976), pp. 21–44. The point of this pivotal essay is to argue that the common distinction between "subjective" and "objective" aspects of experience is merely functional, rather than being a given metaphysical fact.

6 See Dewey's account of "inquiry" in these terms. John Dewey, *Logic: The Theory of Inquiry, The Later Works, 1925–1953*, vol. 12, ed. Jo Ann Boydston (Carbondale, Il: Southern Illinois University, 1986), p. 108.

7 T. S. Eliot, "East Coker," *Collected Poems 1909–1962* (New York: Harcourt Brace Jovanovich, 1970), p. 189.

8 On the notion of vagueness as an objective feature of the world, see Peirce's discussion of vagueness and generality, CP 5.505.

9 CP 1.129.

10 For James' account of radical empiricism, see "A World of Pure Experience," *Essays in Radical Empiricism, Works* (1976), pp. 21–44.

11 That this method is similar to European phenomenology has not gone unnoticed. See, for example, the studies in James Edie, *William James and Phenomenology* (Bloomington, IN: Indiana University Press, 1987). It is worth noting that in 1902, independently of Edmund Husserl, Peirce named his version of this method "phenomenology" (CP 5.121).

12 James provides a similar account of the genesis of value and ethical conflict in Section II of "The Moral Philosopher and the Moral Life," *The Will to Believe and Other Essays in Popular Philosophy, Works* (1979), pp. 141–162.

13 Anthony Weston, "Beyond Intrinsic Value: Pragmatism in Environmental Ethics," *Environmental Ethics* 7 (Winter 1985): 321–339 and "Unfair to Swamps: A Reply to Katz," *Environmental Ethics* 10 (Fall 1988): 285–288 (included with Eric Katz's article and response in this volume).

14 I discuss the notions of adequacy and significance in "The Values of a Habitat," *Environmental Ethics* 12 (Winter 1990): 353–368.

15 I present a more detailed exploration of these two concepts of growth in "Economics, Sustainable Growth, and Community," *Environmental Values* 2 (Autumn 1993): 233–245.

16 Dewey's ideal of participatory democracy is best presented in *The Public and Its Problems, The Later Works, 1925–1953*, vol. 2, ed. Jo Ann Boydston (Carbondale, Il: Southern Illinois University Press, 1988).

17 Maurice Merleau-Ponty, *The Phenomenology of Perception*, trans. Colin Smith (New Jersey: Humanities Press, 1962), p. 203.

18 Anthony Weston has proposed one approach to accomplishing this in "Before Environmental Ethics," *Environmental Ethics* 14 (Winter 1992): 321–338 (included in this volume).

19 On ecofeminism, see Karen J. Warren, "The Power and the Promise of Ecological Feminism," *Environmental Ethics* 12 (Summer 1990): 125–146, and Christine J. Cuomo, "Unravelling the Problems in Ecofeminism," *Environmental Ethics* 14 (Winter 1992): 351–363.

20 Those wishing to explore this connection may wish to start with a reading of Murray Bookchin, *The Philosophy of Social Ecology: Essays on Dialectical Materialism* (Toronto: Black Rose Books, 1990).

21 J. Baird Callicott, "The Case against Moral Pluralism," *Environmental Ethics* 12 (Summer 1990): 99–124.

22 Peter Wenz, "Minimal, Moderate, and Extreme Moral Pluralism," *Environmental Ethics* 15 (Spring 1993): 61–74.

23 This point is developed more fully in Parker, "Economics, Sustainable Growth, and Community", op. cit.

24 The term "anthropometric" (literally "human-measured") is discussed in Alan E. Wittbecker, "Deep Anthropology: Ecology and Human Order," *Environmental Ethics* 8 (Fall 1986): 261–270.

25 Richard Routley, "Is There a Need for a New, an Environmental Ethic?" *Proceedings of the Fifteenth World Congress of Philosophy* (Sophia, 1973), 1: 205–210.

26 J. Baird Callicott, "Intrinsic Value, Quantum Theory, and Environmental Ethics," *Environmental Ethics* 7 (Fall 1985): 262.

2

HOW PRAGMATISM *IS* AN ENVIRONMENTAL ETHIC

Sandra B. Rosenthal and Rogene A. Buchholz

Several recent essays have criticized pragmatism's usefulness for the development of an environmental ethic. Eric Katz has argued that while a pragmatic value theory and ethics may be justifiable in itself, it cannot provide an adequate environmental ethics.[1] Bob Pepperman Taylor has argued that Dewey's position is socio-political at the expense of the environment.[2] And David E. Shaner and R. Shannon Duval have claimed that James' usefulness for environmental ethics is hindered because of his notion of self.[3] This essay defends pragmatism's role in the development of an environmental ethic. It will provide a bird's eye sketch of relevant pragmatic features to show that pragmatic ethics, properly understood, *is* by its very nature an environmental ethics, and that it has implications for some of the dichotomies that appear over and over again in the literature – dichotomies such as anthropocentrism/biocentrism, individualism/holism, and intrinsic/extrinsic values. Pragmatism offers another way to view some of these traditional dichotomies. The pragmatic perspective changes the nature of the issues and can provide an impetus for moving the debates beyond some of their current impasses caused by traditional dichotomies.[4]

Pragmatism arose in part as a reaction against the Modern or Cartesian World View understanding of the nature of science and of the scientific object. This understanding resulted from the general fact that the method of gaining knowledge which was the backbone of the emergence of modern science was confounded with the content of the first "lasting" modern scientific view – the Newtonian mechanistic universe. Such a confusion, based largely on the pre-suppositions of a spectator theory of knowledge, led to a naively realistic philosophic interpretation of scientific content. Scientific knowledge provided the literal description of objective fact, and excluded our lived qualitative experience as providing access to the natural universe. This resulted in a quantitatively characterized universe, a mind–matter dualism, the alienation of humans and nature, and a radical dehumanizing of nature. Nature as objectified justified nature as an object of value-free human manipulation. It is precisely the pragmatic refocus on science in terms of method rather than content which provides the key to its radical correction of modernity.

To bring a pragmatic perspective into focus through its concern with scientific method may strike some as strange, for this scientific focus may well bring to mind echoes of the technological domination of nature by humans which characterized the human–nature split. Yet this is ultimately the key to understanding pragmatism's radical rejection of such a split. While pragmatism is concerned with the findings of science and its import for concrete human existence, its *systematic* focus on science is not on the contents of science but rather on science as *method* or as lived through human activity, on what the scientist *does* to gain knowledge.[5] The stages of this method, as understood by pragmatism, need brief clarification.

The very first stage of scientific inquiry requires human creativity. We are not mere passive spectators gathering ready made data, but rather we bring creative theories which enter into the very character and organization of the data grasped. Second, there is directed or goal-oriented activity dictated by the theory. The theory requires that certain activities be carried out, certain changes brought about in the data to see if anticipated results occur. Finally, the test for truth is in terms of consequences. Does the theory work in guiding us through future experiences in a way anticipated by its claims? Truth is not something passively attained, either by the contemplation of absolutes or by the passive accumulation of data, but by activity shot through with the theory that guides it. This role of purposive activity in thought and the resultant appeal to relevance and selective emphasis which must ultimately be justified by workability are key pragmatic tenets.

Further, such creativity involved in scientific method implies a radical rejection of the "passive-spectator" view of knowledge and an introduction of the active, creative agent who, through meanings, helps structure the objects of knowledge, and who cannot be separated from the world in which such objects emerge.[6] The world of objects with which the scientist deals are not a substitute for, or more real than, our lived qualitative experience, but rather conceptual articulations of ways in which the operations of nature can be understood, products of creative intelligence in its attempt to understand its world, and they are verified in the richness of qualitative, everyday experience.

In this way, a proper understanding of the lessons of scientific method reveals that the nature into which the human organism is placed contains the qualitative fullness revealed in lived experience, and that the grasp of nature within the world is permeated with the action oriented meaning structures by which the human organism and its world are interactionally bound, both at the level of scientific reflection and commonsense experience. The nature of everyday human experience is inherently experimental, reflecting the major features of scientific method developed above.

It is only within this context that the pragmatic focus on the human biological organism and organism-environment adaptation can be understood. The human being is within nature. Neither human activity in general nor human knowledge can be separated from the fact that this being is a natural organism dependent upon a natural environment. But the human organism and the nature within which it is located are both rich with the qualities and values of our everyday experience.[7] Distinctively human traits such as mind, thinking and selfhood are emergent characteristics of nature and part and parcel of its richness. They refer to ways in which the lived body behaves. For none of the pragmatists is the self understood as a self-enclosed entity. Rather, it is viewed as a body-self which is "located," if one speaks of location, throughout the biological organism with its reflexive ability as this emerges from and opens onto the relational contexts in which it functions.[8]

From the backdrop of the non-spectator understanding of human experience, humans and their environment – organic and inorganic – take on an inherently relational aspect. To speak of organism and environment in isolation from each other is never true to the situation, for no organism can exist in isolation from an environment, and an environment is what it is in relation to an organism. The properties attributed to the environment belong to it in the context of that interaction. What we have is interaction as an indivisible whole, and it is only within such an interactional context that experience and its qualities function.

Such a relational view of organism-environment at once has pluralistic dimensions, for environments are contextually located,

40

and significant solutions to problematic situations emerge within such contextually situated environments.[9] Pluralism is not relativism, however, for with the rejection of the spectator theory of knowledge comes the rejection of the correspondence theory of truth and, instead, a view of reality as richer than, or overflowing, our conceptual demarcations. Diverse perspectives grasp the richness of reality in different ways, but must be judged in terms of workability. And workability requires growth, resolution of conflicts in terms of enlargement of context which can adjust or adjudicate the conflicting perspectival claims. Growth cannot be reduced to material growth, but rather is best understood as an increase in the moral-aesthetic richness of experience. Such adjudication must be ongoing, for existence is by its nature filled not just with the stable but the precarious, not just the enduring but the novel, not just conformity but diversity, and it is this dynamic which provides the materials for ongoing growth.

From the backdrop of the above brief sketch of some relevant features of pragmatism, the discussion can turn more directly to general ethical concerns. Value for the pragmatist is not something subjective, housed either as a content of mind or in any other sense within the organism, but neither is it something "there" in an independently ordered universe. Objects and situations, as they emerge in human experience, possess qualities which are as ontologically real in their emergence as the processes within which they emerge. Value and valuings or valuing experiences are traits of nature; novel emergents in the context of organism-environment interaction.[10] Such emergent value in its brute givenness can be positive or negative; one can experience not only value but "disvalue."

The difference between valuings and the experience of the valuable, between valuings and evaluations, is the difference in stages in inquiry, the distinction between experiences which make no future claim and those judgments which make future claims by linking present experience to other experiences in terms of interacting potentialities or causal connections. In brief, valuings are turned into the experience of the valuable by the organizing activity of mind in the ongoing course of experience as experimental. Claims concerning the valuable emerge from the context of conflicting valuings and are dependent for their validity upon their ability to produce a context for harmonious valuing experiences.

Here it becomes important not to fall, within the understanding of value, into what would parallel the remnants of a spectator theory

41

of knowledge. Even the most rudimentary as well as the most casual valuings incorporate not just the fulfillments of anticipatory activity, but also the funded character of past evaluations at least implicitly operative in even the most primitive choices. In brief, the value quality of immediate experience is structured in large part by the context of moral beliefs within which it emerges.[11] The extent of one's awareness of this functional relation between valuings and evaluations or, in other terms, the extent of one's awareness of the moral meanings which enrich experience, determines the extent of morally directed action within experience. Moral action is planned rational action rooted in the awareness of meanings as embodying potentialities for the production of valuing experiences. And when habitual modes of organizing behavior do not work in resolving problematic situations involving conflicting valuings, new moral beliefs embodying new moral standards and new understandings of what is valuable emerge, which in turn give a different funded quality to the immediacy of valuings.[12]

The functional relation between valuing experiences and objective value claims must work as an organic unity in the ongoing course of experience in increasing the value ladenness of experience. Such a workable relationship, to remain workable, requires not stagnation but constant openness to change through intelligent reconstruction incorporating the dynamics of experimental inquiry. Experimental method, as applied in the moral context, is in fact the attempt to increase the value ladenness of a situation through a creative growth of perspective which can incorporate and harmonize conflicting or potentially conflicting values. The expansion of a moral perspective, though not independent of intelligent inquiry, is not merely a change in an intellectual perspective but rather is a change which affects and is affected by the organism in its total concreteness.

At this point it may be objected that an individual can achieve an harmonious life in which valuings are reflective of evaluations in a totally artificial environment, with no concern for nature whatsoever. This, however, is to miss the point that humans are concrete organisms enmeshed in an environment with which they are continuous. Human development is ecologically connected with its biological as well as its cultural world.[13] Growth, for the pragmatist, is inherently moral, and growth involves precisely this deepening and expansion of perspective to include ever widening horizons of the cultural and natural worlds to which we are inseparably bound. This receives its most intense form in Dewey's understanding of experiencing the world religiously as a way of relating one's self with the universe as the totality of conditions with which the self is connected.[14] This unity can be neither apprehended in knowledge nor realized in reflection, for it involves such totality not as a literal content of the intellect, but as an imaginative extension of the self, not an intellectual grasp but a deepened attunement. This is the reason poets get at nature so well.[15]

Such an experience brings about not a change in the intellect alone, but a change in moral consciousness. It allows one to "rise above" the divisiveness we impose through arbitrary and illusory in-group/out-group distinctions by "delving beneath" to the sense of the possibilities of a deep-seated harmonizing of the self with the totality of the conditions to which it relates. And, for all the pragmatists, this involves the entire universe, for their emphasis on continuity reveals that at no time can we separate our developing selves from any part of the universe and claim that it is irrelevant. Indeed, while environmentalists may seek to describe "objective" relationships among interacting individuals – human, non-human, organic, and inorganic that make up the biosphere – yet the properties attributed to the individuals are not possessed by them independently of the interactions in which they exhibit themselves. Nature cannot be dehumanized, nor can humans be denaturalized. Humans exist within and are part of nature, and any part of nature provides a conceivable relational context for the emergence of value. The understanding of "human interests," of what is value*able* for human enrichment, has to be expanded not just in terms of long range vs. short range and conceivable vs. actual, but in terms of a greatly extended notion of human interest or human welfare. Further, to increase the experience of value is not to increase something subjective or within us, but to increase the value ladenness of relational contexts within nature. Dewey's understanding of experiencing the world religiously provides the ultimate context within which pragmatic ethics must be located. While every situation or context is in some sense unique, no situation or context is outside the reaches of moral concern. Pragmatic ethics, properly understood, *is* an environmental ethics.

Such an ethics cannot be called an anthropocentrism. True, only humans can evaluate; and, without evaluation as a judgment concerning what best serves the diversity of valuings, the valuable could not emerge. Further, humans can speak of non-human types of experience only analogically in reference to their own. But, though the concept of the valuable emerges only through judgments involving human intelligence, value emerges – either positively or negatively – at any level of environmental interaction involving sentient organisms. While the value-level emergent in organism-environment contexts increases with the increased capacity of the organism to experience in conscious and self-conscious ways, as long as there are sentient organisms experiencing, value is an emergent contextual property of situations. As James stresses, as moral agents we are forbidden "to be forward in pronouncing on the meaningfulness of forms of existence other than our own." We are commanded to "tolerate, respect, and indulge those whom we see harmlessly interested and happy in their own ways, however unintelligible these may be to us. Hands off: neither the whole of truth nor the whole of good is revealed to any single observer."[16] Although some may question the claim that a distinction in levels of value-emergence can be made, when push comes to shove, when all the abstract arguments are made, is it not the case that claims of the valuable *must* be seen in light of its promotion of or irrepressible harm to human welfare, actual or potential? Does anyone really think that the preservation of the spotted owl and the preservation of the AIDS virus have equal moral claim?[17]

It may be objected that the above evaluation of the relative merits of the AIDS virus and the spotted owl in terms of their promotion of or harm to human welfare is a re-emergence of the anthropocentrism denied above. This objection, however, comes from a failure adequately to cut beneath the "either–or" of anthropocentrism/biocentrism. In fact, "both–and" is closer to the position intended, but even this is inadequate, for it fails to capture the radical conceptual shift which, in making the conjunction, changes the original extremes of the positions brought together. There is no "all or none" involved. It is not the case that all value is such only in relation to humans. Yet neither is it the case that all value has equal claim irrespective of its relation to the welfare of humans. Value is an emergent contextual property of situations as long as and whenever there are sentient organisms experiencing, yet the value-level emergent in organism-environment contexts increases with the increased capacity of the organism to experience in conscious and self-conscious ways. The biological egalitarianism of biocentrism can perhaps be thought consistently, but it cannot be maintained in practice. Surely one is not willing to move from the theoretical egalitarianism of humans and the AIDS virus to an implementation of such theory in practice.

Yet this does not mean that humans can ignore the value contexts of sentient organisms within nature. To do so is not to evaluate in terms of conflicting claims but to exploit through egocentric disregard for the valuings of other organisms. We must make judgments which provide protection for the welfare of humans, yet such judgments must consider the value laden contexts involving other sentient organisms to the largest degree consistent with this goal.

It has been seen that if evaluations are to be about anything, they must be about the way experiences of value, actual or conceivable, are to be organized. And, only contexts involving sentient organisms yield the emergence of value. While this position does not allow for the emergence of value in non-sentient contexts, yet neither does it allow for the exploitation of non-sentient contexts. For is it possible to envision any aspect of nature, any relational context in nature, any thing in nature that cannot be the object of a conceivable experience of sentient organisms?

The problem is not that environments are ultimately valuable in their actual or potential relational contexts of emergent value, but that valuings and the valuable environments which allow for them are taken far too narrowly. At no point can pragmatic ethics draw the line between human welfare and the welfare of the environment of which it is a part. Here it may be objected that to value non-sentient nature in terms of its potentiality for yielding valuing experiences is to say that it has merely instrumental value, and if nature is merely an instrument, then no real environmental ethic is possible. Yet, within the above framework, the entire debate concerning instrumental vs. intrinsic value is wrong-headed from the start. Everything that can conceivably enter into experience has the potential for being a relational aspect of the context within which value emerges, and any value, as well as any aspect of the context within which it emerges, involves consequences and is therefore instrumental in bringing about something further. Thus Dewey holds that no means–end distinction can be made, but rather there is an ongoing continuity in which the character of the means enters into the quality of the end, which in turn becomes a means to something further.[18]

Moreover, evaluations grow, gain novel direction and novel contexts in the resolution of conflicting and novel interests, and it is with choice and creative resolution in these problematic contexts that morality is concerned. If everything has intrinsic value, then decision-making becomes somewhat arbitrary. If, for example, every tree has its own intrinsic value and the right to exist, irrespective of its potential for valuing experiences, how can we choose which trees to cut down? Yet common sense tells us we cannot "save" them all. Arguments must be made, and the literature itself shows that arguments are ultimately made in terms of the potential for valuing experiences and, ultimately, when hard choices must be made, for the valuing experiences of humans.[19]

It may be further questioned as to whether the pragmatic ideal of "fully attained growth" in the union of self and universe merges into an ecocentrism in which value is given to the system rather than to the individual. Here again, these alternatives do not hold within the pragmatic context. Sometimes the system is more important, sometimes the individual, and this is dependent on the contexts in which meaningful moral situations emerge and the conflicting claims at stake. Further, no absolute break can be made between the individual and the system, for each is inextricably linked with the other and gains its significance in terms of the other. The whole notion of an isolated individual is an abstraction, for diversity and continuity have been seen to be inextricably interrelated. Neither individuals nor whole systems are the bearers of value, but rather value emerges in the interactions of individuals, and wholes gain their value through the interactions of individuals, while the value of individuals cannot be understood in isolation from the relationships which constitute their ongoing development. When we slide over the complexities of a problem, we can easily be convinced that categorical moral issues are at stake. And, the complexities of a problem are always context dependent or relational.

While this view cannot tell us what position to take on specific issues, it gives a directive for understanding what is at issue, for making intelligent choices, and for engaging in reasoned debate on the issues. What is needed for responsibility in pragmatic ethics, of which environmental ethics is an essential part, is the development of the reorganizing and ordering capabilities of creative intelligence, the imaginative grasp of authentic possibilities, the vitality of motivation and a deepened attunement to the sense of concrete human existence in its richness, diversity and multiple types of interrelatedness with the natural environments in which it is embedded. The importance of this attunement cannot be over stressed. In Dewey's words, "A problem must be *felt* before it can be stated. If the unique quality of the situation is *had* immediately, then there is something that regulates the selection and the weighing of observed facts and their conceptual ordering."[20] In short, the development of moral respon sibility in general requires education of the whole person.[21] It is education of the whole person which can lead us to extend our valuing experiences sympathetically to the valuings of other sentient organisms, and to make responsible judgments in adjudicating the conflicting claims of diverse valuings both among humans and among humans and other levels of sentient organisms. In brief, it is only education of the whole person which can provide the breadth and depth, the sensitivity and the imagination needed to harmonize conceptual recognition of the valuable and the immediacy of valuing experiences. And it is only such a deep-seated harmonizing which can bring about the change in moral consciousness needed for the implementation of an environmental ethic.

NOTES

1 "Searching for Intrinsic Value: Pragmatism and Despair in Environmental Ethics," *Environmental Ethics*, Vol. 9, Fall, 1987, p. 232 (reprinted in this volume).

2 "John Dewey and Environmental Thought: A Response to Chaloupka," *Environmental Ethics*, Vol. 12, Summer 1990. In "John Dewey's Social Aesthetics as a Precedent for Environmental Thought," *Environmental Ethics*, Vol. 9, Fall 1987, p. 247, William Chaloupka argues that one can find in Dewey's aesthetics what could be called, "with only slight exaggeration," an implicit environmentalism.

3 "Conservation Ethics and the Japanese Intellectual Tradition," *Environmental Ethics*, Vol. 11, Fall 1989.

4 By pragmatism in this essay is intended classical American pragmatism, that movement incorporating the writings of its five major contributors, Charles Peirce, William James, John Dewey, G. H. Mead and C. I. Lewis. That these philosophers provide a unified perspective is assumed in this essay, but this claim is defended at some length in Rosenthal's book, *Speculative Pragmatism* (Amherst, Massachusetts: The University of Massachusetts Press, 1986). Paperback edition (Peru, Illinois: Open Court Publishing Co., 1990).

5 This stress on pure method is not intended to deny that pragmatism is influenced in its philosophical claims by the findings of various sciences. Indeed, it pays careful attention to these findings. However, pragmatic philosophy's inextricable linkage to the model of scientific method as pure method is one thing. Its attention to various findings of various sciences achieved by the general method is something quite different. These two issues should not be conflated, and it is the method of science which provides the key to pragmatism.

6 Dewey expresses this noetic creativity in science in his claim that "What is known is seen to be a product in which the act of observation plays a necessary role. Knowing is seen to be a participant in what is finally

known." As he concludes, both perception and the meaningful backdrop within which it occurs are shot through with the interactional unity between knower and known. Dewey, *The Quest for Certainty*, Vol. 4 (1984), *The Later Works*, ed. Jo Ann Boydston (Carbondale and Edwardsville: University of Southern Illinois Press, 1981–1991), pp. 163–165.

7 Technology, to which the scientific enterprise gives rise, far from being a dominating tool to ravage nature and fragment humans, must be placed in its proper context of the fullness and richness of humans and nature. Technology is needed as the precondition for a full, free, flexible way of life. The problem with technology springs from the ideas, or perhaps better said, the absence of ideas, which too often operate in conjunction with technological factors. Technology can provide the means, but human vision, as creative, as grasping possibilities for the betterment of the human experience in its holistic nature, is required to guide technology. Technology is not over nature, nor can it be set over against humanistic concerns, for it is a part of the natural process by which human beings, who are within and part of nature, alter their environment in order that it better serve the qualitative fullness of their interests.

8 Thus, James' notion of the self, far from being a hindrance to an environmental ethic, as claimed by Shaner and Duval, op. cit., is an essential ingredient for it.

9 For various arguments in the debate on moral pluralism see Christopher Stone, "Moral Pluralism and the Course of Environmental Ethics," *Environmental Ethics*, Vol. 10, Summer 1988, pp. 139–54; Don Marietta, Jr., "Pluralism in Environmental Ethics," *Topoi*, Vol. 12, March 1993), pp. 69–80; J. Baird Callicott, "The Case Against Moral Pluralism," *Environmental Ethics*, Vol. 12, Summer 1990, pp. 99–124; Anthony Weston, "On Callicott's Case Against Moral Pluralism," *Environmental Ethics*, Vol. 13, Fall 1991, pp. 283–286. It should be noted here that what pragmatism is not suggesting is a metaphysical pluralism or a pluralism of "absolute" principles, but rather a metaphysics and epistemology which demand pluralism. This is not a relativism of any type but rather an ontologically grounded perspectivalism or contextualism.

10 The claim by Anthony Weston that pragmatism is a form of subjectivism is misplaced. "Beyond Intrinsic Value: Pragmatism in Environmental Ethics," *Environmental Ethics*, Vol. 7, Winter 1985, p. 321 (reprinted in this volume).

11 As Dewey notes, "Even in the midst of direct enjoyment, there is a sense of validity, of authorization, which intensifies the enjoyment." *The Quest for Certainty*, op. cit., p. 213.

12 For example, awareness of the problems of pollution in terms of potential valuing experiences can lead to a change in the immediate qualitative valuing of throwing one's soft drink can out the window of the car. While some years back this was virtually routinely done, and may even have provided the positive valuing experience of "the throw through the wind," today most morally aware individuals would experience a direct sense of negative valuing were they to even envision

48

doing such a thing. Such an act has a directly felt, qualitative dimension of repulsiveness to it.

13 In "John Dewey and Environmental Thought: A Response to Chaloupka," op. cit., Bob Pepperman Taylor's objections to Dewey as an environmentalist stem from an ongoing illicit abstraction both of the social, cultural, and biological dimensions of the human in Dewey's philosophy from concrete human existence, and of aesthetic sensibility from the very fiber of human life.

14 It should perhaps be pointed out here that this is quite different from theistic beliefs, which often foster environmental indifference.

15 And thus James holds that the broadest forms of moral commitment are held by those who appreciate the religious dimension of existence.

16 "On a Certain Blindness in Human Beings," in *Talks to Teachers* (New York: W. W. Norton, 1958). In "American Pragmatism Reconsidered: William James' Ecological Ethic," *Environmental Ethics*, Vol. 14, Summer 1992, Robert Fuller argues for a possible panpsychism in James such that even inorganic matter has sentience and thus engages in valuing in at least some rudimentary fashion. We are disinclined to accept this interpretation of James or this kind of justification for concern with inorganic nature.

17 Thus we find the characteristic of "harmless" in the above statement by James.

18 See the debate between Anthony Weston "Beyond Intrinsic Value" and Eric Katz, "Searching for Intrinsic Value: Pragmatism and Despair in Environmental Ethics" (reprinted in this volume).

19 Thus, for example, old growth forest is valuable in that it has the potential for yielding valuing experiences for individuals. But here problematic situations emerge. For the old growth forest, as cut down for lumber, has the potential for yielding valuing experiences for humans as they desire more housing. The old growth forest, as a forest, has the potential for providing valuing experiences for individuals as they experience the joys of the outdoors. Further, in these and various other value dimensions of the old growth forest, its potential for the production of valuing experiences extends not just to actual valuings, or even to the valuings of actual individuals, but to its potential for the production of valuing experiences into an indefinite future. These potentialities for future valuing are not something that can be excluded from the present problematic context, for these potentialities to be affected are not in the future; they are there within the present context, to be affected by our present decisions.

20 *Logic: The Theory of Inquiry* (Carbondale: Southern Illinois University Press, 1990), p. 76.

21 Tom Colwell in "The Ecological Perspective in John Dewey's Philosophy of Education," *Educational Theory*, Vol. 35, Summer 1985, views Dewey's entire philosophy of education as a pioneering effort in ecological thought. See also the chapter on Dewey in Arthur Wirth's *Productive Work – In Industry and Schools* (Lanham, Md: University Press of America, 1983) in which he analyzes the social and ecological backdrop of education.

3

NATURE AS CULTURE
John Dewey's pragmatic naturalism

Larry A. Hickman

"... genuine experimental action effects an adjustment *of* conditions, not *to* them: a remaking of existing conditions, not a mere remaking of self and mind to fit into them. Intelligent adaptation is always a *re*adjustment, a re-construction of what exists."

John Dewey (LW 8:98)[1]

"Nature is made better by no mean but nature makes that mean."
William Shakespeare, *A Winter's Tale*, Act 4, Scene 4
(quoted by John Dewey, LW 9:225)

1 THE GUIDING STARS OF DEWEY'S PRAGMATIC NATURALISM

It is as unfortunate as it is unfair that John Dewey has been read as an unabashed apologist for industrial expediency and unhampered business boosterism. One consequence of this has been the assumption that his work has little relevance to current debates regarding the status of non-human nature.[2]

It is true that Dewey was at one time the leader of a school of pragmatism known as "instrumentalism." But his pragmatism was never the vulgar sort that valorizes bald expediency. Nor was his instrumentalism the "straight-line" variety that works towards fixed goals, heedless of the collateral problems and opportunities that arise during the thick of deliberation.

It is also true that Dewey consistently argued that the continued development of experimental science is a necessary condition for the amelioration of the deplorable conditions under which much of the world's human population subsisted during his lifetime (conditions, many of which have since deteriorated). But his notion of experimental science was both more comprehensive and more revolutionary than most of his contemporaries ever grasped, and his conception of its place in human experience was as co-laborer with other forms of inquiry, including the arts, law and politics, and not as their overseer. He consistently held that to view science as tool for the domination of nature is to honor a conception of science, as well as a conception of nature, that has been historically outgrown.

What, specifically, does Dewey have to contribute to the current debates regarding the relations of human beings to non-human nature? Unlike Henry David Thoreau, he did not go to the woods and there articulate an alternative to the stuffy life and genteel transcendentalism of Concord; unlike John Muir he did not develop an evolutionary pantheism in the course of a thousand-mile walk from Indiana to the Gulf of Mexico; and unlike Aldo Leopold he constructed neither land ethic nor land aesthetic based on experiences in the arid American Southwest and the lush farmland of Wisconsin.

In short, Dewey was not a field naturalist. Although his boyhood was spent in small-town and rural Vermont, Dewey's adult home was the city. Apart from his periodic recreational retreats to mountains, seashores and his farm on Long Island, from 1894 until his death in 1952 he lived first in Chicago, then in New York City.

But if Dewey was no field naturalist, he was a naturalist nevertheless. As a committed evolutionary naturalist, Dewey accepted and argued for the view that human beings are in and a part of nature, and not over against it. It was his contention that human life constitutes the cutting edge of evolutionary development (but not its telos), and this because, as he put it, it is only as human beings come to consciousness by means of social intercourse that self-reflection becomes a part of evolutionary history.

For Dewey, the principal difference between human beings and the rest of nature is not that there is no communication elsewhere than within human communities, but that human beings are unique in their ability to exercise control over their own habit-formation and therefore to alter in deliberate ways both the course of their own evolution and the evolution of their environing conditions. In other words, it is only with the advent of human beings that choice, and consequently morality, become a part of life on earth (EW 5:53), and it is only as human beings come to consciousness that nature comes to have "a mind of its own" (MW 4:29).

51

One of the clearest statements of Dewey's naturalism is found in his reply to George Santayana, who had charged him with advancing a "half-hearted" and "short-winded" naturalism. Santayana had argued that Dewey was only interested in "foreground," and that consequently the "rest of nature [in his philosophy] is reputed to be intrinsically remote or dubious or merely ideal."[3] To put a fine point on it, Santayana was accusing Dewey of ignoring, or worse, idealizing, non-human nature.

Dewey had thought his own naturalism such an obvious and fundamental part of his philosophy that he was astounded by Santayana's criticism. His reply was that

> [i]f the things of experience are produced, as they are according to my theory, by interaction of organism and environing conditions, then as Nature's own foreground they are not a barrier mysteriously set up between us and nature. Moreover the organism – the self, the "subject" of action, – is a factor *within* experience and not something outside of it to which experiences are attached as the self's private property.
>
> (LW 14:17)[4]

As further evidence of his naturalism, Dewey cited his appropriation of the radical empiricism of William James:

> My theory of the relation of cognitive experiences to other modes of experience is based upon the fact that *connections* exist in the most immediate non-cognitive experience, and when the experienced situation becomes problematic, the connections are developed into distinctive objects of knowledge, whether of common sense or of science.
>
> (LW 14:18)[5]

Finally, he responded that

> the proof of the fact that *knowledge* of nature, but not nature itself, "emanates"[6] from immediate experience is simply that this is what has actually happened in the history or development of experience, animal or human on this earth – the only alternative to this conclusion being that in addition to experience as a source and test of beliefs, we possess some miraculous power of intuitive insight into remote stellar galaxies and remote geological eons.
>
> (LW 14:19)[7]

In his response to Santayana, then, Dewey reveals the reference points by which the course of his naturalism has been charted. The first is his *instrumentalism*, which is his way of avoiding the traditional problems of both realistic and idealistic views of non-human nature. On the side of ontological realism, for example, seventeenth-century science and philosophy tended to view non-human nature as a clock-like machine, complete in itself. On the side of ontological idealism, some contemporary environmental philosophers have argued for a pantheistic version of the Gaia hypothesis, which in its extreme form holds that not only is the earth a self-regulating superorganism, but it is capable of deliberation in terms of its own ideals.[8] Looked at from a different angle, epistemological realists, including most neo-positivists, have argued that knowledge of nature is secured as its features are "mirrored" in separate human minds; and epistemological idealists, such as Berkeley, have contended that nature is a correlation among ideas.

For Dewey's instrumentalism, however, nature, as a complex of objects of knowledge, is neither complete in itself apart from human interaction, nor the locus of extra-human deliberation. It is neither directly given nor a mental correlation. Nature is instead a multi-faceted construct that has been slowly and laboriously built up over thousands of years of human history by means of various tools of inquiry, including the arts, religion, magic, hunting, manufacture and experimental science, to recall just a few. Nature is a construct, or cultural artifact, but it has not been constructed out of nothing. The raw materials of previous experiences and experiments, unanticipated events, chance insights, moments of aesthetic ecstasy, habits, traditions and institutions have all been continuously reshaped and refined by tools that have included religious rituals, philosophical treatises, novels, poems, scientific hypotheses, television documentaries, and many more.

The instrumentalism that supports Dewey's concept of nature-as-culture bears scant resemblance to the "straight-line" variety of instrumentalism advanced by seventeenth-century philosophers and scientists. His post-Enlightenment instrumentalism calls for careful attention to ends–means relationships at every step of deliberation. This is no less true when the domain of inquiry is non-human nature than when a musician chooses a subject for her song. Tools must be continually revised if they are to be appropriate to new tasks. Tasks must likewise be continually re-evaluated in the light of the tools available for their execution.

Nature-as-cultural-artifact is never finished. Because the rush of time and the jolt of novelty are observable features of experience, nature too, as a complex of objects of knowledge, becomes subject to ongoing re-evaluation and reconstruction in order to effect adjustment to changed and changing conditions. We may be able to get it better and better, truer and truer, but we never get it completely right. This is Dewey's *fallibilism*.

One of the most important features of Dewey's naturalism, so important that it almost becomes synonymous with his larger program, is his distaste of claims to transcendent knowledge. His *anti-transcendentalism* would have led Dewey to reject attempts by some environmental ethicists to "sacralize" nature as a thing-in-itself with values, interests or rights that are purely intrinsic to it and independent of human interests. What would he have made of the view of Carolyn Merchant, for example, that holds that "all living things, as integral parts of a viable ecosystem, . . . have rights"?[9] Should the one remaining sample of smallpox virus be set free from captivity, for example, because of its inherent rights as an integral part of a viable ecosystem? And what would he have thought of the biocentrism of Paul Taylor, with its claim that nature has "intrinsic" value, or value apart from its being valued "either intrinsically or instrumentally, by some human valuer"?[10] Dewey would, I think, have characterized Merchant's "rights" talk and Taylor's suggestion – that an ecological ethic can only be grounded in values never experienced, and perhaps not capable of being experienced, by human beings[11] – as having abandoned naturalism altogether for an excursion into an ideal realm.

Dewey thought it the function of intelligence to expand and enrich experience, and this with a view to the adjustment of experienced situations to new demands. Such adjustment is neither uniquely the alteration of environment for the sake of the experiencing subject, nor the accommodation by the experiencing subject to its environment. Because environments include experiencing subjects as parts, it is both accommodation and alteration.

If Dewey's naturalism eschews the transcendent, it is holistic none the less. Since human beings are a part of nature, their enriched experience of nature enriches nature's experience of itself. This is what Dewey means when he says that the production of the objects of knowledge involves the interaction of one part of the environment with other parts of the environment. At the same time, however, as he argued in his now famous essay "The Reflex Arc Concept in

Psychology" (EW 5:96–109), there is no knowledge without prior interest because it is interest that serves to initiate and focus inquiry. Our knowledge does not come to us fully formed from any region in which we have no interests. Some may wish to call this view "anthropocentric," but it is neither more nor less than a recognition of the fact that human beings transact business within environing conditions beginning where they are, and not where they are not. If "biocentrism" means taking a perspective that is other than human, then Dewey was no biocentrist. If it means, on the other hand, that it is characteristic of human intelligence that it continually broadens its purview, and that its best and most productive perspective is holistic, then Dewey's work from the 1890s onward was "biocentric."

Another component of Dewey's naturalism was his *antifoundationalism*. This is the view, now recognized as one of the central theses of post-modern thought, that the search for epistemic foundations is both futile and unnecessary. One consequence of this is that the individual thinking self is not privileged, as it was for the architects of modern philosophy, Descartes, Locke and Kant. The self is itself a construct, and as such it is experienced neither foundationally, immediately, nor privately, but just as are other parts of the built-up environment of human knowledge. It further follows that there is no objective nature to provide a foundation for knowledge. Nature is not a "thing" but instead a complex and fecund matrix of objects and events, experienced in part as an expanding source of novel facilities and constraints, but nevertheless constructed within the history of human inquiry.

Dewey's *radical empiricism*[12] includes the claim that non-cognitive experience is capable of grasping relations. This is very important for an understanding of nature-as-culture because it means that we can grasp what hangs together in all of nature – human and non-human alike and together – as features of our most immediate and basic aesthetic experiences. In a moment I will suggest that this grasping-of-things-together was also an important stage in the development of the thought of Aldo Leopold.

Dewey recognized that it is notoriously difficult to retain moments of aesthetic insight. Even the most intense delights have a way of turning to dust in our hands. It is at this point that the cognitive portion of experience enters the picture. Cognition develops experienced relations by relating them to one another and making something new and more secure out of them.

But radical empiricism doesn't just say that we experience

relations; it also says that at the edges of the focal points of non-cognitive experience there are unfocused areas, or fringes. This amounts to a powerful antidote to the tendency to go transcendent, to posit a non-human world filled with independent values of its own.

In fact, it turns out that radical empiricism provides the benefits of transcendent views without their disadvantages. It allows us to acknowledge that there is a "beyond" to experience, just as transcendent views do. But it doesn't commit the fallacy of transcendent views, which is their attempt to say something *definitive* about what is experienced only as horizon. Regardless of where the focus of experience moves, according to radical empiricism, it is always fringed by vague areas of which we are only dimly aware but which may provide the opportunity for refocusing. Such refocusing is itself often the occasion for the production of new objects of knowledge.

What all this means is that we can get more and more intimately involved in terms of our experience of non-human nature without having to posit a realm in which animals and plants which are not conscious of themselves or in control of their own behavior have independently inherent "rights," or into which we may only enter provided that we have abandoned the perspective of human beings. The function of cognition is to extend human interest, and therefore human knowing, into areas of experience that had theretofore been no more than fringes or horizons of working knowledge. Properly nurtured, aesthetic delight gives rise to interests, which in turn motivate the kind of inquiry that eventuates in a robust interaction with ever wider dimensions of the human environment.

Radical empiricism does the work of transcendent views of nature, and it does it better. It describes and prescribes ways in which non-human nature can enter into the domain of human concerns, and thus into the widening circle of the moral, without appealing to *a priori* or *ad hoc* devices. Radical empiricism embraces a genuinely evolving naturalism that is rooted in the histories of natural events and that seeks to play a part in their further evolution; it is not, as are some versions of the Gaia hypothesis or most theories of "inherent rights of nature," a short-cut or ersatz naturalism based on the discontinuities of mysticism or logical leaps.

Taken together, radical empiricism and instrumentalism argue that what is cognitive arises out of what is non-cognitive by the intervention of intelligence. But the reverse is not true. Unlike Bertrand Russell, for example, Dewey is no reductionist: he does not claim that the cognitive can be reduced to something primitive and non-cognitive.

Phylogenetically, historically, then, the cognitive emerges from the non-cognitive. This is Dewey's *genetic argument*. Whenever we think seriously about anything, ontogeny recapitulates phylogeny. Cognition is advanced and enhanced whenever it takes its own history into account. The roots of the normative aspects of any discipline grow in the fertile soil of the history of that discipline. By implication, if we are to advance normative claims with respect to non-human nature, then the history of inquiry in that area, including the history of religious, aesthetic, scientific and technological inquiries, must be taken into account.

Dewey's *constructivism* is a thread that runs through each of the other components of his naturalism, and thus deserves emphasis. Nature may be conceptualized in retrospect as nature-as-nature, or what is in fact experienced as immediately and unreflectively valued. In its richer sense, however, nature is also nature-as-culture, an artifact or complex of ideas that has proven valuable and continues to provide grounds for successful action. Nature-as-nature may be and often is the source of romantic or mystical responses that are deeply satisfying in their consummatory moments. But nature-as-nature is nature experienced haphazardly; experienced values have not been secured because their meanings have not been worked out and linked to one another. Nature-as-culture, on the other hand, is the product of conscious attempts to extend and link the meanings of nature in ways that secure experienced values by testing them one against the other in order to determine what can continue to prove valuable.

It is Dewey's constructivism that links these two conceptions of nature. In other words, the functional separation of nature-as-nature, or nature-as-valued, from nature-as-culture, or nature-as-valuable, does not render Dewey vulnerable to the charge of having regressed to a dualism of fact and value, or even a dualism of nature and culture, since what is valuable is a development that grows in ordered richness out of what is valued, and culture is thus continuous with and a part of nature. Nature-as-nature and nature-as-culture are not ontologically separate, but only functionally so. They are phases, earlier and later, of the expansion and extension of the meanings of situated human experience.

Dewey's position avoids the traditional split between facts and values by means of his contention that (a) values and relations are experienced (his radical empiricism), (b) facts are not just given but always selected from a busy and complex environment as facts-of-a-case, that is, always and only in the context of a particular inquiry (his instrumentalism and his anti-transcendentalism), and (c) what is valuable is so only as a result of tests that have proven it to be a reliable basis for further action (his constructivism and his fallibilism).

2   DEWEY'S PRAGMATIC NATURALISM AND LEOPOLD'S LAND ETHIC

If these are the guiding stars of Dewey's naturalism, what course do they indicate for an environmental philosophy? One way of answering this question is to set Dewey's naturalism along side that of Aldo Leopold, allowing each to take the measure of the other.

Chapter 7 of Max Oelschlaeger's *The Idea of Wilderness*[13] provides an excellent guide to Leopold's work. Leopold's land ethic, writes Oelschlaeger,

> which states that humans ought to act to preserve the integrity, stability, and beauty of natural systems, gives Leopoldian ecology an explicitly normative dimension. . . . In Leopold's normative ecology the human species is viewed as a part of rather than apart from nature. Subsequently, the membership of sentient beings in the community of life entails obligations to preserve the land.[14]

This statement sums up in an admirable way the diverse and sometimes conflicting elements within Leopold's work. As a professional scientist, a forester, he had been trained to accept the demands of a *modernist* or "imperial" ecology, an ecology based on a search for epistemological foundations, a faith in quantification, a vision of linear and inevitable progress, an acceptance of the physical sciences as paradigmatic of all rationality, and a conception of nature as machine to be dominated and exploited.

On the other side, however, because of his own profound aesthetic sensibility, Leopold also felt the claim of a *postmodernist* or "arcadian" ecology that had been adumbrated by Thoreau and Muir, an ecology that rejected each of these modernist claims and sought to establish others in their stead. This would be an ecology that emphasizes human situatedness within nature, that holds that science is only one of many productive areas of human experience, that views progress as fragile and attainable only in piecemeal fashion, that treats knowledge as relative because perspectival and fallible and that denies any absolute or final split between fact and value or between culture and nature.

By now it should be obvious that each of these components of postmodernist ecology was also a component within Dewey's naturalism.

In Oelschlaeger's story, Leopold's thinking moves through several developmental stages. From an initial acceptance of the modernist views, learned at Yale and in the employ of the Forestry Service, he moved first to an intuitive, aesthetic appreciation of the connectedness of natural events (bordering on a mystic organicism), and thence to an attempt to construct a land ethic of amelioration that takes into account "the interconnections between the cultural and natural worlds."[15]

Within this later phase, however, Leopold's vocabulary is complex and sometimes conflicted. He variously employed, according to Oelschlaeger's account, (1) an organic model of nature whose key idea is management, (2) a social model of nature whose central idea is community and finally, (3) an enriched organicism that held that "*natural species possessed intrinsic rights* to existence and that these sometimes took precedence over human rights."[16]

Leopold's field naturalism and Dewey's pragmatic naturalism turn out to have a great deal in common. Dewey's radical empiricism, for example, provides a key to understanding the incipient phase of Leopold's shift from modernist to postmodernist ecology. (Conversely, Leopold's shift provides an excellent example of Dewey's radical empiricism.)

Leopold's initial break with Forest Service doctrine was patently non-cognitive. He was profoundly influenced by the relations that he discovered within his aesthetic experience. Tempted to remain within the confines of that experience, he flirted with a transcendent, organic, vitalism. "Possibly, in our intuitive perceptions," he wrote, "which may be truer than our science and less impeded by words than our philosophies, we realize the indivisibility of the earth – its soil, mountains, rivers, forests, climate, plants, and animals, and respect it collectively not only as a useful servant but as a living being."[17]

But Leopold soon realized that it would be impossible to continue indefinitely this celebration of his non-cognitive experience. Mysticism qua mysticism does little work in the public sphere.[18] Beyond the continued celebration of it, the consummatory moment in aesthetic experience can be prolonged only by developing its connections to other experiences. He came to realize that his non-cognitive vision must be reconstructed into an instrument that can function in the sphere of public science and public opinion. His subsequent vocabularies of management, community and the rights of species represent various stages of his attempt to reconstruct his initial experience in ways that would prove sufficiently valuable to have broad appeal and therefore to effect what he took to be ameliorative change.

Leopold never lost sight of the aesthetic dimension of his experience, however. He appealed to both elements within his experience, the non-cognitive and the cognitive alike, in his 1932 remark that a successful ecology must take into account a "residual love of nature, inherent even in 'Rotarians,' [that] may be made to recreate at least a fraction of those values which their love of 'progress' is destroying."[19] On the cognitive side, the term "management" appears as a key word in the title and the chapter headings of *Game Management*, one of his major works during this period.[20]

Leopold's "Rotarian" remark also contains a genetic argument. Evolutionary history equips human beings (even the most ardent land speculator) with a non-cognitive sensibility towards nature that may, if properly managed, provide the springboard for an enriched cognitive response to non-human nature that can take into account what is beyond the narrowly economic and utilitarian.

His remark is also constructivist and anti-foundational. He has recognized that any concept of nature that does real work in the domain of public affairs is a cultural artifact. "Although Leopold never escaped entirely from thinking of ecological facts as 'out there,'" writes Oelschlaeger, "he knew that the objective order of nature was a useful fiction. His research had repeatedly confirmed that Homo sapiens and nature were internally related."[21]

There are numerous parallels to Leopold's conceptual shift in the contemporary literature of environmentalism. Biologist Nathaniel T. Wheelwright, for example, has argued for respect for nature on the grounds of its "resplendence." Contending that it is "poor conservation strategy to bank on the arguments of ecologists or economists alone," Wheelwright has pointed out that the deterioration of natural environments and the loss of species diminishes what is "intricate" and "irreplaceable" and that aesthetic experience is thereby diminished.[22] This is an excellent example of an appeal to what most human beings "feel" about nature, which is something that can be reconstructed in such a way that it performs work in the public sphere.

To what extent is Dewey's pragmatic naturalism consistent with Leopold's environmentalism? Two out of the three metaphors that Oelschlaeger has identified as central to Leopold's thought are also found in Dewey's work. Dewey's pragmatic instrumentalism is an encouragement to "management," in just the Leopoldian sense, that is, as an intelligent reworking of what is unsatisfactory in order to render it more satisfactory. It is true that Dewey utilized the word "control," in connection with his instrumentalism, and that this has been the occasion for some of his critics, especially those of neo-Heideggerian temperament, to dismiss his views as unrepentantly modernist. But Leopold also wrote of "control." What both men mean by "control" is intelligent interaction within a situation in order to effect its improvement.

The second of Leopold's central metaphors, community, also occupies an important place in Dewey's work. There are two important senses in which nature can be understood as "community." In the first, non-human nature would be said to constitute a "community" in the sense of interacting populations, food chains, and so on. "Communication" within nature's community would, on this model, be a way of talking about equilibrating forces within an ecological system that maintain its stability as a whole and with respect to which human beings are either not involved or involved only marginally. This view of communication has the disadvantage of tending towards an idealization of nature that renders it transcendent of human interests.

In the second sense of "community," however, there is no break between human and non-human nature, and human beings themselves are regarded as one of many forces within the larger domain of nature. Communication would then be transaction among all relevant parts of nature, including the human part, that is, the part in which self-conscious intelligence emerges.

The term "management" takes on radically different meanings when applied to these two views of nature's community. The first view presents two scenarios. In the first scenario, that of the onto-logical idealist, the idea of management is replaced by the idea of respect, since nature is something that possesses ideals apart from those of human beings. In the second, the scenario of the ontological realist, management is imposed on a nature no less apart than that of the idealists, but which is, in this case, a machine to be maintained and repaired. Both of these positions have their roots in modernist thought.

In the second view of nature's community, human managerial skills become an active part of the ongoing evolution of a system of which human beings are also a part.

Both Leopold and Dewey understood "community" in the second of these senses, that is, in the postmodernist sense. Civilization is not, Leopold wrote, "the enslavement of a stable and constant earth. It is a state of *mutual and interdependent cooperation* between human animals, other animals, plants, and soils, which may be disrupted at any moment by the failure of any of them."[23] In short, evolution evolves. Continuing communication (ongoing adjustment of the various parts of the entire system to one another) is the condition for the continuing success of the whole.

This is also Dewey's sense of community, and it is his sense of management. In his 1898 essay "Evolution and Ethics," Dewey argued against the position taken five years earlier by Thomas Huxley in his Romanes Lecture. Huxley had taken the view that there has been a radical break in evolutionary history. The rule of the earlier "cosmic" processes had been struggle and strife. This was nature "red in tooth and claw." The rule of the emergent but now radically distinct "ethical" process would be sympathy and cooperation. And whereas the goal of the cosmic process was survival of the fittest, the goal of the ethical process would be that of fitting as many as possible for survival. Huxley had argued that "the ethical progress of society depends, not on imitating the cosmic process, still less in running away from it, but in combating it."[24]

Dewey thought that Huxley had capitulated to an unwarranted and dangerous form of dualism. In his reply he utilizes Huxley's own analogy of a garden in order to undercut his separation of nature from culture. "The ethical process," he writes, "like the activity of the gardener, is one of constant struggle. We can never allow things simply to go on of themselves. If we do, the result is retrogression. Oversight, vigilance, constant interference with conditions as they are, are necessary to maintain the ethical order, as they are to keep up the garden" (EW 5:37).

But what is the relation of the ethical (the cultural) to the process of evolution as a whole (the natural)? Dewey answers

> we do not have here in reality a conflict of man as man with his entire natural environment. We have rather the modification by man of one part of the environment with reference to another part. Man does not set himself against the state of

nature. He utilizes one part of this state in order to control
another part. . . . He introduces and maintains by art conditions of sunlight and moisture to which this particular plot of
ground is unaccustomed; but these conditions fall within the
wont and use of nature as a whole.

(EW 5:37–38)

In other words, human activity, and therefore culture, is one part
of nature. It is one of the ways that nature transacts business with
itself. In intelligent gardening, just as in any other intelligent activity,
one part of the environment is modified with respect to another part
of the environment. Deliberation and intelligent management enter
into the history of evolution.

It might be objected that Dewey has already gone too far; that he
has allowed just anything that human beings happen to do to
amount, by definition, to progress with respect to the whole. But
Dewey meets this objection head-on by means of his pragmatic
instrumentalism. Since the part of human beings within the evolutionary process is intelligent choice, it is not action *simpliciter*, but
intelligent action that produces improved results and that therefore
advances the process of evolution. Doing nothing and doing just
anything are equally unintelligent, since they do not enhance the
adjustment of one part of the environment to another.

Dewey's argument in this essay hinges on his notion of temporality.
"Everyone must have his fitness judged by the whole, including the
anticipated change; not merely by reference to the conditions of
today, because these may be gone tomorrow. If one is fitted simply
to the present, he is not fitted to survive. He is sure to go under"
(EW 5:41). "The past environment," Dewey writes, "is related to the
present as a part to a whole" (EW 5:46). Further, "evolution is a
continued development of new conditions which are better suited to
the needs of organisms than the old. The unwritten chapter in
natural selection is that of the evolution of environments" (EW
5:52).[25]

If Dewey has undercut the grounds for a dualism of evolution and
ethics, nature and culture, we are still left with the question of just
how it is possible for "communication" among the features of the
natural environment to occur. This is a matter that Dewey takes up
in 1925 in Chapter 5 of *Experience and Nature*, where he presents his
theory of communication as an essential ingredient of his naturalism.
"Where communication exists," he writes, "things in acquiring

meaning, thereby acquire representatives, surrogates, signs and implicates which are indefinitely more amenable to management, more permanent and more accommodating, than events in their first state. By this fashion, qualitative immediacies cease to be dumbly rapturous" (LW 1:132).[26]

In other words, communication involves the taking of naturally occurring experiences and making something of them that increases their meaning by relating them to other naturally occurring experiences. This is a type of art that involves, in its turn, careful attention to the qualitative moments of experience in order that their traits may be made manifest, or expressed, by working out their implications.

Communication is a multiplier. It is not a matter of expressing something already there so much as it is a matter of "the cooperation in an activity in which there are partners, and in which the activity of each is modified and regulated by the partnership" (LW 1:141).[27] Communication opens up the doors of perception. We become "capable of perceiving things instead of merely feeling and having them. To *perceive* is to refer the present to consequences, apparition to issue, and thereby to behave in deference to the *connections* of events" (LW 1:143).[28]

Late in his career, and apparently as a reaction to the neo-positivism that was beginning to dominate academic ecology,[29] Leopold seems to have retreated to an organicism that holds that "*natural species possessed intrinsic rights* to existence and that these sometimes took precedence over human rights."[30] This is Leopold's third model of nature, and what he seems to have regarded as the basis for his now famous "Land Ethic." But Dewey's naturalism leads him to rejects this, as well as other varieties of free-standing or transcendent treatments of nature. He rejects foundations in earth as well as sky.

Like many other ethicists, Dewey held that moral rights exist only in the context of a community of moral agents. This is so because of the linkage between rights and obligations. Because there cannot be obligation in the absence of choice, and because it is only with the advent of human life that choice becomes fully a part of evolutionary history, it is a mistake to attribute intrinsic rights either to non-human species or to non-human individuals.[31] To speak of non-human species or non-human individuals as the possessors of intrinsic rights would in Dewey's view amount either to anthropomorphizing non-human nature or to opening up a chasm between human and non-human nature by positing a domain of moral rights that does not involve moral agency and is therefore entirely separate from what human beings understand by the term.[32]

Does this mean that Dewey's naturalism regresses to a modernist anthropocentrism? Does his naturalism open the door to treating non-human species in any way we choose? It does neither. In order to understand why this is so it is necessary to recur to his radical empiricism and his idea that human beings experience nature non-cognitively as well as cognitively.

At the non-cognitive level, nature, both domesticated and wild non-human nature, is a source of intense and immediate aesthetic delight. Because of its immediacy, this type of aesthetic experience requires no warrant. It just is. Hunting, fishing, hiking, boating, bird-watching, celebration of the seasons, and many other forms of interaction with non-human nature, such as the enjoyment of pets, offer the occasion for such delight. The delights of breathing clean air, drinking pure water and the enjoyment of forests untouched by acid rain – all this is valued in its immediacy.

Dewey's radical empiricism also allows for the immediate experience of a "beyond" in the sense that immediately experienced delight possesses sensible fringes. Hints, gaps, leads and clues are experienced on the fringes of focused experiences. In its non-cognitive phase, then, nature is the source of both felt delight and wider expectation. Because of its commitment to radical empiricism, Dewey's naturalism is capable of promoting a piety with respect to non-human nature that is not encumbered by the epistemological problems of transcendent views of nature. A fringe is a vague indication of what may be to come, under the proper circumstances; of what is open to possible development, given sufficient interest.

It is at the cognitive level, however, that appreciation of nature is enlarged. Nature is understood both by means of the arts, as aesthetic experiences are secured and enriched, and by means of the sciences, as experiences are enlarged and related to one another through experimentation, abstraction and quantification. Both the arts and the sciences function in Dewey's work to expand the meanings of experience, and to secure what would otherwise have been immediate and transitory; but they do so in different ways. The arts "express" meanings, as he puts it, and the sciences "state" meanings.

Another way of putting this is to point out that Dewey undercuts the distinction that plagued Leopold throughout his career, namely the distinction between facts and values.[33] In order to do this, Dewey distinguished between what is or has been valued and what has proven or might prove to be valuable. Values in non-human nature, and in human nature as well, are most often just experienced. As such they have not been secured as valuable, that is, they have not been reconstructed as platforms for further action. Dewey thought that values are secured as valuable just as their meanings are developed, enlarged and interrelated. This may be accomplished in the arts, as certain traits and qualities of materials are expressed in ways that single them out from others that are less interesting, less fecund or less evocative of further experience. It may also be accomplished in the sciences by means of experimentation, or the instrumental interaction with natural processes in which mere endings are replaced with consequences and consummations that are worthy of celebration and suggestive of further paths of deliberation.

Dewey's naturalism thus treats non-cognitive nature both as immediately *valued*, and as raw material for the construction of nature as culture, that is, nature as human artifact or nature as *valuable*. Given the complexity of human culture, with its many overlapping and competing interests, including the economic, the artistic, the political and the religious, to name just a few, it is nature-as-human-artifact that enters into public debates regarding the adjudication of conflicting interests. This is because one non-cognitive experience, since it is immediate, has no way of holding its own against the claims of other, potentially competing non-cognitive experiences. Its implications have not been worked out. But in nature-as-culture, implications have been drawn, connections made, and tentative conclusions reached.

3 ENVIRONMENTAL PRAGMATISM AND ENVIRONMENTAL PRESERVATION

The upshot of this is that Dewey's naturalism is capable of supporting Leopold's land ethic, i.e. the view that humans ought to act to preserve the integrity, stability and beauty of natural systems, but without Leopold's occasional lapses into an appeal to a realm of transcendent rights. This can be done by demonstrating that the integrity, stability and beauty of non-human nature is immediately experienced as valued, and further that these factors have proven valuable as a source of continually emerging values, including those that are aesthetic, economic, scientific, technological and religious. Each of Leopold's terms, "integrity," "stability" and "beauty," however, because it is a tool of inquiry and not an absolute whose meanings have been determined for all time, must undergo continual re-evaluation and reconstruction with respect to changing conditions.

Dewey's naturalism is consistent with and anticipatory of at least one current version of the Gaia hypothesis. As Frederic L. Bender characterizes it, Gaia presents four major challenges to traditional thinking about nature. First, traditional notions of individuality are challenged; second, traditional notions of fitness are challenged; third, Gaia intentionally blurs the traditional boundary between life and non-life; and fourth, Gaia's holism rejects the traditional focus on individual ecosystems in favor of attention to global relationships.[34]

Each of these points is also Dewey's. He argued that individuals are only so in the context of environing factors; that the notions of "fitness" must be greatly expanded (see his reply to Huxley); that the difference between life and non-life is primarily a matter of level of organization (LW 1:195);[35] and that intelligent deliberation takes as broad a view as is possible. As I have noted, however, Dewey rejected extreme views of Gaia, which hold that the global ecosystem has intelligence apart from that of human beings.

Dewey's naturalism is also consistent with and anticipatory of some forms of "restoration" ecology, such as that advanced by William R. Jordan.[36] Like Dewey, Jordan's leading metaphor is the garden, with its ancillary metaphors of "maintenance" and "reconstitution." Among the objects of his restoration interest are various portions of the Wisconsin prairie.

Like Dewey, Jordan recognizes that human life is not a "pernicious" factor outside environmental change, but one part of it. His goal is thus not to "protect" nature from human beings, but to "provide the basis for a healthy relationship between nature and culture."[37] He recognizes that restricting human participation in natural events (idealizing nature) is merely another way of fighting nature (the obverse of treating nature as machine), and that the real challenge of restoration ecology is to find ways in which human beings can come to view themselves as participating members of their environments.

Traditional nature activities such as boating, hunting and fishing are consequently parts of his program. "All of these are integrated into an event that is constructive rather than consumptive – as each of these particular activities is in its traditional form."[38] It is by means of these reconstructed activities that Jordan intends to "bring to our attention aspects of our relationship with nature that otherwise we might not recognize."[39]

In short, Jordan thinks that the older versions of environmentalism (which I have argued rest on modernist versions of realism or idealism) have failed because of their fruitless attempts to isolate nature from culture. He thus sees his own restoration model as pragmatic. His intent is to increase the relevance and enlarge the application of Leopold's land ethic.

A key element in Jordan's restoration ecology is ritual celebration. By beginning with the immediate delight afforded by communal and festive (non-cognitive) interaction with natural events, such as programmed prairie burns, he believes that the basis can be laid for an enriched cognitive appreciation of the place of human life within its natural setting, and consequently that restoration will come to be seen as "both an effective [scientific] process and an expressive [artistic] act." "The idea," he continues, "is not merely to *decorate* restoration, but to develop it to enhance its expressive power."[40]

Because of his interest in scientific inquiry, Dewey would have approved of setting aside wilderness areas so that they can serve as laboratories for environmental scientists. But this is not to treat wild nature as apart, ideal, or "untouched." It is instead to preserve it as source of experimental data which would otherwise be lost. As Leopold notes, "A science of land health needs, first of all, a base datum of normality, a picture of how healthy land maintains itself as an organism. . . . Wilderness, then, assumes unexpected importance as a laboratory for the study of land-health."[41]

As a synthesis of the aims of preservationist and restorationists, the work of the Nature Conservancy is also consistent with Deweyan naturalism. As it continues its task of buying up and protecting wildlife habitats checkered within developed areas, both the scientific and the aesthetic dimensions of human experience are served and expanded. Each of these models of naturalism – restoration, preservation and the Nature Conservancy synthesis – can play a part in the wider project of adjusting one part of our environment to other parts in order to effect amelioration of the whole.

If my reading of Dewey is correct, then, his naturalism allows him to accept and defend the central tenets of Leopold's land ethic without the appeal to an idealized non-human nature that sometimes surfaces within his, Leopold's, work. I have argued that Leopold's attempt to provide a foundation for his ethic by this means is the least workable and the least defensible feature of his otherwise excellent project. If I have made my case, then Dewey's work locates itself in the thick of current debates regarding the relations between human beings and non-human nature, and it offers the promise of continuing insights within this arena of experience.

NOTES

1 Standard references to John Dewey's work are to the critical edition, *The Collected Works of John Dewey*, edited by Jo Ann Boydston (Carbondale and Edwardsville: Southern Illinois University Press, 1969–1991), and published as *The Early Works* (EW), *The Middle Works* (MW) and *The Later Works* (LW). These designations are followed by volume and page number.

2 Dewey's British critics including Russell and his German critics, including Adorno and Horkheimer, read, or more properly misread, him in this way. Even George Santayana took up the refrain when he characterized Dewey as "the devoted spokesman of the spirit of enterprise, of experiment, of modern industry" and claimed that his philosophy was "calculated to justify all the assumptions of the American society." See George Santayana, "Dewey's Naturalistic Metaphysics," in *The Philosophy of John Dewey*, 3rd edition, ed. P. A. Schilpp and L. E. Hahn (La Salle, Illinois: Open Court, 1989), p. 247.

3 Schilpp and Hahn, op. cit., p. 251.

4 Ibid., p. 532.

5 Ibid., pp. 532–533.

6 The term "emanates" was Santayana's. Dewey allowed its usage in connection with his own view, provided that the "*aura* that clings to the word" be eliminated. See LW 14:19.

7 Schilpp and Hahn, op. cit., p. 534.

8 This is Frederic L. Bender's reading of James Lovelock. See Frederic L. Bender, "The Gaia Hypothesis: Philosophical Implications," in *Technology and Ecology*, ed. Larry A. Hickman and Elizabeth F. Porter (Carbondale, Illinois: Society for Philosophy and Technology Press, 1993), pp. 64–81.

9 See Carolyn Merchant, *The Death of Nature* (San Francisco: Harper and Row, 1980), p. 293.

10 Paul W. Taylor, *Respect for Nature: A Theory of Environmental Ethics* (Princeton: Princeton University Press, 1986), p. 75. Taylor does not base his biocentrism on the rights of individual non-human organisms, but rather on their status as "teleological centers of life" (p. 122). He does allow that human beings have the right to destroy predatory organisms, such as the smallpox virus, on grounds of self defense (p. 264).

11 Ibid., p. 99.

12 Dewey admits his debt to William James for his notion of radical empiricism. "Long ago I learned from William James that there are immediate experiences of the connections linguistically expressed by

conjunctions and prepositions. My doctrinal position is but a generalization of what is involved in this fact" (LW 14:18, note 16).

13 Max Oelschlaeger, *The Idea of Wilderness* (New Haven: Yale University Press, 1991).

14 Ibid., p. 206.

15 Ibid., p. 216. See Brian G. Norton, *Toward Unity among Environmentalists* (New York: Oxford University Press, 1991), especially pp. 39–60, for an account of Leopold's development that differs slightly from that of Oelschlaeger. Norton thinks that Leopold's position was better integrated than Oelschlaeger's, and I, believe it to have been. Norton does, however, make an interesting comment that relates to the notion of "nature-as-culture" that I am attempting to develop in this essay. He sees Leopold's work as "guiding the search for a *culturally* defined value in nature" (p. 58).

16 Oelschlaeger, op. cit., p. 228.

17 Aldo Leopold, "Some Fundamentals of Conservation in the Southwest," in *The River of the Mother of God and Other Essays*, ed. Susan L. Flader and J. Baird Callicott (Madison: University of Wisconsin Press, 1991), p. 95.

18 Leopold withheld "Some Fundamentals of Conservation in the Southwest" from publication during his lifetime.

19 Aldo Leopold, "Game and Wild Life Conservation," in Flader and Callicott, p. 66.

20 Aldo Leopold, *Game Management* (New York: Charles Scribner's Sons, 1936).

21 Oelschlaeger, op. cit., p. 227.

22 Nathaniel T. Wheelwright, "Enduring Reasons to Preserve Threatened Species," in *The Chronicle of Higher Education*, June 1, 1994, p. B2.

23 Aldo Leopold, "The Conservation Ethic," in Flader and Callicott, p. 183.

24 Thomas H. Huxley, *Evolution and Ethics and Other Essays* (New York: D. Appleton and Co., 1896), pp. 81–83, *et passim*. Quoted in EW 5:36.

25 See, however, the argument of Bob Pepperman Taylor that Dewey's view of nature represents no advance over that of Locke. Bob Pepperman Taylor, "John Dewey and Environmental Thought," in *Environmental Ethics*, Vol. 12, No. 2, Summer 1990, p. 183. Dewey was in fact quite critical of Locke's view of nature. Locke, Dewey writes, "was completely under the domination of the ruling idea of his time: namely, that *Nature* is the norm of truth. . . . Nature is both beneficent and truthful in its work; it retains all the properties of the Supreme Being whose vice-regent it is" (MW 8:59). The irony here is that if Dewey's reading of Locke is correct, then his (Locke's) view of nature is much closer to that of the idealistic environmentalists such as Paul Taylor than it is to Dewey's view. This is a point that Bob Pepperman Taylor apparently misses.

26 John Dewey, *Experience and Nature* (La Salle, Illinois: Open Court, 2nd edition, 1965), pp. 138–139.

27 Ibid., pp. 148–149.

28 Ibid., p. 151.

29 See Oelschlaeger, op. cit., p. 226 for an account of this.
30 Ibid., p. 228.
31 Dewey apparently did not know of Koehler's work with apes, and so he denies choice to non-human animals. Since self-reflexive communication is the basis of his account of responsible action, however, he might well have wished to have included non-human animals, such as chimps who have learned sign language and entered into communication with themselves and humans by its means, as moral agents and thereby the bearers of rights. In any case, Dewey was enough of an evolutionist that he was acutely aware of transitions within nature, and that the history of evolutionary development is more or less continuous, if not in the temporal sense, then at least in the functional sense. For more on Koehler, see W. Koehler, *Mentality of Apes* (London: Kegan Paul, 1924).
32 The question of legal (as opposed to moral) rights is, of course, a different matter. Legislators have in fact given legal rights to entities that are not moral agents. But legal rights are normally extended on the basis of human interests, and not on the basis of some putative status independent of human interests.
33 This tension is clear in Leopold's 1933 essay "The Conservation Ethic," where he writes of ethics as possibly a kind of "advanced social instinct." Whereas Leopold seems to think that there is already an ethics at work at the level of the aesthetic, Dewey would have argued that the primitive aesthetic response furnishes a platform for working out an ethical response to non-human nature. In the next paragraph, however, Leopold takes another tack. He suggests that the ethical dimension of the human relation to the land is still in the formative stage, and that "science cannot escape its part in forming them." Dewey would have argued that if a robust land ethic is to be developed at all, then science will have to play a part. See Flader and Callicott, op. cit., p. 182.
34 See Hickman and Porter, op. cit., pp. 68–71.
35 Dewey (1965), p. 208.
36 See William R. Jordan III, "'Sunflower Forest': Ecological Restoration as the Basis for a New Environmental Paradigm," in *Beyond Preservation: Restoring and Inventing Landscapes*, ed. A. Dwight Baldwin, Jr., Judith De Luce and Carl Pletsch (Minneapolis: University of Minnesota Press, 1994), pp. 17–34. Note that Dewey would have rejected the critique of restoration advanced by Eric Katz and others who have argued that the only real nature is nature that is "permitted to be free, to pursue its own independent course of development." Katz thus valorizes an extreme version of what I have called nature-as-nature, or nature as undisturbed, and he deprecates nature-as-culture, which he treats as a kind of forgery. The problem with this view, from Dewey's point of view, is that nature-as-undisturbed is also nature-as-unknown, since knowledge involves experimental interaction and therefore some measure of "disturbance." I have already addressed the epistemic fallacy that is committed by such views of nature-as-transcendent. The problem is this: what can human beings know of the values of non-human nature with which they, by definition, have had no contact? See

Eric Katz, "The Big Lie: Human Restoration of Nature," *Research in Philosophy and Technology*, Vol. 12, 1992, p. 239.

37 Jordan, op. cit., p. 21.
38 Ibid., p. 24.
39 Ibid. p. 24.
40 Ibid. p. 31.
41 Aldo Leopold, *A Sand County Almanac* (New York: Oxford University Press, 1966), p. 251.

4

THE ENVIRONMENTAL VALUE IN G. H. MEAD'S COSMOLOGY[1]

Ari Santas

"Take any demand, however slight, which any creature, however weak, may make. Ought it not, for its own sole sake to be satisfied? If not, prove why not."

William James[2]

Life is precarious for all living beings, especially for humans, as they find themselves clawing out an existence without the benefit of claws. As long as humans have struggled to survive, they have worked together to insulate themselves from the forces of nature. Although it is true that all beings need some form of insulation from the elements, human beings, being conscious of that need, have the (presumably) distinct capacity to conceive of themselves as separate from the environment in which they live. Such a conception is no doubt important and useful, but it has lent itself to attitudes of antagonism which have created problems of enormous proportions. In the West, such a separation has taken the form of a dualism: a conception of the universe which reifies the difference between humans, *as human*, and the rest of the world. Mind, on such a conception, is one thing, body another; culture is one thing, nature another; humans live by one set of laws, the rest of creation by another.[3] Although dualism may not be the only metaphysical culprit to be identified in the fight against environmental degradation, it is a culprit with which environmentalism needs to contend.[4]

There is clearly work to be done in the area of environmental science and philosophy. Philosophy of nature still tends to maintain the classic dualisms given to us by the Greeks. Our tendency is to see not one, but two distinct worlds – *Nomos* and *Physis* – and to see two fundamentally different kinds of value – intrinsic and extrinsic. We find such dualisms even in environmentalism, a field which is

73

presumably grounded in the life sciences. On questions of value, for instance, we still find a reliance on the Aristotelian distinction between mere means and ends-in-themselves, a distinction which presupposes a metaphysics that has long since been rejected.[5] What is needed is a fundamental shift in worldview, a rethinking of nature and our place therein. What we need, in short, is a philosophy of nature which is in step with the findings and teachings of modern science.

The pragmatic philosophy of George Herbert Mead may seem to be an unlikely starting place for this fundamental shift in worldview. Mead was not an environmentalist, not even a proponent of animal rights.[6] Largely forgotten as a philosopher, he is mostly associated with sociology. Beyond his influence on the school of symbolic interactionism, he is remembered mostly as someone who is nearly impossible to read. Though his reputation as a speaker and teacher was one of brilliance, his writing leaves little trace of the dynamism he is reputed to have had.[7] Nevertheless, there is a great wealth of truth and wisdom in Mead's work, not only in his social psychology, but in his philosophy of nature. From his analysis of the act of reflection, to his treatment of emergence, to his musings on Einsteinian relativity, Mead shows a tremendous depth of understanding in both science and philosophy (which, to his mind, were the same thing). Like so many important thinkers, the beauty of Mead's ideas is under the surface, hidden beneath a tough layer of technical terminology.

There are two aspects of Mead's philosophy of nature that are particularly useful for the present purposes: his conception of the connection between the various living beings, from amoebas to humans, on the one hand, and his analysis of the relation between organism and environment on the other. Together, these aspects of Mead's cosmology should help to break down the radical dualisms between man and nature, mind and body, subject and object, man and "beast." My focus in this work will be Mead's views on the relationships between humans, their environment, and the other organisms within this shared world.

An understanding of Mead's philosophy of nature will generate two results. First, out of Mead's conception of nature will emerge a coherent account of the existence of environmental values, one which avoids the intrinsic/extrinsic dualism and makes ascriptions of environmental value scientifically understandable. Second, Mead's philosophy can provide a general pragmatic framework for environmental ethics, one which would remove the existing tendency to hypostatize distinctions into ontological dualisms and replace it with an understanding of both the differences and continuities between the entities that make up our world.

MEAD'S COSMOLOGY

In the essay entitled "Mechanical and Teleological Objects,"[8] Mead outlines a continuum of existence which ranges from inanimate objects to unicellular organisms, then from unicellular to the multicellular organisms, and then to human beings. Purely mechanical objects, says Mead, are entirely determined by their environment. Having no goals or aims, no need or desire to be fulfilled, they simply react to some precedent action. A rock falls into the water and the water splashes. The relationship is a simple one of cause and effect. Similar is the case for the most simple forms of life. Unicellular organisms resemble mechanical objects in that they interact directly with their surroundings. As nutrients pass through permeable membranes they incorporate themselves into the living form. Yet these organizations of matter are at least partially teleological (i.e. goal directed) in that their matter is organized so as to perpetuate their existence. That is, they select their environment by taking in what they need and expelling what they don't.

This feature of organisms is even more clear in higher forms of life. Multicellular organisms are clearly teleological. Not only do they select their environment to meet their needs, but their responses to the stimuli of their environment are indirect, i.e. they do not follow immediately from environmental stimuli but as a result of complex internal processes. In the case of the multicellular organism, there is no simple and direct passage of nutrients into the organism: they are neither acquired nor prepared for assimilation by the cells requiring nutrition. There is a higher level of organization which makes for a less direct relationship between the interior cells and the larger environment. Muscle cells in need of food, for instance, must communicate that need to the digestive system, which churns and gurgles and pangs so that a desire for consumption emerges. The resulting search for food, if successful, begins another process of digestion and assimilation, and so on. Such organization, for Mead, is a form of micro-society, though the constituent members have no consciousness of their activity.[9] For all their complexity and sophistication, however, multicellular organisms still operate on a level of stimulus and response, since in the absence of self-consciousness, all activity is merely reactionary. There is a stimulus which calls out a response. The organism, ultimately, is reacting to the given stimulus.

The human organism, by contrast, does not simply react to a given stimulus. For Mead, human beings have reached an even higher level of organization. Owing to their capacity for self-consciousness and meaning, they are capable of controlling their stimulus and hence controlling their environment. First, because humans are self-conscious, they are capable of conceiving themselves as something struggling to survive over and against other beings and external forces. They abstract the set of organized strivings that they are from the total set of interactions and strivings. We humans see ourselves, that is, as entities separate from our environment. It is for this reason that people can conceive of pests such as the ants in their yards as a "them" to be exterminated. Had we no self-consciousness, we would only be able to respond to things on the occasions on which they stimulated us to act.

Second, by virtue of this power of abstraction, humans are capable of assigning different meanings to the events in their perception. Depending on the meaning assigned, the response will vary.[10] An ant mound may stimulate us to act in a variety of ways, depending on how we view it. My neighbor sees in ant mounds invaders of his property;[11] ecologists see something quite different. As the perception varies, so does the response. To take another example, if a dish of food is placed before someone, and that someone, though hungry, knows that it is intended as a test of willpower, it's likely that he won't eat it; but if a hungry dog is given such a dish she will not search for a significance, a way of looking at it: she will only sniff for edibility before she eats. We may choose, Mead would argue, to respond to the "test stimulus" by attending to that character of the event, but dogs cannot.

Notice that although humans can be distinguished from other life forms by virtue of this capacity for self-reflection, there are no grounds here for conceiving of humans as ontologically different from the rest of creation. Thinking, on this account, is not some metaphysical quality given to humans from on high; it is a natural product of evolution.[12] Mind, according to Mead, is a form of consciousness emerging out of the peculiar conditions of the human environment.[13]

Whether the life form is human or not, the relationship it holds with its environment is a dynamic one, having a much greater degree of complexity than we are apt to presume. In the essay "Form and

76

Environment,"[14] Mead explains that a living form, human or otherwise, must not be construed in complete abstraction from its environment. Although it is useful for many purposes to consider an organism as an entity distinct from the elements surrounding it, strictly speaking, no such distinction exists. First of all, there is always a flow of material in and out of the membranes of the organism as it reconstructs and replenishes itself with the stuff of the "outside world." In unicellular organisms such as amoebas, the flow is direct and easy to understand; in the multicellular the flow is less direct and correspondingly more difficult to keep in mind. Ants, cats, frogs and lizards all breathe in and out, eat and excrete. In so doing they recreate themselves and their environment. Moreover, what they create themselves out of is their environment and that with which they recreate their environment is themselves.

Second, the organism is part of what *constitutes* its environment in the obvious sense that it is a member of an ecosystem. An ecosystem, after all, is but a series of relations between constituent members. An environment, accordingly, must include the living things which interact within it. Part of most people's environment (like it or not) is the ants in their yard, and vice versa. The ants, were they capable of self-consciousness, would be just as inclined to be rid of us as many of us are of them.

Third, an organism is part of its own environment in so far as its activities determine future states of those surroundings in which it acts, so that its present environment is a function of its past activities and its future is a function of the present. Every life form determines its environment in this way, though it is clear that larger and more complex beings often have greater determining power than their less complex counterparts. Aerobic bacteria take gas from the atmosphere; beavers build dams; humans clear forests and spray pesticides. All this is simply to say that organisms have an impact on their environment, and that impact has a role to play in the organism's future character.

In this interactive process between living form and environment, the lines separating organism from environment blur into areas of interchange, of reciprocal adaptation and readaptation. Unlike the case of purely mechanical interaction (where one event simply determines the next), the interaction between organism and environment is one that is truly *interactive* or dynamic. It is here that one may draw the distinction between mechanism and life form – along the lines of the difference in their modes of interaction. The processes of the living, therefore, are irreducible to those of the non-living. It would be a mistake, however, to attribute to Mead a *dualistic* conception of living vs. non-living. For Mead, the living emerged out of the non-living, just as molecules emerged from atoms and conscious beings from the non-conscious.[15] Although there are distinctions to be drawn between the various beings in the world, distinctions by themselves are not ontological chasms. Ironically, in the case of mechanical and teleological interaction, it is this very difference between the interactive modes of these objects which constitutes the blurring of the lines of separation between form and environment.

Putting the various forms of life on a scale from the simple to the more complex, we find varying levels of both determining power and dynamism on the part of the organism. The greater the complexity and organization, the greater the determining power. In the case of human beings, the power has taken on frightening proportions. Mead remarks in *Mind, Self and Society* that what is striking about the human organism is the level of control that it commands over its environment. He writes:

> The human form establishes its own home where it wishes; builds cities; brings its water from great distances; establishes the vegetation which shall grow about it; determines the animals that will exist; gets into that struggle which is now going on with insect life, determining what insects will continue to live; is attempting to determine what micro-organisms shall remain in its environment.[16]

Written in the early part of this century, this passage anticipates some of our current dilemmas. The fact that we have such control is a great part of our ecological problems, although it can be argued that our control is ultimately a false control, one that put us in a precariously fragile position on the planet. Mead can be criticized here for a certain amount of *naïveté*, but his analysis shows great insight, none the less, into the purpose behind our conduct: we are, in the end, organisms seeking control. What appears today is that much of the control we have been taught to accept as necessary and good is illusory.

PRAGMATIC ENVIRONMENTALISM

Mead's philosophy of nature is not simply a descriptive environmental philosophy; it provides a framework for developing an environmental ethic by making room for a coherent conception of environmental values. Unlike traditional theories which construe values as either metaphysical properties of things (attached to them somehow by divine design) or as the mere preferences of the human animal, Mead's view conceives of values as functional properties which emerged along with the needs and wants of living things. In a brief moment of eloquence, Mead writes: "Even the scientifically tough-minded men must recognize that values have arisen in the universe as genuinely as molecules or galaxies or the tides of the seas."[17] Values, for Mead, are intrinsic[18] properties of things in the world. With the emergence of living things emerge functional needs; values are nothing more than potential or actual fulfillments of those needs. Grass, for instance, came to have the emergent property of value as soon as there were animals who used it for food. The dung of those animals, by the same token, came to have value in so far as it provided nutrition for plants. Value, on this view, is a triadic relation between organism, the thing it needs, and the specific function which that thing fulfills.[19] Given such a conception, it is a mistake to relegate values to the realm of human subjectivity. Mead continues:

> [Values] belong, to be sure, to the perspectives of human society, but there is no aspect of the universe that is not a perspective, and the tough-minded scientist is the last person who can deny that the human perspective belongs to the universe of science, for his toughness consists in denying any transcendence of man above his habitat.

The subjectivity of value judgments, therefore, is no greater in principle than that of any other set of judgments. Like John Dewey,[20] Mead believes that values can be postulated to exist and tested with the same methodological rigor as any other postulated qualities.

There are further conclusions to be drawn from Mead's philosophy of nature, religious and ethical. A religious attitude, according to Mead, is the extension of the self beyond the ordinary boundaries of ethics in the moral community. "[T]he attitude [of religion] has to be one that carries this particular extension of the social attitude to the universe at large."[21] Getting people to adopt this attitude, of course, is no small task – having them identify with large groups of human beings is difficult enough. Even Mead himself saw the consideration of non-human interests as going beyond the call of duty; yet effecting such an identification is not an insurmountable obstacle. There was a time when the "civilized" world saw the members of tribal societies (not to mention members of other races) as literally subhuman. These others were seen as radically other, their resemblance to ourselves being hidden by our ignorance and fear. Although there are remaining rags and remnants of this attitude, our basic worldview no longer ignores the continuity between all human beings.

The problem – perhaps the main problem – with which environmentalism has to deal is getting people to see the connections between themselves and the rest of the world. It is here, I think, that Mead's pragmatism is particularly useful. Focusing on the connectedness outlined in Mead's cosmology, one can create a basis for what Mead called a religious (and we should call an ethical) attitude towards our environment. A self-conscious feeling of connectedness, of being at home in the world – so long as it is tempered with a respect for the processes which comprise and sustain the home – has for a long time been seen as a condition of religiosity and the apogee of moral life. The time has come, perhaps, given the current state of the globe, to stop conceiving of such a feeling as supererogatory, and treat it as a subject of moral education.[22]

The ironic fact of our ecological crisis is that self-consciousness – the precondition for this feeling of connectedness – is precisely what makes it possible for us to perceive ourselves as radically distinct from our surroundings. Mead's model explains the irony: because we are self-conscious, we are capable of conceiving of ourselves as distinct beings; and as we develop this capacity of distinguishing, we run the risk of confusing our abstractions with ontological categories. Yet Mead also explains that the very capacity which makes us liable to this error is what makes the greatest understanding of our true place possible. *Conscious* activity is a higher level of connectedness. When the awareness of such connection is greatly heightened, a feeling of exhilaration ensues. We may even deem it a religious experience, though the connotations of this expression are too complex to introduce here. Suffice it to say that so far as a moral agent endeavors to find her true place in the nature of things, she is in need of cosmology; and in as much as environmental ethics concerns itself with seeking a place in a larger whole, Mead's cosmology, which emphasizes continuity between the various elements of the world, lends itself to the environmental point of view. Mead's cosmology is certainly not a panacea, but it gives us a pragmatic framework for effecting environmental reform. At the very least, his philosophy provides a foundation for a sorely needed shift in attitude.

NOTES

1 An earlier version of this paper was delivered to the 1994 North American Interdisciplinary Wilderness Conference, Weber State University, Ogden, Utah.

2 From William James, *Essays in Pragmatism*, Alburey Castell, ed. (New York: Hafner, 1948), p. 73.

3 See, for example, Lynn White's essay, "The Historical Roots of Our Ecologic Crisis," *Science*, Vol. 155, No. 3767 (March 1967), pp. 1203–1207, for a discussion of the role of dualism in environmental degradation.

4 Reductionism (which arguably is a child of dualism) can also lead to environmental recklessness. Just as one may be abusive to those with whom one shares no commonality, she can also rationalize degradation of another by claiming the other to be merely a part of his or her self – to be used in whatever way is seen fit. Complete identification can be just as problematic as complete isolation. Reductionism, which lends itself to identification, can be as much a source of rationalized selfishness as dualism. (I say "lends itself" because there is no necessary connection between reductionism and identification, any more than there is an infallible link between ontological dualisms and environmental degradation. Nevertheless, there are points of view which make the rationalizations flow more freely. Both dualism and reductionism, to my mind, encourage the sort of simple-mindedness to which foolish persons continually submit.)

5 Aristotle's distinction between intrinsic and extrinsic goods can be found in the *Nicomachean Ethics*, Bk I. Keeping in mind that he identified "The Good" with "that towards which all things aim," we can see the connection of the doctrine of ends-in-themselves (for him the same as intrinsic goods) to that of final causes and his teleological metaphysics. Dewey traces this connection in Chapter III of *Reconstruction in Philosophy* (Boston: Beacon, 1948), and in the chapter of *Human Nature and Conduct* (New York: Henry Holt, 1922) entitled "The Nature of Aims."

6 Mead writes: "We put personalities into animals, but they do not belong to them; and ultimately we realize that those animals have no rights. We are at liberty to cut off their lives; there is no wrong committed when an animal's life is taken away." His reasoning is that since they have no self-consciousness, they have neither a past nor a future to be taken away. "They have not that future and past which gives them, so to speak, any rights as such." *Mind, Self and Society*, Charles Morris, ed. (Chicago: University of Chicago, 1934), (hereafter, MSS) p. 183. Mead is following Kant's moral rationalism here, which makes reason the fundamental moral criterion. Had he followed Bentham, Mill and James and adopted sentience as his criterion instead, his cosmology might easily have led him to a different position on the moral status of non-human animals. For James' discussion of sentience and obligation, see his essay "The Moral Philosopher and the Moral Life," in William James, *Essays in Pragmatism*, pp. 65–87.

81

7 I am reminded of a story of an aspiring graduate student proposing to his major professor that he translate Mead's *MSS* into French, only to evoke the response "Hadn't you ought to first translate into English?!"

8 G. H. Mead, *Philosophy of the Act*, Charles Morris, ed. (Chicago: University of Chicago, 1938) (hereafter, *PA*), Essay XVII, pp. 301–307.

9 Such a conception is implicit in his discussion of want in the essay "Mechanical and Teleological Objects," in *PA*, pp. 301–307. See esp. pp. 304–305. Compare this to his discussion of human versus insect societies. (He argues that the main difference is that insect societies differentiate roles on a physiological basis, making its members more like physiological parts of an organism than individual members of a group.) See *MSS*, pp. 227–237. It also works the other way around: a society is a highly developed form of organism. When a society reaches *conscious* organization, there emerges a fifth dimension in the universe – the institution.

10 See *MSS*, Part II, esp. pp. 117–125.

11 My neighbor informs me almost daily as he spreads poison on the fire ant mounds in his yard that there is absolutely nothing valuable about ants. Our government, moreover, is cheating us, he says, by not allowing us to use stronger chemicals to get rid of them. As certain as I am that the global ecosystem could not survive without ants, I am even more certain that telling him so would have no effect on his convictions. Unfortunately, he is not alone in his views.

12 Mind, however, is not reducible to merely mechanical phenomena. In explaining his behavioristic point of view, Mead sharply criticizes Watson's reductionist behaviorism for failing to take note of the fact that thinking and all that it entails – attitudes, language, significance – exists as a dynamic functionality which cannot be explained away in mechanistic terms. See *MSS*, Part I, for a criticism of reductionist behaviorism.

13 See *MSS*, Part II.

14 Mead, *PA*, Essay XVIII, pp. 308–312.

15 For a concise discussion of emergence, see *PA*, pp. 640–643.

16 Mead, *MSS*, pp. 249–250.

17 Mead, *PA*, p. 495.

18 The word "intrinsic" may bring to mind the sort of metaphysics Dewey and Mead were trying to dismantle, but it should not. Dewey, who spent most of his career criticizing the doctrine of ends-in-themselves, did not argue against intrinsic value per se. All he meant by intrinsic value was something one prizes for its own sake. Traditional metaphysics offered a "grounding" for Intrinsic Value, while Dewey offered a means of evaluating an act of prizing by appeal to the means required to bring them about. (See his "Theory of Valuation," in *International Encyclopedia of Unified Science* (Chicago: University of Chicago, 1939), Vol. II, No. IV, esp. Secs. V and VI.) Intrinsic value for Dewey was no more absolute than instrumental value was "merely relative." Part of the point of *Art As Experience* (New York: Perigee, 1934), after all, was to show that we need to see the instrumental nature of intrinsic value and vice versa. What Mead's philosophy adds to Dewey's conception of

value is the idea that all values are intrinsic to (i.e. inseparable from) some functional system.

19 Mead never speaks of value as a triadic relation, but following C. S. Peirce, he identifies meaning as such. A sign, according to both Mead and Peirce, is a triadic relation between the signifier, the interpreter and the thing signified. A significant symbol, that is, means some*thing* to some*one*. For Mead's specific version of this general pragmatic conception of meaning, see *MSS*, pp. 75–82. For Peirce's discussion, see "The Peirce/Welby Correspondence," in C. Hardwick, ed., *Semiotic and Significs*, Oct. 12, 1904, pp. 22–36. See also "Logic as Semiotic: The Theory of Signs," in Justus Buchler, ed. *Philosophical Writings of Peirce* (New York: Dover, 1955), pp. 98–119.

20 For Dewey's discussion, see "Theory of Valuation."

21 Mead, *MSS*, p. 275. See also p. 289.

22 I shudder to think of the totalitarian sound of this sentence and anticipate the label "environmental fascism." I do not intend here a program of rigorous inculcation, pledges of allegiance to Mother Nature and rote repetition of some environmental creed. Besides being overly intrusive on individuality, such measures wouldn't work. What I have in mind is the sort of educational environment espoused by Dewey in *Democracy and Education* (New York: Macmillan, 1916). The whole point of education, for him, was to get the pupil to see and understand her connection to the world in which she lives. There is no reason why the study of our "natural" environment should not be part of that process.

5

THE CONSTANCY OF LEOPOLD'S LAND ETHIC

Bryan G. Norton

In 1920 Aldo Leopold enthusiastically described his predator eradication program. He had formed a coalition of sportsmen and stockmen to eliminate wolves, mountain lions and other large predators from Arizona and New Mexico: "But the last one must be caught before the job can be called fully successful," he said.[1] Twenty-four years later Leopold repented his war on wolves in a graceful and humble essay, "Thinking Like a Mountain," which was drafted in 1944 and published in *A Sand County Almanac and Sketches Here and There*. What happened in the meantime?

It is tempting to believe that, during this period, Leopold discovered his revolutionary land ethic, that his thinking underwent a profound religious-metaphysical-moral change, and that his about-face on predator control programs was a direct result of this profound philosophical conversion. Since Leopold was acting, in 1920, as a representative of the US Forest Service, which remained under the philosophical domination of Gifford Pinchot's humanistic utilitarianism, this interpretation sees Leopold as later rejecting utilitarian management because he came to espouse "a right to exist" for all members of the land community.[2]

This essay seeks to show that Leopold's intellectual odyssey during this period was more complex than this straightforward account would suggest. In particular, Leopold had embraced the main philosophical elements of his land ethic early in his career, even while he was advocating predator eradication. These main elements include important influences, hitherto unnoticed, derived from American pragmatism, a philosophical approach that Leopold borrowed from Arthur Twining Hadley who was president of Yale University when Leopold was a student there. Leopold never abandoned the main elements of his early philosophy, and I will argue that his shift from predator eradication to predator protection was motivated, not by a shift in religious, metaphysical or moral views, but rather by a recognition that scientific knowledge is inadequate to guide gross manipulations of ecosystems and by an increasingly pessimistic view of the prospects of environmental management. I conclude that, while Leopold was fascinated by organicism and its metaphysical and moral implications, these abstract views had little direct impact on his managerial style.

I

In 1923 Leopold drafted an essay, "Some Fundamentals of Conservation in the Southwest."[3] He included, as a final section, some brief remarks that he called, "Conservation as a Moral Issue." This essay remained unpublished until thirty years after Leopold's death; commentators have treated the final section as an immature draft of Leopold's conservation ethic, and some have suggested that Leopold later abandoned significant elements of the philosophy expressed there.[4]

It cannot be denied that Leopold's brief 1923 discussion of conservation morality is confusing. In only three pages Leopold stated that "economic determinism" is insufficient to understand land conservation; invoked the prophet Ezekiel; considered the Russian organicist philosopher P. D. Ouspensky's view that the world is a "living thing" with "a soul, or consciousness"; questioned whether the world "was made for man's use, or has man merely the privilege of temporarily possessing an earth made for inscrutable purposes"; approvingly quoted John Muir on the rights of rattlesnakes but decided "I will not dispute the point"; and finally concluded that we have an obligation to future generations to prove ourselves "capable of inhabiting the earth without defiling it." Along the way, he admitted that most scientists and laymen hold "anthropomorphic" views.[5] He also considered the effect that Ouspensky's organicism would have on "most men of affairs" and observed that for them "this reason is too intangible to accept or reject as a guide to human conduct."[6]

It is difficult to see, at first glance, a unifying principle in this densely packed presentation of so many grand ideas. It seems clear that, after introducing non-anthropocentric ideas, Leopold opted in the end for a conservation ethic based on our obligations to future generations of humans – a forward-looking anthropocentrism. But the reasoning by which he shifted from his prior discussion of organicism and non-anthropocentrism to long-sighted anthropo-centrism is so compact as to nearly defy understanding. Fortunately, a clue to Leopold's thinking appears in a parenthetical comment, which is embedded in his discussion of our obligations to future generations: "How happy a definition is that one of Hadley's which states, Truth is that which prevails in the long run."

Since the essay was never published, Leopold prepared no notes or list of references. This definition, however, clearly derives from the American pragmatists.[7] Arthur Twining Hadley, a child prodigy in Greek, graduated at the head of his class at Yale in 1876. He studied political economy at the University of Berlin and returned to become a tutor and later a professor at his undergraduate institution. Noted for the breadth of his knowledge, his classes on economics and political ethics were extremely popular. Hadley became the first lay president of Yale in 1899.[8] He described himself as a "thoroughgoing pragmatist"[9] and generally quoted William James' work as representative of modern philosophical thinking.

In his most thoroughly philosophical book, *Some Influences in Modern Philosophic Thought*, Hadley said: "The criterion which shows whether a thing is right or wrong is its permanence. Survival is not merely the characteristic of right; it is the test of right."[10] These views he characterized as the views of pragmatists and in the next paragraph he discussed James' view, which he stated as "We hold the beliefs which have preserved our fathers" and accepted this view while changing its emphasis somewhat:

> I do not mean that we should consciously adopt a belief because it is useful to us, as James seems to imply. I would rather take the ground that we hold the belief that has preserved our fathers as an intuition and act on it as an instinct.[11]

Leopold began "Conservation as a Moral Issue" with a quotation from Ezekiel:

> Seemeth it a small thing unto you to have fed upon good pasture, but you must tread down the residue of your pasture? And to have drunk of the clear water, but ye must foul the residue with your feet?[12]

When Leopold invoked Ezekiel, he was invoking the "beliefs that have preserved our fathers." As understood by Hadley, the pragmatists' notion of truth amounted to a recommendation that we respect the wisdom of our ancestors. Hadley said:

86

The moral and religious instincts that bind the group together, which some men, not so many years ago, were condemning as outworn prejudices, count for even more than individual intelligence. In our practical philosophy, of politics and of life, we are reverting to the words of Edmund Burke: "We are afraid to put men to live and trade each on his own private stock of reason, because we suspect that this stock in each man is small, and that the individuals would do better to avail themselves of the general bank and capital of nations and of ages."[13]

These ideas (due to Hadley and the pragmatists) provide the key to understanding an important passage in Leopold's "Conservation as a Moral Issue." Leopold said, "Possibly, in our intuitive perceptions, which may be truer than our science and less impeded by our words than our philosophies, we realize the indivisibility of the earth." This brief passage shows a connection between three important and related ideas in Leopold's thought. He was referring to Hadley's conception of "truth" or "rightness" of social practices interpreted as the "intuitive perceptions" we have inherited from Ezekiel and others who have counseled protection of resources. Second, he implied that our science is fallible, and perhaps less reliable than these intuitive perceptions. Finally, he suggested that our intuitions are more helpful than philosophies, because the latter are "impeded by language."

In a discussion of Darwin's influence on history and political science, Hadley explained that Darwin's idea of survival of the fittest was readily accepted by historians, who had long recognized that survival of a culture depended on the adaptability of its institutions to situation.[14] He criticized Herbert Spencer and others for trying to apply the Darwinian criterion over too brief time periods and to individual behaviors.[15] Hadley concluded: "It is the institution even more than the man that has been marked out for survival by the process of natural selection."[16]

I have found no evidence that Leopold read or studied the American pragmatists in detail, although he no doubt knew about the publication of Hadley's book, as it was given a major review in the *Yale Review*.[17] Apparently, Leopold read this magazine regularly.[18] Further, Leopold was clearly familiar with, and impressed by, Hadley's conception of truth, because he referred to it on several occasions in the journals he kept during this early period of his career.[19] It may, at this point, be impossible to determine the extent of Leopold's study of Hadley, but it is clear that Leopold absorbed and applied the basic elements of Hadley's approach to evaluating cultures.

When Leopold mentioned intuitive perceptions, Ezekiel's admonition to treat the land with respect, he was invoking Hadley's intuitions of our fathers. In this same passage Leopold compared this intuition favorably with both science and philosophy. Since Leopold was here striving for a very large-scale understanding, a kind of worldview, to support his conservation goals, it is important to examine the treatment he gave both philosophy and science in this large perspective.

Leopold is obviously wary of philosophical pronouncements. In only three pages "Conservation as a Moral Issue" contains no less than five cautions against the "pitfalls of language," and three of them are related to doubts about the efficacy of philosophical ideas. While on first reading these passages may appear as expressions of humility and little more, I believe they also provide a clue to the philosophical approach underlying Leopold's conservation ethic. These references to language derive from the linguistic pluralism of Ouspensky's *Tertium Organum*, which Leopold here blended with American pragmatism as a justification for a long sighted anthropocentrism to support his conservation ethic.

In his discussion of Ouspensky, Leopold characterized two "conceptions of the earth." There is a "mechanistic conception of the earth as our physical provider and abiding place." This conception was opposed to another: the world is a living organism and the "soil, mountains, rivers, atmosphere, etc. [are] organs, or parts of organs, of a coordinated whole, each part with a definite function." On this view of the world, which "many of the world's most penetrating minds" (he cited Ouspensky) have found compelling, the earth has "a soul or consciousness." It begins even to make sense to respect it as a living thing and to relate to it morally. In this context of competing worldviews, Leopold's references to the importance of language appear more significant:

> There is not much discrepancy, except in language, between this conception of a living earth, and the conception of a dead earth, with enormously slow, intricate, and interrelated functions among its parts, as given by physics, chemistry, and geology.[20]

A similar form of conceptual pluralism, also linked with cautions about the inadequacy of language, is a major theme of Ouspensky's *Tertium Organum*.[21]

I am suggesting that, in "Conservation as a Moral Issue," Leopold amalgamated ideas from the American pragmatists with the organicism of Ouspensky, concluding that the most effective conservation ethic represents a concern that we pass on, to future generations, a world not despoiled by our current activities. The relationships between these ideas become more explicit if one distinguishes several levels of discourse. When Leopold compared different conceptions of the earth, he was speaking of what might be called second-order beliefs. The facts about the world around us, the facts of physics, chemistry, and geology, are first-order beliefs about the way the world is. Organicism and mechanism, two alternative conceptions of the world in Leopold's terminology, are second-order beliefs about how to interpret the first-order facts of the particular sciences. Leopold was arguing that organicism and mechanism can accept the same first-order facts about the world and that the choice between these two interpretations is mainly a difference of language. "The essential thing for present purposes is that both admit the interdependent functions of the elements."[22] Thus, the choice of a second-order interpretation of scientific facts, the choice between organicism and mechanism as alternative conceptions of the world, is essentially a linguistic choice of how to conceptualize this interrelatedness.

Leopold placed so much emphasis on the pitfalls of language because he believed that empirical data from the natural sciences will never determine our second-order, linguistic choices as to how to conceptualize that data. These questions depend on human perception, not reality: "The very words living thing have an inherited and arbitrary meaning derived not from reality, but from human perceptions of human affairs."[23] Further, he recognized that the choice of a conception of the world, an interpretation of the data, will be closely related to the way we think and act. Here we see the amalgamation of Ouspensky's organicism with Hadley's pragmatism. It is a central idea of American pragmatism that linguistic forms depend upon perception and perception depends in turn upon human affairs. What we do determines what we say and think as much as vice versa. Leopold combined pragmatism with Ouspensky's organicism and arrived at a form of homegrown perspectivism, a view that metaphysical conceptions of the world are projections of human perceptions which depend, in turn, upon cultural practices.

From this amalgamation, Leopold concluded that different cultures with radically different practices will have different characteristic vocabularies and therefore different conceptions of the world.

But since these different conceptions amount only to different conceptualizations of the same hard, scientific data, Leopold was cautious whenever he compared ideas from one conception with those from another. Leopold thus shied away from metaphysical and theological pronouncements: "It is just barely possible that God himself likes to hear birds sing and see flowers grow. But here again we encounter the insufficiency of words as symbols for reality." Similarly, his concern that philosophical and theological ideas are artifacts of conceptions of the world led Leopold to express a deep distrust of pronouncements about the "truth" of anthropocentrism or non-anthropocentrism:

> Probably many of us who have neither the time nor the ability to reason out conclusions on such matters by logical processes have felt intuitively that there existed between man and the earth a closer and deeper relation than would follow the mechanistic conception of the earth. . . . Of course, in discussing such matters we are beset on all sides with the pitfalls of language.[24]

So far, we have seen that Leopold understood Ouspensky's organicism and modern, atomistic science as two alternative conceptions of the world, the choice between which is a second-order choice among vocabularies for presenting first-order data, and unlikely to be fully decided by that data. At the same time, he recognized that the choice between these conceptions has profound consequences for the way we treat the earth. On a mechanistic worldview, the earth is dead and we would have no moral concern for it. This viewpoint has led to "economic determinism," which Leopold disparagingly characterized as the "language of compound interest." On the organicist view, the earth is alive and worthy of our respect and moral concern. But the choice between the two conceptions of the world is mainly a linguistic choice, underdetermined by the scientific data available.

This combination of views, without supplementation, would result in a radical relativism: we either speak as mechanistic scientists or as organicists and, depending on this arbitrary choice, we will either be economic determinists or we will react morally to the land. It is at this crucial juncture that Leopold introduced Hadley's definition of truth as that which prevails in the long run. Following Hadley, Leopold chose not to dispute the metaphysical and theological issues of anthropocentrism, but rather to rely on the intuitive perceptions of the tradition – Ezekiel. Or, to put the point in more

Hadleian terms, the test of rightness of cultural practices is their long-term survivability. Hadley's pragmatic definition of truth therefore functioned, in Leopold's early philosophy, as a third-order principle, as a means to judge second-order conceptions of the world and to provide a criterion for distinguishing acceptable cultural practices from unacceptable ones.

This was surely an attractive idea to Leopold for several reasons. First, it provided a unification of his philosophical thought with his biological belief in the Darwinian principle of "survival of the fittest." Second, it allowed him to relate, in a larger perspective, scientific knowledge with his speculations about Ouspensky's organicism – these are two differing linguistic approaches to the same "earth." Third, it provided him an ideal platform from which to denounce the practices, described in detail in the first two sections of "Some Fundamentals of Conservation in the Southwest," which he believed were destroying the land. Our treatment of the land is wrong because it is not sustainable, as Ezekiel said.

Immediately after he decided not to dispute anthropocentrism, Leopold granted "that the earth is for man." He then said, "there is still a question: what man? The cliff dwellers, the Pueblos, the Spaniards, and now the Americans all believed the earth was their possession." But the prior cultures "left the earth alive, undamaged." If we are "logically anthropomorphic," he said, we must consider what the next civilization will say of us. If there is, indeed, "a special nobility inherent in the human race – a special cosmic value, distinctive from and superior to all other life," then it must manifest itself in "a society decently respectful of its own and all other life, capable of inhabiting the earth without defiling it." If we do not manifest that nobility, we shall "be judged in the derisive silence of eternity."[25] Leopold was, ultimately, basing his argument for conservation on the fact that the land in the Southwest was going through a series of less and less productive cycles owing mainly to overgrazing. This factual basis, outlined in detail in earlier sections of the draft article, seemed to him an adequate basis for conservation when combined with Hadley's pragmatic definition of truth as survivability. This argument could be made independent of appeals to non-anthropocentrism, which is a second-order belief not possible to establish conclusively. By applying Hadley's definition of truth to the cultural practices that were causing deterioration of the land in the Southwest, Leopold could sidestep the issue of anthropocentrism and declare those practices "false." He relied not upon Ouspensky (in spite of his obviously

91

deep attraction to organicism) but upon Ezekiel buttressed by Hadley's definition of truth.

Leopold's philosophical pragmatism also provides an explanation of his comment that "most men of affairs" will find organicism and other forms of non-anthropocentrism "too intangible to either accept or reject as a guide to human conduct."[26] Leopold recognized that both modern science and traditional Judaeo-Christian religion are anthropocentric in their conception of the world and, given his admission that conceptions of the world involve unresolvable differences in linguistic forms, he recognized that he would be unsuccessful in preaching non-anthropocentrism to them, at least in the short run. He resolved, instead, to argue for a long sighted anthropocentrism based on the intuitive perceptions of Ezekiel: we ought not to "tread down the residue of [our] pasture." Leopold therefore resolved, early in his career, to enter the policy arena armed only with arguments based on long sighted anthropocentrism, rather than basing his moral strictures on non-anthropocentrism.

II

Leopold's approach to environmental management underwent a profound change between 1920 and 1944. The question I am posing is whether this change resulted from a change in his religious-metaphysical-moral views, or whether the change was motivated by expanding scientific information and hands-on management experience. Having sketched a unified philosophy underlying Leopold's thinking in 1923, it is now possible to ask whether he changed this philosophy in subsequent years. How significantly do Leopold's views, as expressed in the final version of "The Land Ethic" (which dates from 1947), differ from his early views as stated in "Conservation as a Moral Issue"?

"The Land Ethic" begins with the story of Odysseus' arbitrary hanging of a dozen slave-girls on mere suspicion of misbehavior, a historical example of the changing nature of moral judgments. This example, which was first introduced in 1933, evokes his earlier view that moral ideas are tied closely to changing patterns of behavior and to the conceptions of thought associated with them.[27] These, in turn, are likely to differ across epochs. Subsequent developments have seen the extension of moral concepts to human individuals previously treated as mere property. This much-discussed example stands as an analogy for the eventual emergence of a full-blown land ethic: "The land-relation is still strictly economic, entailing privileges but not obligations." The emergence of a land ethic

> is actually a process in ecological evolution. An ethic, ecologically, is a limitation on freedom of action in the struggle for existence. An ethic, philosophically, is a differentiation of social from anti-social conduct. These are two definitions of the same thing.[28]

Here we see a later representation of Leopold's view that there are multiple conceptions of the world (here represented as two different fields of study, philosophy and ecology) and that these conceptions, which are associated with differing practices, are winnowed by the forces of competition and evolution. An ethic "has its origin in the tendency of interdependent individuals or groups to evolve modes of cooperation. The original free-for-all competition has been replaced, in part, by cooperative mechanisms with an ethical content."[29]

In 1923 Leopold was drawn to organicism and the ethic associated with it, but he was wary of trying to establish these views rationally because an ethic is so tied up with its characteristic vocabulary and world conceptions. An ecologically sensitive conception of the world will emerge only gradually as we learn, in practice, the extent of our interdependence. In the meantime we must rely on the intuitive wisdom of Ezekiel and our forefathers. In the 1947 version, Leopold was looking forward, predicting that, in the face of greater recognition of our mutual dependencies with other species, we will eventually develop an ethic consonant with the ecological conception of the world. Assuming we survive at all, we will have discovered a worldview that is adapted to the modern world, a new set of Hadleian intuitions that will promote survival: "Ethics are possibly a kind of community instinct-in-the making."[30] Leopold's temporary reliance on the traditional strictures of Ezekiel, as well as his faith that a new ethical age will dawn (if we do not destroy ourselves first) are both manifestations of the pluralistic view of world conceptions and the pragmatic conception of truth as survivability.

Near the end of "The Land Ethic" Leopold discussed "Land-Health and the A–B Cleavage":

> Conservationists are notorious for their dissensions. Superficially these seem to add up to mere confusion but a more careful scrutiny reveals a single plane of cleavage common to many specialized fields. In each field one group (A) regards the

land as soil, and its function as commodity-production; another group (B) regards the land as a biota, and its function as something broader.

Here, again, we see the conception of a "dead earth" contrasted with that of a "living earth" and Leopold understood the dissensions among conservationists as based, ultimately, on the acceptance of one or the other of these conceptions. References to Ouspensky do not occur, but Leopold's words indicate that he was still fascinated with second-order systems that he called in 1923 "conceptions of the earth." "In all of these cleavages, we see repeated the same basic paradoxes: man the conqueror versus man the biotic citizen; science the sharpener of his sword versus science the search-light on his universe; land the slave and servant versus land the collective organism."[31]

"The Land Ethic" differs from "Conservation as a Moral Issue" in lacking the latter's numerous cautions about language and its pitfalls. This is indicative of a deeper shift in Leopold's strategy – he apparently decided not to emphasize philosophical theory (which he saw as raising issues that can be settled only in the slowheating crucible of evolutionary selection), but to rely instead on his vast experience as an environmental manager. But the pragmatism and conceptual relativity are still there in muted form, as evidenced by the references, quoted above, to different fields (ethics and ecology) as using alternative vocabularies to describe similar processes and in the discussion of the A–B cleavage.

Thus, while Leopold's philosophical scaffolding is less explicit in his later work, traces of his theoretical commitments remain. The strategic decision to emphasize his experiences as a forester and wildlife manager is deeply pragmatic in spirit. Theory, according to the pragmatist, must ultimately be tested against experience.

Leopold was admittedly tentative in his 1923 pronouncements, perhaps because he had to rely on the philosophical and linguistically relative concepts that he cautioned against. As he matured, gaining experience and replacing philosophical speculation with knowledge of ecological science and the consequences of management strategies, he relied more confidently on his managerial experience.[32] If, therefore, we use the term "philosophy" to refer to a basic worldview including a metaphysics and value system, we can conclude that Leopold acted, throughout this period, against the backdrop of a consistent and unified philosophical approach. If, however, we use the term "philosophy" more broadly, in a sense in which one might say that Leopold had a "philosophy of environmental management," his philosophy did indeed change. Why did Leopold, acting against a backdrop of unchanging metaphysical and moral worldviews, change his approach to predator control and other management strategies? How could Leopold, in the early 1920s, have approved organicism while eradicating wolves and mountain lions?

There is a ready explanation for Leopold's early attempts at predator eradication: trained in the Pinchot mold of utilitarian forestry, Leopold set out to maximize resources for human use.[33] When he became interested in game management in 1915, he simply transferred forestry management practices to a new resource, fish and game. If a little game is valuable, more is correspondingly so. Predators compete for game, so he set out to eliminate them, as well as to enforce hunting laws and to stock streams – all were management methodologies designed to maximize sport resources. What requires an explanation is not Leopold's use of these management practices (which were assumed as part of his job), but his doing so while approving organicism and questioning the adequacy of "economic determinism." Leopold already believed in this early stage of his career that the health of the land community is important; but he had not yet realized that all species are necessary to promote that health. He believed that resource managers, provided they are scientific in their management practices, can manipulate populations and he seems to have believed that predator control would enhance the overall health and productivity of biological systems.

With this view as a starting point, Leopold's views of management began a slow evolution towards less radical interventions. By 1925 he believed that wolves and mountain lions contributed to the diversity of an area and he retreated from a goal of eradication to one of control.[34] When Leopold met Charles Elton, who had initiated the transformation of ecology from a purely descriptive to a more functionally oriented science with the publication of *Animal Ecology* in 1927, he integrated these ideas into his own 1933 text, *Game Management*.[35] Defining "management" as "the coordination of science and use," Leopold stated that "the central thesis of game management is this: game can be restored by the creative use of the same tools which have heretofore destroyed it – axe, plow, cow, fire, and gun."[36]

In the 1933 essay "The Conservation Ethic" Leopold noted that there was emerging a new and more positive approach to conservation of the land and living things. The means to the goal of protection was biological research:

The duty of the individual is to apply its findings to the land
. . . [because] the soil and plant succession are recognized as
the basic variables which determine plant and animal life and
likewise the quality of human satisfactions.[37]

Leopold applied this strategy to species preservation, arguing that
species become rare and extinct because their habitats have shrunk.
He asks:

Can such shrinkage be controlled? Yes, once the specifications
are known. How known? Through ecological research. How
controlled? By modifying the environment with those same
tools and skills already used in agriculture and forestry.[38]

In 1933, then, Leopold optimistically believed that ecological
research would usher in a new era of plenty based on a positive
program of environmental management.

Given, then, the knowledge and the desire, this idea of
controlled wild culture or "management" can be applied not
only to quail and trout, but to any living thing from bloodroots
to Bell's vireos.[39]

He as yet saw no inherent contradiction between conservation and
intense management for production of resources. Given ecological
knowledge, populations could be managed for the good of humans.
The reduction of predators would be unproblematic: human hunters
would simply absorb the ecological function of wolves and mountain
lions. While accepting organicism, he did not yet see that organicism
implied a goal of saving all species. He believed that, given sufficient
knowledge and sensitivity in management, the living organism (the
land) could be kept thriving, while some of its less desirable organs
were removed. Faith in ecological technique, therefore, shielded him
from the conclusion that destruction of wolves and mountain lions
would cause serious illness in the organic system of nature.

By 1939, however, Leopold's view of the role of ecology had
changed drastically. He still believed that ecology "is the new fusion
point for all the natural sciences," but its results were not those he
had hoped for:

The emergence of ecology has placed the economic biologist
in a peculiar dilemma: with one hand he points to the accumu-
lated findings of his search for utility, or lack of utility, in this

or that species; with the other he lifts the veil from a biota so complex, so conditioned by interwoven cooperations and competitions, that no man can say where utility begins or ends . . . the old categories of "useful" and "harmful" have validity only as conditioned by time, place, circumstance. The only sure conclusion is that the biota as a whole is useful, and biota includes not only plants and animals, but soils and waters as well.[40]

Leopold concluded that human management activities do not successfully mimic nature in the creation of habitats because "evolutionary changes are slow and local," while human use of tools "has enabled him to make changes of unprecedented violence, rapidity, and scope." Unforeseen changes in species composition result as "larger predators are lopped off the cap of the pyramid." The effects of such changes "are seldom foreseen; they represent unpredicted and often untraceable readjustments in the structure."[41]

Leopold was now regretting his war on wolves. His predator control program and the restrictions on hunting he began to enforce in 1915 in the South-west had resulted in huge, but starving, herds of deer. The deer overbrowsed the land and caused yet one more downward turn in the cycle of succession. Weedy, brushy species replaced more useful trees and shrubs, and the diversity of the area diminished.[42]

The 1939 essay "A Biotic View of Land" was in effect a retrospective view of what Leopold had learned as a forester and wildlife manager. He referred to the German experience in tree farming: "Thus the Germans, who taught the world to plant trees like cabbages, have scrapped their own teachings and gone back to mixed woods of native species."[43] Similarly, he came out definitively against predator control as a "highly artificial (i.e. violent)" method of management.[44] Additionally, he had observed the extent and seriousness of the dust bowl phenomenon, of what pervasive "economic management" could do to fragile ecosystems.

Leopold therefore changed his views of management because:

In short, economic biology assumed that the biotic function and economic utility of a species was partly known and the rest could shortly be found out. That assumption no longer holds good; the process of finding out added new questions faster than new answers. The function of species is largely inscrutable, and may remain so.[45]

We need not posit any shift in Leopold's metaphysical or moral views to explain the changes in his views on wildlife management. He learned through practice that "violent" methods of management and control are inappropriate because they also cause unforeseen effects and damage the biotic community. This is an insight that was implicit in his belief in the importance of ecology; but it was obscured by his initial faith that ecology would teach us enough about ecological interactions among species to allow manipulation of populations for utilitarian purposes. He underestimated the complexity of systems and overestimated our ability to control them; he consequently failed to see that predator protection was one of the principles implied by the holistic approach that he advocated in opposition to the economic determinism he rejected. In the face of practical evidence, the pest problems of monocultural forestry, and deer starving on overgrazed reserves, Leopold eventually adopted a less violent and disruptive approach towards management.

III

Should Leopold be classed as an anthropocentrist? Yes and no.

Yes, in the sense that he believed that, for better or worse, humans must and should manage the natural world. Given that, and current attitudes in society, arguments based on the good of the human species will carry more weight in policy debates. After summarizing the structure of the land pyramid and describing the land as an "energy circuit," Leopold summarized his land ethic in three basic ideas:

> (1) That land is not merely soil. (2) That the native plants and animals kept the energy circuit open; others may not. (3) That manmade changes are of a different order than evolutionary changes, and have effects more comprehensive than is intended or foreseen. These ideas, collectively, raise two basic issues: Can the land adjust itself to the new order? Can the desired alterations be accomplished with less violence?[46]

This is anthropocentrism of sorts. Leopold accepted that humans will alter the biota. Their management will be successful if they protect life and if the human race survives. It will be a failure if the human race, "like . . . John Burrough's potato bug, which exterminated the potato, . . . exterminates itself."[47] Leopold never questioned the right of humans to alter nature, provided these alterations were consistent with ecological knowledge and would protect, in the long run, human life and the living land on which it depends.

But Leopold regarded both anthropocentrism and its denial as representing only human conceptions, as artifacts of human perceptions rather than reality. To try to determine the truth of these speculative pronouncements without reference to the systems in which they are embodied is to go beyond the possibilities of language. And yet there is a legitimate sense in which Leopold was a non-anthropocentrist. He saw organicism as an alternative to mechanism, one that would carry with it a deeper, even moral, reaction to the land. This view, while it can hardly be expressed in our current vocabulary, is true in the pragmatic sense – it has survival value. Leopold's dream that someday our culture will evolve a more sensitive reaction to the land explains the central role he always placed, as an environmental professional, on developing public perception. He believed that, as Americans become more aware of their interdependence with the rest of the biotic world, they will gradually develop a new conception of the world, including a moral reaction to the community of life. This development will, he thought, improve the survival chances of our culture and he therefore devoted his career to improving the perception of the American public.[48]

Leopold's actual target was not anthropocentrism, however. He concluded that non-anthropocentrism raises issues too intractable to make it useful in management discussions. Instead, he attacked shortsighted economic reasoning that ignores the scientific evidence that intense management often leads to gradual decline in productive systems.[49] Leopold recognized, in the degeneration of vegetative systems in the South-west, in German forestry, and in the dust bowl phenomenon, the inadequacy of management practices based solely on Pinchot's utilitarian criterion. The search for profit, "economic determinism," leads inevitably to an undervaluing of future resources. But shortsighted, destructive practices are wrong even if we are logically anthropocentric. Anthropocentrism itself should imply a concern for future generations.

Leopold acted upon what I call the "convergence hypothesis."[50] According to that hypothesis, the interests of humans and the interests of nature differ only in the short run. If we recognize the extent to which the human species is an integral part of the community of life, long-term human interests coincide with the "interests" of nature. To protect the fullness of life is to protect the far-distant future of the human species and its evolutionary successors; and vice versa. Since the survival of our culture depends upon the survival of the ecosystems on which we, in turn, depend, the conception of the world one adopts is less important than the longsightedness with which it is applied in environmental management.

ACKNOWLEDGMENT

"The Constancy of Leopold's Land Ethic" originally appeared in *Conservation Biology* Vol. 2, No. 1 (1988). I gratefully acknowledge the help of Curt Meine, J. Baird Callicott, Sara Ebenreck and two very helpful anonymous reviewers who read and commented on earlier drafts of this paper.

NOTES

1 S. L. Flader, *Thinking Like a Mountain* (Lincoln, Nebraska: University of Nebraska Press, 1974), p. 3.
2 J. Petulla, *American Environmentalism: Values, Tactics, Priorities* (College Station, Texas: Texas A & M University Press, 1980), pp. 16, 20.
3 A. Leopold, "Some Fundamentals of Conservation in the Southwest," *Environmental Ethics* 8 (1979) 195–220.
4 S. L. Flader, "Leopold's 'Some Fundamentals of Conservation': A Commentary," *Environmental Ethics* 1 (1979): 143–144; J. B. Callicott, "The Conceptual Foundations of the Land Ethic," in *A Companion to a Sand County Almanac: Interpretive and Critical Essays*, ed. J. B. Callicott (Madison, Wisconsin: University of Wisconsin Press, 1987); H. Rolston, III, "Duties to Ecosystems," in *A Companion to a Sand County Almanac: Interpretive and Critical Essays*, ed. J. B. Callicott (Madison, Wisconsin: University of Wisconsin Press, 1987).
5 Leopold used the term "anthropomorphic" as we currently use "anthropocentric" to refer to a value system that bases all value in human motives. Except in quotations of Leopold, I will follow current practice and use "anthropocentric."
6 Leopold, "Some Fundamentals of Conservation in the Southwest," pp. 138–141.
7 C. S. Peirce, "Pragmatism in Retrospect: A Last Formulation" in *The Philosophy of Peirce*, ed. J. Buchler (New York: AMS Press, Inc., 1978), p. 288.
8 M. Hadley, *Arthur Twining Hadley* (New Haven, Connecticut: Yale University Press, 1948).
9 Ibid., p. 197.
10 Pragmatists described their conception of truth in several ways. James

tended to emphasize the "effectiveness" of true ideas. (See, for example, W. James, "Pragmatism's Conception of Truth," in *Essay in Pragmatism*, ed. A. Castell (New York: Hafner Publishing Company, 1948), p. 162.) Peirce emphasized that truth is that which will last through an indefinite number of experiments and actions. He said that truth is "the predestined result to which sufficient inquiry *would* lead" (Pierce, "Pragmatism in Retrospect: A Last Formulation," p. 288). The reader should not be disconcerted by Leopold's shift from Hadley's discussion of "right" to a definition of "truth" – the pragmatists drew no sharp distinction between facts and value and therefore treated "truth" and "right" as largely interchangeable. Leopold's application of "truth" to cultural practices would be acceptable to pragmatists such as Hadley.

11 A. T. Hadley, *Some Influences on Modern Philosophy* (New Haven, Connecticut: Yale University Press, 1913), p. 73.
12 Leopold, "Some Fundamentals of Conservation in the Southwest," p. 138.
13 Hadley, *Some Influences on Modern Philosophy*, p. 75.
14 Ibid., pp. 121–126.
15 Ibid., p. 130.
16 Ibid., p. 127.
17 S. P. Sherman, "Review of Some Influences in Modern Philosophy Thought," *Yale Review* 3: 383–385.
18 Curt Meine, who has just completed a biography of Leopold based on the collection of Leopold's papers, informs me that he found evidence that Leopold read *Yale Review* regularly.
19 Again, Meine is my source of information here.
20 Leopold, "Some Fundamentals of Conservation in the Southwest," pp. 139–140.
21 P. D. Ouspensky, *Tertium Organum* (New York: Alfred Knopf, 1968), p. 222.
22 Leopold, "Some Fundamentals of Conservation in the Southwest," pp. 139–140.
23 Ibid., p. 139.
24 Ibid., p. 139.
25 Ibid., p. 141.
26 Ibid., p. 140.
27 A. Leopold, "The Conservation Ethic," *Journal of Forestry* 31 (1933): 634–643.
28 A. Leopold, A *Sand County Almanac and Sketches Here and There* (Oxford: Oxford University Press, 1949), pp. 201–203.
29 Ibid., p. 202.
30 Ibid., p. 203.
31 Ibid., pp. 221–223.
32 Flader, *Thinking Like a Mountain*, p. 18.
33 Ibid., p. 25.
34 Ibid., p. 154.
35 Ibid., pp. 24–25.
36 Ibid, p. 25, quoted from A. Leopold, *Game Management* (New York: Scribner Publishing, 1933).

37 Leopold, "The Conservation Ethic," p. 641.
38 Ibid.
39 Ibid.
40 A. Leopold, "A Biotic View of Land," *Journal of Forestry* 37 (1939): 727.
41 Ibid., p.728.
42 Flader, *Thinking Like a Mountain,* p. 117.
43 Leopold, "A Biotic View of Land," p. 730; see also Flader, *Thinking Like a Mountain,* p. 139.
44 Leopold, "A Biotic View of Land," p. 729.
45 Ibid., p. 727.
46 Leopold, *A Sand County Almanac and Sketches Here and There,* p. 218.
47 Leopold, "Some Fundamentals of Conservation in the Southwest," p. 141.
48 For a fuller discussion of the central role of "transformative values" in environmental ethics, see B. Norton, *Why Preserve Natural Variety?* (Princeton, New Jersey: Princeton University Press, 1987), Chapter 10.
49 For a discussion of Leopold's use of the term "economic," see B. Norton, "Conservation and Preservation: a Conceptual Rehabilitation," *Environmental Ethics* 8 (1986): 208.
50 B. Norton, *Toward Unity Among Environmentalists* (New York: Oxford University Press, 1991).

Part 2

PRAGMATIST THEORY AND ENVIRONMENTAL PHILOSPHY

6

INTEGRATION OR REDUCTION

Two approaches to environmental values

Bryan G. Norton

INTRODUCTION: THE ROLE OF ENVIRONMENTAL ETHICISTS IN POLICY PROCESS

Environmental ethics has been dominated in its first twenty years by questions of axiology, as practitioners have mainly searched for a small set of coherent principles to guide environmental action. In axiological studies, a premium is placed on the systematization of moral intuitions, which is achieved when all moral judgments are shown to be derivable from a few central principles. The goal of these studies is to propose and defend a set of first principles that is (1) *complete* in the sense that this small set of principles can generate a single correct answer for every moral quandary and (2) *jointly justifiable* in the sense that, once the principles are warranted, then every particular moral directive derived from the principles must also be warranted.

The limiting case of axiological simplification is *moral monism*, the view that a single principle suffices to support a uniquely correct moral judgment in every situation.[1] Monism represents to some philosophers an ideal because, provided the adopted principle is self-consistent, problems of coherence and consistency are resolved once and for all – there is no need to worry about what to do if two principles imply differing actions in a given situation, no worry that there will be irresoluble conflicts among competing and equally worthy moral claims. This reasoning motivates the drive towards unification.[2] The goal of environmental ethics as a discipline, in keeping with this ideal, has most centrally been to offer a unified and monistic account of our moral obligations. The adoption of this goal is what has given environmental ethics its axiological character.

105

What is curious is that this axiological approach rests on an assumption that is common to both sides in what has become a polarized debate: both neoclassical welfare economists – who believe that all value is expressible in units of individual, human welfare – and advocates of attributing inherent value to non-humans – who argue that the moral force of environmental principles derive from the moral considerability of natural objects – are unyieldingly monistic in their approaches. The adoption of the monistic view-point and the associated goal of developing a universal moral theory applicable in all cases is inevitably "reductionistic." Because all values, which are experienced in multiple modes and contexts, must on the monistic approach be accounted for under a single theory, the basic strategy must be to reduce all moral concerns to a unified analytic vernacular in which solutions to specific moral quandaries are generated, by unavoidable inferences, from a single theory.

This shared assumption of monism has, I believe, locked environmental ethicists into a paralyzing dilemma, a dilemma that lies at the heart of most discussions of environmental values. Most participants in these discussions have subscribed to a crucial alternation in the theory of environmental valuation: either the value of nature is entirely instrumental to human objectives, or elements of nature have a "good of their own" – value not dependent on human valuations.[3] Could it be that the polarized thinking that paralyzes environmental policy today results from false alternatives forced upon us by the assumption, unquestioned by neoclassical economists and by most of their opponents among environmental ethicists alike, that whatever the units of environmental value turn out to be, there will be only one kind of them?

The thesis of this paper is that the goal of seeking a unified, monistic theory of environmental ethics represents a misguided mission, a mission that was formulated under a set of epistemological and moral assumptions that harks back to Descartes and Newton. An assessment of the contribution of environmental ethics to environmental policy in its first two decades is accordingly bleak. The search for a "Holy Grail" of unified theory in environmental values has not progressed towards any consensus regarding what inherent value in nature is, what objects have it, or what it means to have such value. Nor have environmental ethicists been able to offer useful practical advice by providing clear management directives regarding difficult and controversial problems in environmental planning and management.[4] One very practical effect of the monistic assumption is that the range of topics open for discussion in environmental ethics has been narrowed, and opportunities for building bridges with other, more practice-oriented disiplines have been lost. Another effect has been to define an often unhelpful role for environmental ethicists in environmental policy debates.

In order to emphasize these practical implications of the issues raised in this paper, I complete this introduction by drawing a distinction between "applied" and "practical" philosophy as representing two somewhat different roles for environmental ethicists in the process of developing and implementing environmental policy. After this practice-oriented introduction, the remainder of the paper falls into two parts, one destructive and critical, and the other positive and speculative. Part 1 illustrates the problems of formulating a monistic environmental ethic by exploring the evolution of the monistic, ecocentric theory of J. Baird Callicott. This retrospective of Callicott's position, and a criticism of his current position, provide reasons to be very skeptical both of Callicott's specific monistic theory and also of his mission as he understands it.[5] As counterpoint to this negative argument, I briefly present a pluralist conception of the role and possible content of an environmental ethic in Part 2. Influenced by a pragmatist attitude towards social problems, I will sketch an environmental ethic that applies multiple principles, but one which seeks integration of these principles in a way that is sensitive to place orientation and to temporal and spatial scales.

Let me explain my basic methodology and advocate the importance of practice by reference to a distinction between two kinds of non-theoretical philosophy: I call them "applied" philosophy and "practical" philosophy, despite the fact that these terms are sometimes used interchangeably. I use them here to correspond to two somewhat different roles for philosophers in the process of public policy formation. Applied philosophy refers to the application of general philosophical principles in adjudications among policy goals and options. Applied philosophy's method is usually to develop very general and abstract principles and then to illustrate their use by discussing a few, carefully circumscribed hypothetical cases. This conception of the role of environmental ethicists has encouraged the confinement of philosophers, in their day-to-day work, within their traditional academic roles of teaching and writing. The actual applications of these principles is usually left to others such as environmental managers or environmental groups.[6] Moral monism and applied philosophy are naturally complementary – a single principle agreed upon by all disputants provides just the sort of moral guidance that applied philosophers would like to give. They want to furnish a universal principle from which actual decision-makers can derive moral directives and then apply them to the cases they encounter in the day-to-day process of setting policy. Since the universal principle functions as an essential premise in an argument that one or another policy is justified, agreement on a policy option will emerge only if the general principle is accepted by all parties to the dispute.[7] Philosophers' contributions, given the role envisaged by applied philosophers, can only be as strong and decisive as the case for one universal principle. If some disputants do not accept the unitary moral principle proposed by applied philosophers, or if applied philosophers cannot agree among themselves regarding the formulation of the universal principle, they must retreat to theoretical arguments and attempt to establish more definitively the universal, monistic principle/premise before returning to applications. I therefore turn to a discussion of the assumptions that shape environmental ethicists' view of what they can offer in the policy process.

Practical philosophy, as I am defining it here in contrast to applied philosophy, is more problem-oriented; its chief characteristic is an emphasis on theories as tools of the understanding, tools that are developed to resolve specific policy controversies. It shares with applied philosophy the goal of contributing to problem solution; but practical philosophy does not assume that useful theoretical principles will be developed and established independent of the policy process and then applied within that process. It works towards theoretical principles by struggling with real cases, appealing to less sweeping rules of thumb that can be argued to be appropriate in a particular context, rather than establishing a universal theory and "applying" it to real cases. Practice is prior to theory in the sense that principles are ultimately generated from practice, not vice versa.

I do not mean to claim that theorizing is worthless; on the contrary, theory-building that addresses real-world problems, in the spirit of John Dewey and Aldo Leopold – the forester-philosopher – is absolutely essential if the environmental movement is to develop a vision for the future.[8] In the meantime, however, theoretical differences often need not impede progress in developing current policy; if all disputants agree on central management principles, even without agreeing on ultimate values, management can proceed on these principles.[9] And philosophers have a lot to offer policy-makers in specific, complex situations in which they face many conflicting moral directives, even though it has proved impossible for them to deliver the Holy Grail of monism as promised.

What sets practical philosophy in contrast to applied philosophy is the differing practices and impacts they envision for philosophers in the processes of policy articulation, evaluation and implementation. Not surprisingly, such deep differences in conceiving the role of environmental ethics are associated with differences of philosophical theory. This paper explores the philosophical beliefs and assumptions that shape the thinking of avowed applied philosophers.

PART 1: MONISM AND THE MISSION OF ENVIRONMENTAL ETHICS

Having noted that moral monism and applied philosophy tend to go together, I now explain the reasons that have convinced me that the search for a monistic ethic is intellectually, as well as practically, misguided in a profound sense. Monists are not simply wrong in that they have not yet proposed the correct universal principle, or because they have not quite successfully specified the precise boundaries of moral considerability in nature. No, I believe that the entire project of shoehorning all of our obligations regarding other humans and nature into a "monistic" system of analysis is the wrong strategy at the wrong time, given that it allows decisive intervention in public policy formation only after a single, unified moral principle is articulated and agreed upon, an outcome that seems unlikely in the foreseeable future.

As noted above, monism is embraced by both environmental ethicists and economic theorists – both are equally "reductionistic" in this sense. In this paper I will focus on the dominant form of monism in environmental ethics – the large and diverse collection of theories that assert nature has value independent of humans in some sense.[10] It may not at first be clear how such theories achieve monism, so I begin by tracing briefly the development of the idea of human-independent value in the writings of J. Baird Callicott. Callicott provides an excellent case study for several reasons. First, his position claims less than other non-anthropocentric theories in the sense that Callicott does not assert that human-independent values in nature are independent of human consciousness; he claims only that value in nature is independent of human *valuations*. So criticisms of this moderate view may apply equally to theorists who defend a more

109

radical independence of values in nature.[11] Second, Callicott has experimented with several different versions, or at least formulations, over the past fifteen years, and has readily noted shifts in his own thinking. By tracing Callicott's changing formulations of inherent value we can better understand the dynamic of a complex argument. Finally, Callicott has explicitly embraced monism, explained explicitly the sense in which he considers himself a monist and criticized pluralistic alternatives, providing us with greater insight into the nature and implications of monism as a moral mission.[12]

As a preliminary to this brief historical examination, it must be noted that Callicott and his monistic colleagues never question an underlying conceptual assumption, an assumption that lies at the heart of the assumed mission of applied environmental ethics: success in the axiology of environmental ethics must include as a centerpiece an answer to the question of *moral standing*: "What beings are morally considerable?" Given the project of applied philosophy, it is not surprising that non-anthropocentrists believe that, whatever monistic principle or theory turns out to be the correct one, this principle will fulfill two conditions: (1) The principle/theory must specify what objects in nature are morally considerable. Interestingly, success in this specification has been identified with the task of identifying which objects in nature "own" their own inherent worth, of which more below. (2) The principle/theory must also provide some *motivation* for moral beings to protect natural objects. The universal, underlying principle is that moral individuals act to protect inherent value, wherever it is determined to reside. Condition (1) ensures environmental activists can identify which objects deserve moral consideration in any given situation. Condition (2) ensures that goals to protect inherent value are invested with moral gravity.[13] Any morally committed environmentalist ought always to act so as to maximize the protection of inherent value, wherever it occurs. Monistic, non-anthropocentric theory can on these conditions rival economists in universalism. This monistic principle is attractive to philosophers who hope to resolve environmental problems by throwing fully formed, general principles over the edge of the ivory tower, to be used as intellectual armaments by the currently outgunned environmental activists, to aid them against the economic philistines in the political street wars that determine the fate of natural environments.

Callicott's dilemma

This heroic version of applied philosophy's role in the policy process can only be realized if environmental ethicists, laboring in the tower, can agree on which principles to throw down to the streetfighters. And, if these principles are to exert moral force to protect the environment, they must be "objectively" supportable.

The measure of objectivity, on Callicott's view, is the extent to which the central theory of environmental values succeeds in attributing human-independent value to natural objects themselves. In the words of Callicott, the blue whale and the Bridger wilderness "may therefore, be said in a quite definite, straightforward sense to own inherent value, that is to be valued *for themselves*." Callicott goes on in the same paragraph to state that the institution of a "genuine" environmental ethic – one that recognizes inherent value in nature – provides the only defensible basis for the environmentalists' platform of social reforms: "Environmental policy decisions, because they may thus be based upon a genuine environmental ethic, may thus be rescued from reduction to cost-benefit analyses in which valued natural aesthetic, religious, and epistemic experiences are shadow priced and weighed against the usually overwhelming material and economic benefits of development and exploitation."[14] In this passage, Callicott commits himself to a good, old-fashioned realist interpretation of the problem of objectivity: "the very sense of the hypothesis that inherent or intrinsic value [sic: exists?] in nature seems to be that value *inheres* in natural objects as an intrinsic characteristic. To assert that something is inherently or intrinsically valuable seems, indeed, to entail that its value is objective."[15] What is interesting is that Callicott, having formulated the problem of objectivity in terms of representational realism, immediately retreats from asserting an objectivist solution. Instead he argues that his own Humean subjectivist solution asserts as much objectivity for claims that objects in nature own their inherent values as exists for scientific claims. So, Callicott's "ownership" theory of inherent value, which attributes to ecosystems their own inherent value, is offered to environmental activists as the fruits of his search for the Holy Grail of monistic ecocentrism.

Three comments are necessary. First, the general principle of "ecocentrism," so defined, hardly resolves the question of what beings in nature are proper owners of inherent value. The Bridger Wilderness and the blue whale are given as examples, but they themselves represent different scales on the biological hierarchy, and Callicott owes his readers an account of the breadth to which he would generalize these examples. Second, as long as the first comment remains unanswered, Callicott cannot claim to have provided any definitive policy direction to activists because they can only know what they are obliged under the universal principles of non-anthropocentrism to protect after they know what particular entities in nature have inherent value. Third, Callicott, who wishes to interpret the land ethic as a moral theory, betrays an underlying commitment to moral individualism. He interprets Leopold's holism as attributing inherent value to ecosystems *as individuals* who can "own" their own goodness. But this conclusion only brings us to the heart of the matter – Callicott's original assumption that the land ethic is to be interpreted as monistic and holistic. It will therefore be necessary to look briefly at Callicott's changing definitions of ecocentric holism, and to question whether they express the kernel ideas of Leopold's land ethic.

In an important 1980 article, Callicott established himself as the leading interpreter and proponent of Aldo Leopold's land ethic, and also caused an important alteration in the intellectual terrain on which the principles of environmental ethics were to be debated in the subsequent decade.[16] Callicott showed that, if one took Leopold's holistic pronouncements and arguments seriously, the land ethic was logically incompatible with extensionist ethics of animal rights advocates whose individualistic ethics are based in utilitarianism or rights theory.

Callicott's initial interpretation of Leopold's land ethic was therefore boldly holistic and strongly non-anthropocentric. He argued that ecological communities, not individuals, are the real locus of values in nature (as we have learned by drawing out the metaphysical implications of ecology) and that individuals have value in so far as they contribute to ecosystematic processes that support the community, leaving the clear implication that protecting inherent value in nature might involve sacrificing individual specimens of *any* species – including, presumably, human individuals – if those individuals threaten "ecological integrity" of the biotic community. Not surprisingly, his position was attacked as brutal towards individual animals and apparently misanthropic. In particular, it was pointed out by critics that this reduction of all individual value to functional value in a larger whole smacks of fascism.[17]

Subsequently, admitting that he had left his holism unqualified in order to be provocative, Callicott offered a much more conventional view of our moral obligations to human individuals, other species, and ecosystems.[18] But Callicott revised the apparent implications of the bold holism of the 1980 paper without so much as questioning his earlier conclusions that the land ethic establishes the whole biotic community as a morally considerable being. What Callicott did instead was simply to specify how we should rank our obligations to various objects that have inherent value, theoretically and practically, in accord with the "communitarian" principles of the land ethic, taking communitarianism to imply that humans and other elements of nature make up one moral community among others. Nonhuman elements of nature, including species and biotic communities, have inherent value because they are morally considerable "owners" of their own value as members, along with humans, in the land community. He therefore explains why we may give precedence to obligations to family members or our human community over obligations to ecosystems: "we are members of nested communities each of which has a different structure and therefore different moral requirements . . . I have obligations to my fellow citizens which I do not have to human beings in general *and* I have obligations to human beings in general which I do not have towards animals in general."[19] Building on this gradation of obligations, Callicott extricates himself from charges of fascism thus: "our holistic environmental obligations are not preemptive. We are still subject to all the other more particular and individually oriented duties to the members of our various more circumscribed and intimate communities. And since they are closer to home, they come first."[20]

It is significant that Callicott shifts the grounds of the debate over fascism from obligations flowing from attributions of inherent value – the central source of normative obligation in his monistic theory – to the origins of special obligations that emerge in specific communities in biology, culture and ecology; he differentiates the obligations according to the intimacy of the community. There can be multiple criteria of right action – one stringent criterion of protectionism applicable in cases of parents to children, and a less stringent moral criterion of protectionism that applies to the broader, ecological community. The commitment to moral monism recedes into the theoretical background as these special, community-based (and presumably not universal) obligations do the hard work of resolving conflicts that were introduced by generalizations of inherent value to species and ecosystems as well as individuals in human communities.

Callicott's 1980 formulation, which alarmed readers who wondered if persons and individual animals would be sacrificed to a single principle that apparently followed from his theory that all value originates in ecosystems, has given way more recently to an endorsement of theoretical monism rather than a monism of principles (as explained in note 1). Callicott therefore adopted a qualified version of holism that recognizes a plurality of rules applicable according to specific circumstances, explaining that pluralism on the level of principles of action is not inconsistent with monism on a more general, theoretical level in which various practical rules are unified and related to a single moral ontology.

Callicott's version of monism allows, he thinks, *both* unification under a single theory of value *and* flexibility in the formulation of rules. He claims a unified theory because he relates all obligations to a moral ontology in inherent value. Recognizing the subjective source of inherent value in human consciousness and in cultural ideas and institutions, however, the theory can nevertheless be elaborated in ways that are appropriate, depending on the special circumstances of the communities in which the obligations arose. But this solution to the charge of fascism – and one assumes Callicott must answer this charge if his theory is to be taken seriously – surely taxes the semantic elasticity of the concept of "inherent" value. Inherent value of natural objects, on this account, is due to "virtual" characteristics of objects of value, even though the specific, practical implications of the evaluations of these characteristics is ultimately determined by individual actors, according to the moral sensibilities of particular, independent, moral communities.

Ignoring these semantic difficulties, the upshot is that Callicott advocates allegiance to monistic inherentism in theory, but recognizes that the more intimate obligations of kinship and culture will usually outrank obligations to protect species and ecosystems. If this seems a capitulation to business as usual in environmental affairs, with inherent value reduced to a meaningless slogan, it must in fairness be said that Callicott faces a difficult and apparently destructive theoretical dilemma, for which he has thus far offered no resolution.[21] If moral inherentism is to provide the unified foundation for an environmental ethic, inherent value must be protected wherever it occurs. This apparently implies that any conflicts between a person's obligations to protect her children, for example, and her obligations to protect

some inherently valuable biotic community, should strictly be deter
mined by the obligation to maximize the protection of inherent value.
But this would leave the theory open to charges of fascism, unless
Callicott can prove that our obligations to persons could never, in
principle, conflict with our obligations to ecosystems. Such a proof
seems highly unlikely, so he chooses the other horn of the dilemma.
While we have obligations to ecosystems as owners of inherent value
and obligations to persons as owners of inherent value, the latter can
override the former because of the special circumstances of the moral
agent within the specific community in which those obligations arose.

But what are we to make of these auxiliary rules that resolve
disputes when the interests of inherently valuable and morally
considerable entities conflict? Can they, or can they not, be derived
from the central, monistic theory of ecocentrism? If they can, it
would seem that inherent value must come in grades, providing
objective resolution of conflicts in interest among inherently valuable
entities. But Callicott has never, to my knowledge, offered even a
sketch of the required theory of gradations of inherent value or an
explanation of how such gradations should inform our choices of
what protectionist goals to give priority in action. If, on the other
hand, these auxiliary rules cannot be derived from the central theory,
we apparently have uncovered a most important class of moral
quandaries that require we step outside Callicott's complete and
unified theory, negating any claim to monism and universality. Until
this dilemma has been resolved, it appears that Callicott's modified
monistic ecocentrism can tell us nothing about what we are morally
obligated to protect.

While this criticism of Callicott's adventures in ecocentric holism
has proceeded on a theoretical level, it is important to note
that Callicott's decision to interpret the land ethic as monistic and
ecocentric has had at least two important practical consequences for
the development of environmental ethics. Callicott's early comments
on the land ethic established the non-anthropocentric interpretation
of the land ethic as the standard interpretation of Leopold's mature
thought. The criterion has not, accordingly, been operationalized
because, on this standard interpretation it embodies all of the ambiguities of non-anthropocentrism, as just listed; worse, the criterion
has been used as a shibboleth by one side in the polarized situation
described above, rather than as the powerful practical guide it could
be, because this interpretation identifies the criterion with the
idiosyncratic views of a small subset of scientists and the public.

Second, Callicott and other non-anthropocentrists have used this interpretation to support a highly tendentious, and narrowing, definition of the field of study of environmental ethics. These two issues are discussed in the following two subsections.

Non-anthropocentrism and the land ethic

Callicott has interpreted what is perhaps the most important passage in the history of conservation thought, Leopold's famous "criterion" of right management, as monistic, holistic and non-anthropocentric in its philosophical commitments. Leopold said: "A thing is right when it tends to preserve the integrity, stability and beauty of the biotic community. It is wrong when it tends otherwise."[22] Callicott and the Deep Ecologists have taken the two sentences of Leopold's criterion to imply that the community or ecosystem is the *object of value* which conservationists should be attempting to protect. They have assumed, accordingly, that the ecosystem/community must, for Leopold, be *an object of value independent of human values.* This passage is read as an endorsement of the view that the ecological community has "integrity" understood as wholeness, *and therefore that ecosystems are moral subjects.* To be a moral patient, however, requires, given the objectivity requirement, non-human ownership. And ownership requires a moral subject. Under the influence of his commitment to ownership as the basis of moral considerability, and implicitly unliberated from the subject–object dichotomies of Cartesian dualism, Callicott reifies biotic communities as "moral subjects" who can own human-independent value. Callicott and his followers therefore interpret this passage as Leopold's definitive statement that communities themselves are loci of inherent, human-independent value that can be considered in competition with human values. To the extent that this has become the standard interpretation of Leopold's land ethic, environmental ethicists have encouraged environmentalists to understand this passage as an assertion of non-anthropocentrism.[23]

While I agree with Callicott's identification of "integrity" as the key concept of environmental ethics and management, we nevertheless differ strongly regarding how to interpret this conceptual centerpiece of the land ethic. Callicott believes that, by attributing integrity to the biotic community, *taken as a whole,* Leopold stepped across the line to non-anthropocentrism and declared his moral allegiance to the hypothesis that nature has inherent value. Our obligations to protect this integrity are "objective" in the sense that they originate in the integrity of whole agent/object which are morally considerable owners of their own value. Being constantly tempted to think of ecosystems as persons by the requirement that they are "owners" of their own value, it is natural also to think of them as objects capable of strategizing and having "interests" and "strategies" of their own.[24] Philosophers and ecologists, under the influence of this misguided, morally driven holism, have unfortunately failed to confront another ancient philosophical problem – how an organ can behave relatively independently, as an object in its own right, while at the same time being an organ that functions as a part of a larger organism. In philosophy, it can be called the problem of parts and wholes. In ecology, it can be called the problem of scalar dynamics. It is a problem that cannot get a hearing on the current assumption that the objects of moral attention in nature are necessarily "wholes."

Consider an alternative interpretation of Leopold's famous remark; consider it as a *practical* remark on the *proper focus of conservation management*, not as a philosophical statement of what objects in nature are of ultimate value. On this view, Leopold is in this passage summarizing his wisdom, drawing a broad, inductive generalization from the experiences of his long and varied career in environmental management, rather than asserting a moral, "first principle." On this reading Leopold was making the ontologically less committed, but none the less insightful point that, because of the complexity of the interrelationships in nature, and because there are so many different values exemplified in nature, the only way to manage to protect *all* of these diverse and pluralistic values is to protect the integrity of community processes (which supports and sustains the individuals and species of which it is composed). The latter two guidelines – stability and beauty – are then interpreted as glosses on, or specific criteria to employ – in our search for the sometimes elusive analogy of integrity. On this interpretation Leopold is not telling us *what to value* in nature, but rather telling us *what to protect* in our practical environmental management (given the diversity of values and scales involved).

On this view, then, Leopold is proposing an approach to management: he is defining right action in environmental use and management, rather than addressing the problem of standing. Further, he is strongly endorsing an integrative, systems approach to environmental management as the only way to encompass multifarious human goals as we manage a many-level, complex system which is our habitat. An advantage of this interpretation is that it unites the land ethic with Leopold's seminal managerial metaphor of "thinking like a mountain."[25] As I have argued elsewhere, the key idea behind the admonition to think like a mountain is the recognition of multi-scalar relationships in time and space and a prescription that citizens and environmental managers in a technological age must pay attention to longer-scale values embodied in the structures and processes of slow-changing systems, as well as the immediate and short-term values of economics.[26] But on my interpretation of Leopold, he is emphatically *not* committing himself to a moral ontology of inherent value; nor is he expressing fealty to a single, monistic principle or theory of value. I see Leopold as a moral pluralist who was struggling to integrate multiple values rather than as an axiomatic deducer of applications from some universal theory.

My interpretation, it can be argued, fits much better into the context in which the criterion is found in the essay, "The Land Ethic." In the section immediately preceding the one in which the criterion is stated, Leopold outlined a systematic practical difference, cutting across all fields of resource management, between Group A and Group B conservationists.[27] Group A conservationists are primarily concerned with commodity production, whereas Group B conservationists, among whom Leopold counted himself, regard "the land as a biota, and its function as something broader. How much broader is admittedly in a state of doubt and confusion." Now Leopold often used the term "broader" to apply to the philosophical aspects of a problem, so I interpret this passage as explicitly choosing *not* to embrace any particular moral ideology. Leopold confidently made the separation between the two activist groups but, faced with an opportunity to make a pronouncement on the philosophical principles of the Group B movement, Leopold deferred and turned humble, admitting the question of what their principles meant philosophically is "in a state of doubt and confusion." It would be rather odd, would it not, if he had changed moods in the very next section of the essay and explicitly endorsed a full-blown theory of inherent moral goodness which implies that ecological communities are morally considerable beings who own their own good?

The scope of environmental ethics

Non-anthropocentrists such as Callicott and Tom Regan, though they disagree strongly on what non-anthropocentric principle to apply, and which beings/subjects have it, agree that, if environmental ethics is to have a distinctive subject matter, the field must embrace some form of non-anthropocentrism. Regan asserts that, lacking a commitment to biocentrism, environmental ethics "collapses into an ethic for managing the environment [for human purposes]."[28] Remarkably, this definition excludes problems of environmental justice within generations and problems of intergenerational fairness from the discipline of environmental ethics. Callicott nevertheless quotes Regan approvingly and, seeing non-anthropocentrism as the only route to rationally defensible values, he accepts Regan's tendentious definition of environmental ethics as the search for a non-anthropocentrc ethic.[29] One suspects that Callicott's belief that the land ethic is non-anthropocentric must be at work in predisposing him to accept Regan's otherwise implausible definitional narrowing; one also suspects that this definition is encouraged by the unquestioned goal of monistic theory-building and by the mission of applied philosophy as it is understood by Callicott and other monists. Non-anthropocentrism is not, apparently, an empirical hypothesis; it is a principle that is established "philosophically" – independent of experience, because that is what is required if philosophers are to operate independently of real environmental problems and constraints to establish moral principles independent of management science.

Ontology and epistemology

Having examined a number of characteristics of applied philosophy by examining the ideas and practices of a leading proponent of that approach, and generalizing from them, I conclude this first part with one last look at the standard problematic of environmental ethics since its beginning as a recognized sub-discipline having a "distinctive" subject matter. Because of his prior commitment to monism, Callicott assumed that the discipline of environmental ethics must achieve two goals with a single theory. There is, of course, the problem of determining who, or what, has *moral standing*, as noted above. And there is the quite distinct problem of the *warranted assertibility* of environmentalists' pronouncements that certain policies should be instituted because they are morally required and trump mere preferences of consumers. His solution is to provide a realistic moral ontology in which there exist moral objects in nature as well as among human individuals, subjects capable of "owning"

119

inherent value. But whereas the first problem might (although I doubt it) be usefully addressed by an ontological theory of moral monism, the second problem is essentially a question of warranted assertibility of environmentalists' claims to priority in certain cases. This is an *epistemological* problem; we require no *ontological* solution to it. Once we avoid Callicott's conflation and problematize monism, there is no reason to consider it obligatory that our theory of value will be (1) realistic and (2) designed to provide *both* a theory of value *and* an epistemological warrant for assertions of priority.

The single ontological solution to two distinct problems seems plausible to Callicott, I submit, because he is operating on a fundamentally Cartesian conception of knowledge and reality, a conception in which epistemological justification requires location of a causal antecedent of perception and knowledge that can be located in reality, independent of human perception. He accordingly attempts to establish independent existence of environmental values by reifiying ecosystems and making them owners of value. His single solution to the two problems is to recognize Cartesian moral subjects in nature as the owners of objective moral characteristics.

It is especially puzzling that Callicott attempts to reify ecosystems in this way, given that he clearly rejects the description of ecosystems as tending towards an inevitable and stable context. Callicott brilliantly shows that the lesson from ecology is that ecosystems are multi-layered, dynamic processes, not self-like, organismic wholes that seek a stable equilibrium.[30] Is it even plausible to say that multi-layered dynamic processes are owners of inherent value?

We can now see why Callicott's ontological, monistic interpretation of the integrity criterion and of the land ethic has led to ambiguities and paralysis. If the land ethic must be morally monistic, then to fulfill its unifying role in the non-anthropocentrists' moral ontology, it must consider ecosystems analogous to human organisms or at least Cartesian subjects who can "own" inherent value. But to answer the epistemological demand of activists for warranted assertibility, Callicott must claim that these owned values must have their existence independently of human valuation – they must be "objective" and exist independently of human evaluations. The theoretical dragon, "inherent value in nature," is necessarily two-headed, because it was created to slay two logically distinct monsters: ontological pluralism in value theory and moral skepticism. But the ontological solution of positing independent moral subjects in nature has unfortunately encouraged the organicist interpretation of ecological assemblages and led to the fascist tendencies of reified holism.

The difficulty in all of this, of course, is that organicism does have an important point to make. Organicists are correct that mechanistic models do not explain the ability of ecological processes to create, sustain and to heal themselves. Ecological management requires that we accept two elements of organicism – the idea that the whole is more than the sum of its parts and the related idea that relationships among multi-scalar processes – not the static characteristics of objects – provide the key to understanding ecosystems. The problem is to express this idea in a way that does not carry us all the way to teleology and personalism. We must emphasize the creative nature of environmental processes and the key role of energy flows in those processes, without personalizing them.[31]

My concern in this paper is not so much with epistemology[32] as with the ontology that should unify the underlying theory of environmental ethics. Callicott's overall position represents a curious mixture of Cartesian modernism and postmodernism. He attributes to Leopold, and seems to accept himself, a Darwinian, dynamic, particularistic and postmodern viewpoint in *ethics*, while at the same time addressing the problem of warranted assertibility within a distinctly modern epistemology. Despite direct evidence that Leopold employed a pragmatist theory of truth,[33] Callicott assumes Leopold would have given a *realist* answer to the question: how can we justify our pursuit of environmental protection? I, on the other hand, see Leopold's many cautions about specifying the *purposes* behind creation and his numerous remarks that the exact interpretation of reality is "beyond language" as indicating that Leopold had at least the glimmerings of a postmodern conception of knowledge and objectivity, as well as morals.[34] Had Callicott placed Leopold's remarks regarding good environmental management in the context of Darwinian epistemology as well as Darwinian ethics, he would have conceived the "objectivity" problem very differently.[35]

If we focus for a moment on the problem of warranted assertibility of environmentalists' goals, it seems likely that environmentalists will achieve more by appealing to the relatively non-controversial and intuitive idea that the use of natural resources implies an obligation to protect them for future users[36] – a sustainability theory based in intergenerational equity – rather than exotic appeals to hitherto unnoticed inherent values in nature. Callicott argues it is an advantage of intrinsic value in nature that, if it can be shown to exist, then it would shift the burden of proof in environmental arguments from environmental protectors to the despoilers.[37] But from the fact that such value might be *sufficient* to shift the burden of proof, it does not follow that it is the only, nor the best, means available to environmentalists to shift the burden.

The epistemological problem is that environmentalists need to be able to enter the public arena armed with genuine and defensible moral principles so that they can assert the priority of their goals over the mere preferences of the consumer society. As long as we can assert *other* morally binding obligations – such as an obligation to sustain the integrity and health of the ecological systems we are now damaging so that future generations can enjoy the bounties of intact ecological communities – we have a basis warrantedly to assert obligations to protect biodiversity over many generations. But these obligations are anthropocentric and cannot, apparently, be comprehended in a monistic non-anthropocentrism, even though abiding by these less controversial obligations would lead to most of the environmental protections favored by inherent value theorists.

PART 2: AN ALTERNATIVE TO ONTOLOGY

In our search for an environmental ethic we will never, I submit, find any environmental values or goals more defensible than the sustainability principle, which asserts that each generation has an obligation to protect productive ecological and physical processes necessary to support options necessary for future human freedom and welfare. The normative force supporting the protection of the environment for future generations should be based on a commitment to building just, well-adapted and sustainable human communities. Accepting responsibility for our expanding numbers and for the power of our technologies follows simply from the recognition that we now affect the productivity of the human habitat and the very survival of the human community. This responsibility becomes less and less escapable as we learn the many consequences, expected and unexpected, of our increasingly violent and pervasive alteration of natural systems.[38] This principle is consistent with a Darwinian emphasis on survival and complements a pragmatic conception of truth. The acceptance of both the facts of human impacts and the associated *moral responsibility* to protect the integrity of ecological communities as repositories of many human options and values in the future is destined, in the terms of Peirce, to be adopted as the conclusion of all rational inquirers, as they struggle through many experiments to make coherent sense of human experience. I believe that *both* the descriptive problem of understanding the impacts of our actions on future generations *and* our resulting responsibilities as moral beings must be addressed within processes of inquiry constitutive of the Peirceian community of inquirers/actors. For example, considering species threatened with extinction to represent "books" of information – information that may be essential to future generations in their struggle to understand and act within a changing environment – seems to entail that the obligation to contribute to the process of inquiry requires protection of the sources of information and knowledge for future inquirers.

While the sustainability principle can give a certain unity to environmental action, this unity represents an open-ended direction, an appeal to learning and small-scale adaptation, community-building, and experimentation, not a slavish commitment to a priori principles of value such as monism and anthropocentrism. Thus this principle is unifying without being monistic – it sees the task of living sustainably as a problem of social learning, guided by a method that is socially open and scientifically experimental. This means that diverse stakeholder groups will be encouraged to assert and defend their own values and interests and to participate in social experiments in search of solutions that allow diverse users to fulfill their diverse needs with minimal disruption of the interests of other groups. The goal of value and policy discussions in a democracy should be inclusive, recognizing that the diversity of values humans and communities find in nature is only the first step towards an integrative policy that preserves differences; in this sense it differs from monisms, which demand that diverse groups formulate their values in a single vernacular or have their considerations excluded from discussion and possible consensus.

I propose a new beginning for environmental philosophy – both for environmental ethics and environmental epistemology – a beginning based broadly in the pragmatic epistemology of Charles Sanders Peirce, who replaced the failed project of representational and foundational realism with a constructivist method that recognizes that the correctibility of scientific inquiry must be fully characterized within human experience, not by reference to "external objects" that exist beyond experience.[39] Truth and objectivity must be sought in the specific characteristics of specific situations in which action is required. If environmental values can justifiably be asserted to have priority in some situations, the mark of this will be their eventual emergence in a complex process of inquiry, including diverse groups with diverse interests and viewpoints, that will submit both values and scientific hypotheses to discussion and testing. The relevant intellectual community is not philosophers with a distinctive subject matter, but the activist community that is committed to human survival and improvement. Knowledge and moral discussion must be understood as a part of the struggle to determine adaptable policies, rather than as a distinct "field" of theoretical morality. The inclusion of values as well as information in the process of inquiry can be traced to the experimental approach to social activism of John Dewey, who clearly rejected as folly the search for certainty and deductivism in moral and social matters.[40]

I therefore place my work in the tradition of "adaptive management,"[41] first introduced by C. S. Holling and developed by his colleagues in the Pacific Northwest and in Northern Canada. Adaptive management has recently been given a political formulation by Kai Lee, who explicitly and concretely bases his political analysis in the philosophy of John Dewey.[42] Lee also shows that the Deweyan approach is usefully complemented by the creative work of policy analysts on "bounded rationality," who have recognized that arriving at improved policies is often a matter of "muddling through" rather than a matter of establishing idealistic goals and instituting decisive and abrupt changes to achieve those goals.[43] The pragmatic approach recognizes that there is great uncertainty in both human knowledge and human valuations and attempts to nurture processes and institutions that seek compromise and incremental change and improvement of understanding and goals. In the process, both information and values will be adjusted to become more appropriate and adaptive to particular situations.

Nature as a multi-scalar, open system

Modeling a natural system is not a simple matter of choosing the hierarchy which best represents natural processes because, for any given ecological system, there are many – no doubt an infinite number – of models which will correctly *describe* some aspect of ecosystem functioning.[44] Further, because ecological systems are irreducibly complex, it follows that these models are not reducible to a single model without loss of descriptive content. I represent this essential complexity by using *hierarchy theory* (HT) – a theoretical approach to multiscalar systems of analysis developed by theoretical ecologists such as Holling, T. F. H. Allen, and Robert O'Neill. Hierarchy theory represents a scalar application of general systems theory.[45] Systems are best understood as open, multi-scaled processes in which each system is viewed also as a subsystem on a more encompassing level of the system. Multi-scalar analysis has just as important a positive role in understanding environmental values and management goals as it does in describing natural systems; it is the scientific cornerstone of the adaptive management approach. Multi-scalar analysis is based on the neutral, model-constituting assumption that smaller sub-systems function on faster dynamics, and that all models that capture the multi-scaled nature of ecological and physical systems should embody this spatio-temporal correlation between greater extent in physical space and slower dynamics of change. The central idea of a multi-scalar, contextual approach to analysis and management is therefore the very simple idea that human activities, economic and otherwise, take place within, and affect, larger systems that should be interpreted as multi-scaled ecological processes. As noted above and elsewhere, this multi-scalar approach represents a more precise and operational formalization of Aldo Leopold's insight that we must learn to "think like a mountain," and therefore provides both a connection backward to Leopold's seminal managerial insights and also points forward to new directions for scientific and policy research that proceed by organizing information and value studies into a spatio-temporally structured system of analysis.[46]

Most uses of HT so far have been purely descriptive. T. F. H. Allen and his co-authors, for example, argue that the only type of reason that can be advanced to prefer one scalar model of an ecological phenomenon over another is that the preferred model *enhances our understanding of the system* more extensively than do less preferred models. While they refer to this justification as "utilitarian," it is clear from the context that utility is understood by Allen and Starr as "scientific" or methodological utility, not general social utility.[47] By contrast, I also apply a normative filter of *usefulness in understanding and protecting social values* as well as a descriptive filter, reducing drastically the class of descriptive models that must be considered as relevant to policy. The goal should be to develop scalar models that improve our understanding in the specific sense that they illuminate environmental problems and allow us to focus on those natural dynamics that are causally related to important social values.

Hopefully, these models also help us to define our management goals clearly and to measure success and failure in attempts to achieve those goals.

Conservation biology – and also the "pure" disciplines of zoology, botany, etc. – must adopt the twin goals of understanding *and protecting* ecological systems just as human and veterinary medicine have adopted the twin, descriptive-and-normative goals of understanding *and healing* their patients. Commitment to the Peircean ideal of eventually understanding nature apparently carries with it the obligation to protect the sources of biological knowledge, living organisms, especially from irreversible losses such as species extinctions. The descriptive models chosen in normative sciences must therefore pass a double criterion. They must help us to understand nature, but they must also encourage us to understand nature in a way that will help us to formulate and measure environmental goals effectively and to propose and implement policies to achieve those goals. The problem in linking biological science, social science and value theory into an adequate plan for environmental management is to choose from among this multitude of *descriptively* adequate models a few models that truly help us to understand *and to integrate* human activities into the landscape.

This *multi-scalar and biogeographic* approach to environmental values assumes at the outset that management will proceed from a human perspective and also, that human values quite legitimately shape the modeling decisions of ecological and physical scientists. This latter point is deserving of further explanation: the decisions of biological and physical scientists have an unavoidable normative component. The point is not to *purge* science of those values, which is both impossible and undesirable, the point is to *understand and justify* those values in specific contexts requiring action, and to attempt to adjust them through public discussion and education when they become maladaptive.

A successful integrative ethic for the environment must be morally pluralistic, but it must also be contextual, rather than either objectivist or subjectivist. Good environmental decisions are ones that take into account likely impacts on a number of spatio-temporal scales in specific contexts. As the world becomes more full of humans and as technology becomes more powerful, there will be more and more cases in which there will be spill-over impacts from one level of hierarchical organization to another, especially from our expanding economic and social systems to the natural systems that form their ecological context. Environmental policy and action must do more than enhance values in one dynamic, such as the dynamic driving the economic decision of individual farmers; it is necessary also to examine the impacts on the larger- and usually slower-changing dynamic that determines the structure and diversity of the landscape. Here the focus of moral analysis turns to multiple generations and to the landscape scale.

The goal of an integrative ethic should be to sort the many and various values that humans derive from their environment and to associate these variables with real dynamic processes unfolding on the various levels and scales of the physical and ecological context of our activities. Environmental problems are in this sense essentially scalar problems and I seek to define models that illuminate the dynamics which support human values.

A tri-scalar model

Since the practical philosopher and the adaptive manager set as a goal of all model-building that the models inform environmental decision-making, and since we firmly believe that environmental decision-making must be democratic, any model for this purpose must be fairly simple in structure. The model must be a simple enough representation of multi-scaled natural processes to serve as an aid in public discussion of the goals of a forest management plan or a plan for ecological restoration of a river system. Prescriptive, multi-scalar models must therefore provide a publicly useful vocabulary for discussing environmental goals; they must shape our models of management by associating them with the temporal and spatial scales of the natural dynamics that generate the values guiding our choice of goals. The goal is to develop a spatio-temporally organized and ecologically informed phenomenology of the moral space in which individuals formulate and pursue personal and environmental values.[48] The problem, on this pragmatic approach, is to characterize and categorize pluralistic values in a way that is sensitive to spatio-temporal features – what I call the "scale" – of human interactions with nature. It is to design a method of inquiry, which includes both a descriptive and evaluative component so organized as to increase the likelihood of both achieving truth (defined in a broadly Darwinian sense) and sustainability of fair and equitable communities, and of traditional values and cultural diversity for the benefit of humans today and in the future.

127

To initiate discussion, let me suggest three basic scales, each of which corresponds to a temporally distinct *policy* horizon:

1 Locally developed values that express the preferences of individuals, given the established limits and "rules" – laws, physical laws, governmental laws, and market conditions, for example – within which individual transactions take place;
2 A longer and larger community-oriented scale on which we hope to protect and contribute to our community which might be taken to include the entire *ecological* community;
3 A global scale with essentially indefinite time scales on which humans express a hope that their own species, even beyond current cultures, will survive and thrive. (See Figure 1.)

On the first scale, which unfolds in the relatively short term and local space in which individuals make economic choices, an economics of cost-benefit, if supplemented with a sense of individual justice and equity, can provide useful decision models. The middle scale, on which we feel concern for our cultural connection to the past and future, is especially important for two reasons. Viewed socially, this multiple-generational level is the one on which we protect, develop and nurture our sense of who we are as a culture. It is on this level that we "decide" what kind of a society we want to be. These "decisions" are expressed in our art, in our religion and spirituality, and in our governing political institutions such as the constitution. It is on this scale that we feel concern about the culture's interaction with the ecological communities that form its context. This second scale is doubly important because it corresponds roughly to the ecological time scale on which multiple generations of human individuals, organized into communities, must relate to populations of other species that share their habitat. Thinking intergenerationally apparently requires that we pay special attention to large-scale aspects of the landscape.

Modern ecological knowledge has forced upon us the conclusion that we must act as members of the natural as well as the human social community; it follows also that we must pay attention to the context in which our values are formulated and acted upon, and that context is the interaction between a culture and its habitat that is described in the "natural history" of a place. That natural history must reach back into time, and project itself in a creative way into the future.[49] According to the multi-scalar, contextual view, humans necessarily understand their world from a given local perspective. A

Temporal Horizon of Human Concern	Time Scales	Temporal Dynamics in Nature
Individual/Economic	0–5 years	Human economies
Community, intergenerational bequests	up to 200 years	Ecological dynamics/ Interaction of species in communities
Species survival and our genetic successors	indefinite time	Global physical systems

Figure 1 Correlation of human concerns and natural system dynamics at different temporal scales

preference for localism is really a preference over preferences – a favoring of values that emerge from experience of one's home place. This home place locates the perspective from which one understands and values elements and processes in the natural context of our actions. Localism is represented in the proposed theory of environmental values by an endorsement of the importance of *sense of place values*. I therefore advocate many different locally originating sustainability ethics, each of which is anchored to a particular place by a strong sense of the history and the future of the place. In particular, a sense of place manifests itself, for example, in the defense of local determination and in an inclination of citizens to fight and defeat solutions imposed by centralized and authoritarian institutions.[50]

But a full-blown sense of place must include also a sense of the space around that place.[51] And here we invoke our multi-scalar phenomenology of environmental concerns as the evaluative space in which each citizen, as an individual, as a member of an ongoing, cultural community, and as a member of the global community, must seek local solutions which reduce impacts on larger systems.

We can represent action-decision aspects of the phenomenological models I am describing as in Figure 2. Imagine, conceptually, that the surface of the earth is represented as many points, or individual perspectives, each of which is tied by a cultural history to a human community and by a "natural" history to the land community. These individuals are understood as representative individuals who live in a community – their personal identity is therefore associated with that

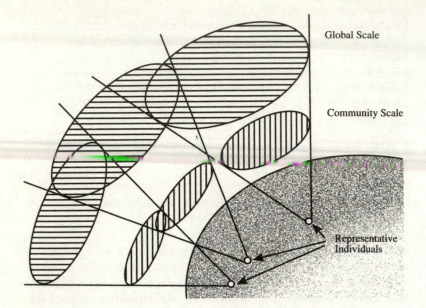

Figure 2 Multi-scalar relationship of individual, community and
global scales

community. They are therefore not selfish only – they seek to project
their culture by affecting future generations, though they unavoid-
ably conceive the world from their own, local perspective. On these
assumptions, our individuals must view themselves also as members
of a community of plants and animals, as well as a community of
humans. They therefore experience, articulate and defend environmental values from a local perspective and from the present point in
time. But impacts on the larger and long-term, intergenerational
scale can also impact local and personal values if the ecological
context changes so rapidly that traditional values and practices
become meaningless in a few years. While individuals perceive values
from a local perspective, those values are also shaped within a larger
space in which there can be impacts on larger, physical systems,
which in turn constrain future choices. Note that greater population
density and expansions in the scale of a community increase the likelihood of impacts from one community to another horizontally,
across communities, and also vertically, across generations.

130

An implication of the multi-scalar approach outlined here is that the search for an acceptable environmental policy will not be a search for the policy that maximizes benefits to costs as measured in present dollars. A good environmental policy will be one that has positive implications for values associated with the various scales on which humans *are in fact* concerned, and also on the scales on which environmentalists think we *should* be concerned if we accept responsibility for the impacts of our current activities on the life prospects and options – the "freedom" of future generations.[52]

In a situation that recognizes multiple human values and associates these with various natural dynamics, it is possible to conceive, describe and seek policies that will protect or even enhance the processes maintaining ecological structure and processes that are crucial in future interactions of human and natural communities. The goal of policy is not to analyze and rank various policies with respect to how well they score on a single criterion, but rather to devise win-win policies that are robust enough to score highly on a number of relevant criteria of good management. Win-win policies, on the approach proposed here, are policies that have positive (or, failing that, neutral, results on all three "scales" of human concern – the individual welfare level, the community level and on the emerging values of the global community).

Consider an example which fulfills this robustness condition. In many deforested developing countries poor families must expend considerable effort to gather firewood for cooking and heating as the scarcity of firewood can cause extended searches consuming many person-days per week per household. In such areas a creative environmental policy would institute many locally based tree-planting programs. Successful programs begin with small loans to private entrepreneurs who use the loan to purchase seedlings and for other start-up costs. Full payment to the entrepreneurs will occur when the trees reach a certain age or height, encouraging the entrepreneurs' clan to protect the trees so that they can reach the pay-off goal. If the trees are planted close together, they will provide increases to economic welfare within a few years as culled trees provide firewood nearer home. Meanwhile, local ecological processes should become more healthy as eroding land is replaced by forests, or at least small and diverse tree farms, improving water retention and improving streamwater quality. Finally, on the largest scale, the impact on current choices on the functioning of global physical systems, we can expect that birth rates will be reduced because peasant families will have less incentive to have children to help with household chores.

Figure 2 helps us to conceptualize a new criterion for acceptable and appropriate environmental action. I refer to the proposed criterion as advocating actions and policies that conform to the *scalar Pareto criterion*. The scalar Pareto criterion represents a multi-scalar application of the Pareto optimality criterion, which was originally stated on the individual level as the requirement that all actions have a positive impact on some individuals and negative impacts on nobody. The scalar application of the Pareto criterion is stated as follows: choose policies that, from the viewpoint of a representative individual in each community, the policy will have positive (or at least non-negative) impacts on goals formulated by that person on the individual level, on the community level, and on the global level. While the scalar Pareto approach retains an individualist perspective (it is human individuals who formulate, discuss and defend values on all levels), it does not seek reduction of all values to economic preferences or to some generic form of "inherent value." It is *pluralistic* in the sense that the value of ecosystems is understood on a community, not an individual, scale and no reduction of community-level values to individual values is attempted. But the pluralistic ethic is also *integrative* in the sense that we seek actions that will have positive (or at least non-negative) impacts on the relatively distinct dynamics that produce and support human values that are expressed on multiple scales (here hypothesized as three).

CONCLUSION

Thus ends my explanation of, and plea for, a practical environmental ethic that seeks to *integrate* pluralistic principles across multiple levels/dynamics. Rather than *reducing* pluralistic principles by relating them to an underlying value theory that recognizes only economic preferences or "inherent" value as the ontological stuff that unifies all moral judgments, I have sought integration of multiple values on three irreducible scales of human concern and valuation, choosing pluralism over monism, and attempting to integrate values within an ecologically informed, multi-scalar model of the human habitat. I believe that the non-ontological, pluralistic approach to values can better express the inductively based values and management approach of Leopold's land ethic, which can be seen as a precursor to the tradition of adaptive management. And, if the problem of environmentalism is the need to support rationally the goals of environmental protection – the problem Callicott misconceived as the need for a realist moral ontology to establish the "objectivity" of environmental goals – then I endorse the broadly Darwinian approach to both epistemology and morals proposed by the American pragmatists. The environmental community *is* the community of inquirers; it is the community of inquirers that, for better or worse, must struggle, immediately as individuals and indefinitely as a community, both to survive and to know. In this struggle useful knowledge will be information about how to survive in a rapidly evolving culture and habitat. It is in this sense that human actors are a part of multilayered nature; our actions have impacts on multiple dynamics and multiple scales. We humans will understand our moral responsibilities only if we understand the consequences of our action as they unfold on multiple scales; and the human community will only survive to further evolve and adapt if we learn to achieve individual welfare and justice in the present in ways that are less disruptive of the processes, evolving on larger spatio-temporal scales, essential to human and ecological communities.[53]

ACKNOWLEDGMENT

A slightly altered version of the first part of this paper appears in *Environmental Ethics* Vol. 17, under the title "Why I am Not a Nonanthropocentrist."

NOTES

1 It is interesting, and perhaps not accidental, that the power and importance of the assumption that environmental ethics should be monistic in form was pointed out not by an environmental ethicist or philosopher, but by a legal scholar. See Christopher Stone, *Earth and Other Ethics* (New York: Harper and Row, 1987). Environmental ethicists, who responded directly to the charge by historian Lynn White, Jr. that Western thought has led to environmental degradation because it is anthropocentric, have for the most part never questioned that anthropocentrism must be replaced with a non-anthropocentric form of monism. White's argument appeared in *Science* 155 (1967): 1203–1207.

More recently, it has been noted that monisms come in several versions. See Peter Wenz, "Minimal, Moderate, and Extreme Moral Pluralism," *Environmental Ethics* 15 (1993): 61–74. Following J. Baird Callicott, I will restrict my discussion to "principles monism" and "theoretical monism," as explained in Callicott's "The Case Against Moral Pluralism," *Environmental Ethics* 12 (1990): 99–124. According to principles monism, there is a single principle that covers all moral quandaries, with principle understood as a moral standard sufficiently practical to imply a single correct action in every situation. Theoretical monism, by contrast, might employ more than one principle in different situations, but achieves monism on a theoretical level by providing an over-arching theory that explains and unifies the use of divergent principles in terms of a monistic theory. It seems reasonable to consider principles monism to be a special case of theoretical monism because, as in the case of some simple versions of utilitarianism, a single theory of value justifies a single principle applicable to all cases. This paper is directed at both principle and theoretical versions because of the reductionistic tendencies they share. For simplicity, I will refer only to the general form, theoretical monism, because arguments applicable to it will apply also to the special case of principles monism as well.

2 See Callicott, "The Case Against Moral Pluralism," op. cit., for a discussion of the strengths of monism as opposed to pluralism in environmental ethics.

3 I have discussed the "environmentalists' dilemma" in detail in *Toward Unity Among Environmentalists* (New York: Oxford University Press, 1991).

4 See B. Norton, "Applied Philosophy vs. Practical Philosophy: Toward an Environmental Policy Integrated According to Scale," in Donald Marietta and Lester Embree, eds, *Environmental Philosophy and Environmental Activism* (Totowa, NJ: Rowman and Littlefield, forthcoming), for a more detailed discussion of the failure of non-anthropocentrism to address real management problems.

5 It should be noted that Callicott has on occasion suggested an alternative. See note 53.

6 See J. Baird Callicott, "Environmental Philosophy *Is* Environmental Activism: The Most Radical and Effective Kind," in Marietta and Embree, *Environmental Philosophy and Environmental Activism*, for a brief exposition and aggressive defense of applied philosophy and the role he sees for it.

7 Interestingly, Callicott asserts (ibid., p. 25) that applied philosophy does "rather little deducing of specific rules of conduct." Rather, he sees the role of philosophy as one of "articulating and thus helping to effect . . . a radical change in outlook. The specific ethical norms of environmental conduct remain for the most part implicit – a project postponed to the future or something left for ecologically informed people to work out for themselves." Despite this well-placed skepticism regarding whether environmental ethicists can get much mileage out of abstract, philosophical theories, Callicott never doubts that his monistic theories of value are important in effecting a change in consciousness of citizens and capable of bringing about a new, environmental era: "Therefore, since human actions are carried out and find their meaning and significance in a cultural ambience of ideas, we speculative environmental philosophers are inescapably environmental activists" (p. 25). So, whether deduction is involved or not, the applied philosopher's contribution is to provide thoughtful persons with reasons or some form of motivation

to adopt a new "cultural ambience," which includes his, particular, monistic principle. Only after persons (at some time in the future) agree on the existence and meaning of non-anthropocentric monistic value can the principle begin to provide guidance in specific situations.

8 This point shows how badly Callicott misunderstands the anti-monistic position, which he refers to as "anti-philosophical." "Environmental Philosophy Is Environmental Activism" (p. 5). I, at least, am not anti-philosophical; I am against philosophical theory which is developed independently of real-world problems and I reject the role for theory assumed by Callicott and other monists and applied philosophers. Callicott mistakes an attack on *outmoded, modernist theory* for an attack on all theorizing.

9 See Bryan G. Norton, *Toward Unity Among Environmentalists* (New York: Oxford University Press, 1991).

10 See Bryan G. Norton, "Evaluating Ecosystem States: Two Competing Paradigms," *Ecological Economics* (1995, forthcoming); "Economists' Preferences and the Preferences of Economists," *Environmental Values* 3 (1994): 311–332; and "Thoreau's Insect Analogies: Or, Why Environmentalists Hate Mainstream Economists," *Environmental Ethics* 13 (1991): 235–261 for criticisms of welfare economics as a similarly inadequate monistic theory of environmental valuation.

11 Such as Holmes Rolston, III, *Environmental Ethics: Duties to and Values in the Natural World*, (Philadelphia: Temple University Press, 1988) and Paul Taylor, *Respect for Nature* (Princeton, NJ: Princeton University Press, 1986).

12 Mainly in "The Case Against Moral Pluralism" and "Moral Monism in Environmental Ethics Defended," *Journal of Philosophical Research* XIX (1994): 51–60.

13 I use this vague phrase, "moral gravity," because Callicott has recently admitted that, because of the subjectivist foundations of his philosophy in David Hume, he cannot claim to offer a theory that exerts "moral force" (the more usual formulation), but only a "moral dimension." See "Can a Theory of Moral Sentiments Support a Genuinely Normative Environmental Ethic?" *Inquiry* 35 (1992): 183–198.

14 J. Baird Callicott, *In Defense of the Land Ethic* (Albany, NY: State University of New York Press, 1989), p. 163.

15 Ibid., p. 160.

16 J. Baird Callicott, "Animal Liberation: A Triangular Affair," *Environmental Ethics* 2 (1980): 311–228. Reprinted in Callicot, *In Defense*.

17 Tom Regan, *The Case for Animal Rights* (Berkeley, CA: University of California Press, 1983), p. 362.

18 Callicott, *In Defense*, pp. 55–59; 93–94.

19 Ibid., pp. 55–56.

20 Ibid., p. 58.

21 I first pointed out the difficulty of this dilemma in "Review of *In Defense of the Land Ethic*," *Environmental Ethics* 13 (1991): 185. As far as I know, Callicott has not responded in print or verbally to this dilemma, which apparently requires some response if we are to understand what is meant by monistic inherentism.

22 Aldo Leopold, *A Sand County Almanac and Sketches Here and There* (London: Oxford University Press, 1949), pp. 224–225.

23 There have, however, been dissenting opinions regarding the non-anthropocentrism of the land ethic. See Scott Lehmann, "Do Wildernesses Have Rights? *Environmental Ethics* 3 (1981): 129–146; Bryan G. Norton, "The Constancy of Leopold's Land Ethic," *Conservation Biology* 2 (1988): 93–102 (reprinted in this volume); and Norton, *Toward Unity*.

24 See, for example, Eugene Odum, "The Strategy of Ecosystem Development," *Science* 164 (1969): 262–270. I have criticized this form of strong organicism in "Should Environmentalists be Organicists?" *Topoi* 12 (1991): 21–30.

25 Leopold, *A Sand County Almanac*, pp. 129–133.

26 See Bryan G. Norton, "Context and Hierarchy in Aldo Leopold's Theory of Environmental Management," *Ecological Economics* 2 (1990): 119–127; Norton, *Toward Unity*.

27 Leopold, *A Sand County Almanac*, pp. 221–222.

28 Tom Regan, "The Nature and Possibility of an Environmental Ethic," *Environmental Ethics* 3 (1981): 20.

29 Callicott, *In Defense*, p. 157; J. Baird Callicott, "Rolston on Intrinsic Value: A Deconstruction," *Environmental Ethics* 14 (1992): 130–131.

30 Ibid.: 107–112.

31 See Norton, "Should Environmentalists be Organicists?"

32 See Norton, "Epistemology and Environmental Values," *Monist* 75(1992): 208–226, for a detailed criticism of the epistemology of intrinsic value theory.

33 See Norton, "The Constancy of Aldo Leopold's Land Ethic" (this volume) for an argument that the land ethic has explicit roots in pragmatism.

34 See A. Leopold, "Some Fundamentals of Conservation in the Southwest," *Environmental Ethics* 1 (1979): 131–141; also see Norton, "The Constancy of Leopold's Land Ethic" (this volume).

35 See Michael Ruse, *Taking Darwin Seriously: A Naturalist Approach to Philosophy* (Oxford: Blackwell, 1987), for a survey and convincing defense of Darwinian epistemology.

36 See E. B. Weiss, *In Fairness to Future Generations* (Tokyo, Japan and Dobbs Ferry, NY: The United Nations University and Transnational Publishers, Inc., 1989). Weiss shows that every major world religion, and many minor ones as well, assert that the use of resources obligates current generations to an obligation to pass the resources on to future generations.

37 Callicott, "Environmental Philosophy Is Environmental Activism."

38 See A. Leopold, "A Biotic View of Land," *Journal of Forestry* 37 (1939): 727–730.

39 C. S. Peirce, *The Philosophical Writings of Peirce* (New York: Dover, 1955), especially pp. 21, 39.

40 See, especially, John Dewey, *The Public and Its Problems*, in *John Dewey: The Later Works, Volume 2: 1925–1927*, edited by Jo Ann Boydston (Carbondale, IL: Southern Illinois University Press, 1984), pp. 235–372.

41 The adaptive management approach was introduced by C. S. Holling, ed. *Adaptive Environmental Assessment and Management* (New York: John Wiley & Sons, 1978). Also see Carl J. Walters, *Adaptive Management of Renewable Resources* (New York: Macmillan, 1986) and, especially, Kai N. Lee, *Compass and Gyroscope: Integrating Science and Politics for the Environment* (Covelo, CA: Island Press, 1993).

42 Lee, *Compass and Gyroscope*, pp. 91–115.

43 Herbert Simon, *Administrative Behavior* (New York, Macmillan, 1954), and Charles Lindblom, "The Science of Muddling Through," *Public Administration Review* (1959): 79–88. The present volume at last provides a discussion of the philosophical principles of pragmatism in connection with environmental policies and values.

44 See Simon Levin, "The Problem of Pattern and Scale in Ecology," *Ecology* 73(6): pp. 1943–1967 for the scientific argument for this conclusion.

45 T. F. H. Allen and T. B. Starr, *Hierarchy: Perspectives for Ecological Complexity* (Chicago: University of Chicago Press, 1982); and R. V. O'Neill, D. L. DeAngelis, J. B. Waide and T. F. H. Allen, *A Hierarchical Concept of Ecosystems* (Princeton, NJ: Princeton University Press, 1986).

   Despite connotations sometimes associated with the term "hierarchy", it is important to understand that there is no implication that higher levels of the system dominate, or should dominate, lower levels. In fact, the processes described in hierarchical systems analysis exhibit communication both upward and downward through the hierarchy, and for this reason we prefer the more neutral terminology, "multiscalar analysis."

46 This use of hierarchy theory differs dramatically from recent applications of the theory by environmental ethicists Cheney and Warren, who have used the theory negatively to attack Callicott's uses of ecological theory. They argue that, because hierarchy theory does not claim ontological priority for any of the hierarchical models it proposes, the efforts of Callicott and others to use ecological theories to support "moral ontologies" is a doomed project. See Karen J. Warren and Jim Cheney, "Ecosystem Ecology and Metaphysical Ecology: A Case Study," *Environmental Ethics* 15 (1993): 99–116. But I believe hierarchy has much stronger potential as a useful method for dealing with troubling scalar issues in the discussion of environmental information, values and goals positively. See Norton, "Scale and Hierarchy in Aldo Leopold's Land Ethic."

47 See T. F. H. Allen and T. B. Starr, *Hierarchy: Perspectives for Ecological Complexity* op. cit., p. 6. See also T. F. H. Allen and Thomas W. Hoekstra, *Toward a Unified Ecology* (New York: Columbia University Press, 1992), pp. 31–35.

48 B. Norton and Bruce Hannon, "A Biogeographical Theory of Environmental Values," submitted for publication.

49 Holmes Rolston, III, *Environmental Ethics*.

50 B. Norton and Bruce Hannon, "Democracy and Sense of Place Values," submitted for publication.

51 Yi–Fu Tuan, *Space and Place: The Perspective of Experience* (Minneapolis, MN: University of Minnesota Press, 1977); Tuan, "Man and Nature," Commission on College Geography, Resource Paper 10, Association of American Geographers, Washington, DC, 1971; and Norton and Hannon, "Democracy and Sense of Place Values."

52 See T. F. H. Allen and T. B. Starr, *Hierarchy: Perspectives for Ecological Complexity* op. cit., p. 15, for a definition of "freedom" in this sense, a sense that can be understood within hierarchical models.

53 I gratefully acknowledge that my understanding of these problems has been improved by countless discussions, in formal and informal situations, with Baird Callicott. I have learned much from our discussions, and have concluded that Callicott and I agree on most practical issues of management, but that we cannot agree regarding the theoretical foundations of environmental ethics. I do not, of course, expect that Callicott will agree with my arguments or my approach, but I hope this opposing viewpoint will carry forward the discussion of the proper role and mission of environmental ethics into the larger community of scholars.

   It should also be said that Callicott has, on occasion, recognized the insidious associations of monistic inherentism when associated with a modernist conception of scientific objectivity with Cartesianism and he has even suggested that achievement of a truly adequate, postmodern conceptualization of nature and value would require a different formulation. If I understand this post-modern version of Callicott's philosophy, it still includes a central role for inherent value. See *In Defense*, especially pp. 165–174 for a tentative trying-out of this alternative position. The arguments of this paper, however, have not been directed at the postmodern, non-Cartesian version of inherent value, but at the version defended in the vast majority of Callicott's published writings.

7

BEFORE ENVIRONMENTAL ETHICS

Anthony Weston

I INTRODUCTION

To think "ecologically," in a broad sense, is to think in terms of the evolution of an interlinked system over time rather than in terms of separate and one-way causal interactions. It is a general habit of mind. Ideas, for example, not just ecosystems, can be viewed in this way. Ethical ideas, in particular, are deeply interwoven with and dependent upon multiple contexts: other prevailing ideas and values, cultural institutions and practices, a vast range of experiences, and natural settings as well. An enormous body of work, stretching from history through the "sociology of knowledge" and back into philosophy, now supports this point.[1]

It is curious that environmental ethics has not yet viewed itself in this way. Or perhaps not so curious, for the results are unsettling. Some theories, in particular, claim to have transcended anthropocentrism in thought. Yet these theories arise within a world that is profoundly and beguilingly anthropocentrized.[2] From an "ecological" point of view, transcending this context so easily seems improbable. In part two of this chapter, I argue that even the best non-anthropocentric theories in contemporary environmental ethics are still profoundly shaped by and indebted to the anthropocentrism that they officially oppose.

I do not mean that anthropocentrism is inevitable, or even that non-anthropocentric speculation has no place in current thinking. Rather, as I argue in part three, the aim of my critique is to bring into focus the slow process of culturally constituting and consolidating values that underlies philosophical ethics as we know it. My purpose is to broaden our conception of the nature and tasks of ethics, so that we can begin to recognize the "ecology," so to speak, of environmental

139

ethics itself, and thus begin to recognize the true conditions under which anthropocentrism might be overcome.

One implication is that we must rethink the practice of environmental ethics. In part four, I ask how ethics should comport itself at early stages of the process of constituting and consolidating new values. I then apply the conclusions directly to environmental ethics. In particular, the co-evolution of values with cultural institutions, practices and experience emerges as an appropriately "ecological" alternative to the project of somehow trying to leapfrog the entire culture in thought. In part five, finally, I offer one model of a co-evolutionary approach to environmental philosophy: what I call "enabling environmental practice."

II CONTEMPORARY NON-ANTHROPOCENTRISM

I begin by arguing that contemporary non-anthropocentric environmental ethics remains deeply dependent upon the thoroughly anthropocentrized setting in which it arises. Elsewhere I develop this argument in detail.[3] Here there is only room to sketch some highlights.

For a first example, consider the very phrasing of the question that most contemporary environmental philosophers take as basic: whether "we" should open the gates of moral considerability to "other" animals (sometimes just: "animals"), and/or to such things as rivers and mountains. The opening line of Paul Taylor's *Respect for Nature*, for example, invokes such a model. Environmental ethics, Taylor writes, "is concerned with the moral relations that hold between humans and the natural world."[4]

Taylor's phrasing of "the" question may seem neutral and unexceptionable. Actually, however, it is not neutral at all. The called-for arguments address humans universally and exclusively on behalf of "the natural world." Environmental ethics, therefore, is invited to begin by *positing*, not by questioning, a sharp divide that "we" must somehow cross, taking that "we" unproblematically to denote all humans. To invoke such a divide, however, is already to take one ethical position among others. For one thing, it is largely peculiar to modern Western cultures. Historically, when humans said "we," they hardly ever meant to include all other humans. Moreover, they often meant to include some individuals of other species. Mary Midgley emphasizes that almost all of the ancient life patterns were "mixed communities," involving humans and an enormous variety of other creatures, from dogs (with whom, she says, we have a "symbiotic" relationship) to reindeer, weasels, elephants, shags, horses and pigs.[5] One's identifications and loyalties lay not with the extended human species, but with a local and concretely realized network of relationships involving many different species.

Taylor might respond that his question is at least *our* question: the urbanized, modern, Westerner's question. So it is. But it is precisely this recognition of cultural relativity that is crucial. "The" very question that frames contemporary environmental ethics appears to presuppose a particular cultural and historical situation – which is not the only human possibility, and which may itself be the problem. Cross-species identifications, or a more variegated sense of "the natural world," fit in awkwardly, or not at all.

Consider a second example. A defining feature of almost all recent non-anthropocentrism is some appeal to "intrinsic values" in nature. Once again, however, this kind of appeal is actually no more neutral or timelessly relevant than an appeal to all and only to humans on behalf of the rest of the world. Intrinsic values in nature are so urgently sought at precisely the moment that the *instrumentalization* of the world – at least according to a certain sociological tradition[6] – has reached a fever pitch. It is because we now perceive nature as thoroughly reduced to a set of "means" to human ends that an insistence on nature as an "end in itself" seems the only possible response. We may even be right. Still, under other cultural conditions, unthreatened by such a relentless reduction of everything to "mere means," it at least might not seem so *obvious* that we must aspire to a kind of healing that salvages a few non-traditional sorts of ends while consigning everything else to mere resourcehood. Instead, we might challenge the underlying means–ends divide itself, turning toward a more pragmatic sense of the interconnectedness of all of our values.[7]

Also, unthreatened in this way, we might not be tempted to metaphysical turns in defense of the values we cherish. Jim Cheney has suggested that the turn to metaphysics in some varieties of contemporary environmental ethics represents, like the ancient Stoics' turn to metaphysics, a desperate self-defense rather than a revelation of a genuine non-anthropocentrism. Cheney charges in particular that a certain kind of radical environmentalism, which he dubs "Ecosophy S," has been tempted into a "neo-Stoic" philosophy – an identification with nature on the level of the universe as a whole

– because neo-Stoicism offers a way to identify with nature without actually giving up control. In this way, abstract arguments become a kind of philosophical substitute for "real encounter" with nature.[8]

Cheney argues that Ecosophy S reflects a profoundly contemporary psychological dynamic. I want to suggest that it also reflects the diminished character of the world in which we live. The experiences for which Ecosophy S is trying to speak are inevitably marginalized in a thoroughly anthropocentrized culture. They are simply not accessible to most people or even understandable to many. Although wild experience may actually *be* the starting point for Ecosophy S, there are only a few, ritualized, and hackneyed ways to actually speak for it in a culture that does not share it. Thus – again, under present circumstances – environmental ethics may be literally driven to abstraction.

Once again it may even be true that abstraction is our only option. Nonetheless, in a different world, truly beyond anthropocentrism, we might hope for a much less abstract way of speaking of and for wild experience – for enough sharing of at least the glimmers of wild experience that we can speak of it directly, even perhaps invoking a kind of love. But such a change, once again, would leave contemporary non-anthropocentric environmental ethics – whether neo-Stoic or just theoretical – far behind.

As a third and final example, consider the apparently simple matter of what sorts of criticism are generally regarded as "responsible" and what sorts of alternatives are generally regarded as "realistic." The contemporary anthropocentrized world, which is, in fact, the product of an immense project of world reconstruction that has reached a frenzy in the modern age, has become simply the taken-for-granted reference point for what is "real," for what must be accepted by any responsible criticism. The absolute pervasiveness of internal combustion engines, for example, is utterly new, confined to the last century and mostly to the last generation. By now very few Westerners ever get out of earshot of internal combustion engines for more than a few hours at a time. The environmental consequences are staggering, the long-term effects of constant noise on "mental health" are clearly worrisome, and so on. Yet this technology has so thoroughly embedded itself in our lives that even mild proposals to restrict internal combustion engines seem impossibly radical. This suddenly transmuted world, the stuff of science fiction only fifty years ago, now just as suddenly defines the very limits of imagination. When we think of "alternatives" all we can imagine are car pools and buses.

Something similar occurs in philosophical contexts. Many of our philosophical colleagues have developed a careful, neutral, critical style as a point of pride. But in actual practice this style is only careful, neutral, critical in certain directions. It is not possible to suggest anything *different*, for the project of going beyond anthropocentrism still looks wild, incautious, intellectually overexcited. Anthropocentrism itself, however, is almost never scrutinized in the same way. Apparently, it just forms part of the "neutral" background: it seems to be no more than what the careful, critical thinker can *presuppose*. Thus it is the slow excavation and the logical "refutation" of anthropocentrism that, perforce, occupy our time – rather than, for one example, a much less encumbered, more imaginative exploration of other possibilities, less fearful of the disapproval of the guardians of Reason, or, for another example, a psychological exploration of anthropocentrism itself, taking it to be more like a kind of lovelessness or blindness than a serious philosophical position. Anthropocentrism still fills the screen, still dominates our energies. It delimits what is "realistic" because in many ways it determines what "reality" itself is.

III   ETHICS IN SOCIAL CONTEXT

The conclusion of the argument so far might only seem to be that we need better non-anthropocentrisms: theories that rethink Taylor's basic question, theories that are not so easily seduced by intrinsic values, and so on. Although such theories would be useful changes, the argument just offered also points towards a much more fundamental conclusion, one upon which very large questions of method depend. If the most rigorous and sustained attempts to transcend anthropocentrism still end up in its orbit, profoundly shaped by the thought and practices of the anthropocentrized culture within which they arise, then we may begin to wonder whether the project of transcending culture in ethical thought is, in fact, workable *at all*. Perhaps ethics requires a very different self-conception.

Here, moreover, is a surprising fact: ethics generally *has* a very different self-conception. Most "mainstream" ethical philosophers now readily acknowledge that the values they attempt to systematize are indeed deeply embedded in and co-evolved with social institutions and practices. John Rawls, for example, who at earlier moments appeared to be the very incarnation of the philosophical drive towards what he himself called an "Archimedean point" beyond culture, now explicitly justifies his theory only by reference to its "congruence with our deeper understanding of ourselves and our aspirations, and our realization that, given our history and the traditions embedded in our public life, it is the most reasonable doctrine for us." For *us*, culture answers "our" questions. "We are not," he says, "trying to find a conception of justice suitable for all societies regardless of their social or historical circumstances." Instead, the theory "is intended simply as a useful basis of agreement in our society."[9] The same conclusion is also the burden, of course, of an enormous body of criticism supposing Rawls to be making a less culturally dependent claim. Rawls, thus, does not transcend his social context at all. His theory is, rather, in a Nietzschean phrase, a particularly scholarly way of *expressing* an already established set of values. That contemporary non-anthropocentric environmental ethics does not transcend *its* social context, therefore, becomes much less surprising. At least it is in good company.

Similarly, John Arras, in an article surveying Jonsen and Toulmin's revival of casuistry, as well as the Rawls–Walzer debate, remarks almost in passing that all of these philosophers agree that "there is no escape from the task of interpreting the meanings embedded in our social practices, institutions, and history."[10] Michael Walzer argues for a plurality of justice values rooted in the varied "cultural meanings" of different goods.[11] Alasdair MacIntyre makes the rootedness of values in "traditions" and "practice" central to his reconception of ethics.[12] Charles Taylor localizes the appeal to rights within philosophical, theological and even aesthetic movements in the modern West.[13] Sabina Lovibond updates Wittgensteinian "form of life" ethics along sociologically informed "expressivist" lines.[14]

It may seem shocking that the "Archimedean" aspirations for ethics have been abandoned with so little fanfare. From the point of view of what we might call the "theology of ethics," it probably is. Day to day, however, and within the familiar ethics of persons, justice and rights practiced by most of the philosophers just cited, it is less surprising. Operating within a culture in which certain basic values are acknowledged, at least verbally, by nearly everyone, there is little practical need to raise the question of the ultimate origins or warrants of values. Because the issue remains metaphilosophical and marginal to what are supposed to be the more systematic tasks of ethics, we can acquiesce in a convenient division of labor with the social sciences, ceding to them most of the historical and cultural questions about the evolution of values, while keeping the project of systematizing and applying values for our own. "Scholarly forms of expression" of those values – or at least systematic forms of expression, "rules to live by" – are then precisely what we want.

It now seems entirely natural, for example, to view persons as "centers of autonomous choice and valuation," in Taylor's words, "giving direction to their lives on the basis of their own values," having a sense of identity over time, and so on. It also seems natural to point to this "belief system" to ground respect for persons, as Taylor also points out. He does *not* ask how such a belief system came into being and managed to rearrange human lives around itself. He does not *need* to ask. But we need at least to remember that these are real and complex questions. It is only such processes, finally running their courses, that make possible the consensus behind the contemporary values in the first place. Weber traces our belief system about persons, in part, to Calvinist notions about the inscrutability of fate, paradoxically leading to an outwardly calculating possessiveness coupled with rigid "inner asceticism," both self-preoccupied in a fundamentally new way. In addition, he traces it to the development of a system of increasingly impersonal commercial transactions that disabled and disconnected older, more communal ties between people.[15] The cultural relativity of the notion of persons is highlighted meanwhile, by its derivation from the Greek dramatic "personae," perhaps the first emergence of the idea of a unique and irreplaceable individual. A tribal African or Native American would never think of him or herself in this way.[16]

It may be objected that to stress the interdependence of ethical ideas with cultural institutions, practices and experience simply reduces ethical ideas to epiphenomena of such factors. However, the actual result is quite different. The flaw lies with the objection's crude (indeed, truly "vulgar," as in "vulgar Marxist") model of causation. Simple, mechanical, one-way linkages between clearly demarcated "causes" and "effects" do not characterize cultural phenomena (or, for that matter, *any* phenomena). Thus the question is emphatically *not* whether ethical ideas are "cause" or "effect" in cultural systems, as if the only alternative to being purely a cause is to be purely an effect. Causation in complex, interdependent, and evolving systems with multiple feedback loops – that is, an "ecological" conception of causation – is a far better model.[17]

One implication of such a model, moreover, is that fundamental change (at least constructive, non-catastrophic change) is likely to be slow. Practices, habits, institutions, arts and ideas all must evolve in some coordinated way. Even the physical structure of the world changes. Individualism and its associated idea of privacy, for example, developed alongside a revolution in home and furniture design.[18] Thus it may not even be that visionary ethical ideas (or anything else visionary, e.g., revolutionary architecture) are impossible at any given cultural stage, but rather that such ideas simply cannot be recognized or understood, given all of the practices, experiences, etc. alongside of which they have to be placed, and given the fact that they cannot be immediately applied in ways that will contribute to their development and improvement.[19] To use a Darwinian metaphor, all manner of "mutations" may be produced at any evolutionary stage, but conditions will be favorable for only a few of them to be "selected" and passed on.[20]

It may also be objected that any such view is hopelessly "relativistic." Although the term *relativism* now seems to be confused and ambiguous, there is at least one genuine concern here: if values are thoroughly relativized to culture, rational criticism of values may become impossible. In fact, however, rational criticism remains entirely possible – only its "standpoint" is internal to the culture it challenges, rather than (as in the Archimedean image) external to it. Much of what we tend to regard as radical social criticism reinvokes old, even central, values of a culture rather than requiring us to transcend somehow the culture in thought. Weber, for example, reread Luther's conception of the individual's relation to God as an extension of the already old and even revered monastic ideal to society at large. Likewise, the challenges of the 1960s in the US arguably appealed not to new values but to some of the oldest and most deeply embedded values of our culture. The Students for Democratic Society's "Port Huron Statement" persistently speaks in biblical language; the Black Panthers invoked the Declaration of Independence; the Civil Rights Movement was firmly grounded in Christianity. In his 1981 encyclical "Laborem Exercens," Pope John Paul II appealed to Genesis to ground a stunning critique of work in industrial societies reminiscent of the early Marx.[21]

In general, those who worry about the implications of social-scientific "relativism" for the rationality of ethics should be reassured by Richard Bernstein's delineation of a kind of rationality "beyond objectivism and relativism," a much more pragmatic and processual model of reason built upon the historical and social embeddedness and evolution of ideas.[22] Those who worry that "relative" values will be less serious than values that can claim absolute allegiance might be reassured by the argument that it is precisely the profound embeddedness of our ethical ideas within their cultural contexts that marks their seriousness. For *us*, of course. Nevertheless, that is whom we speak of and to.

Although these last remarks are very sketchy, they at least serve to suggest that a sociological or "evolutionary" view of values is not somehow the death knell of ethics. Instead, such a view seems to be almost an enabling condition of modern philosophical ethics. At the same time, however, "mainstream" ethics does not need to be, and certainly *has* not been, explicit on this point. The actual origins of values are seldom mentioned at all, and the usual labels – for example, Lovibond's "expressivism" and even MacIntyre's "traditions" – only indirectly suggest any social-scientific provenance. But it is time to be more explicit. As I argue below, large issues outside the "mainstream" may depend upon it.

IV THE PRACTICE OF ETHICS AT ORIGINARY STAGES

In order to begin to draw some of the necessary conclusions from this "evolutionary" view of values, let us turn our attention to the appropriate comportment for ethics at what we might call the "originary stages" of the development of values: stages at which new values are only beginning to be constituted and consolidated. In the case of the ethics of persons, for example, we must try to place ourselves back in the time when respect for persons, and persons themselves, were far less secure – not fixed, secure or "natural" as they now seem, but rather strange, forced, truncated, the way they must have seemed to, say, Calvin's contemporaries. How then should – how *could* – a proto-ethics of persons proceed in such a situation?

First, such early stages in the development of a new set of values require a great deal of exploration and metaphor. Only later do the new ethical notions harden into analytic categories. For example, although the concept of the "rights" of persons now may be invoked with a fair degree of rigor, throughout most of its history it played a much more open-ended role, encouraging the treatment of whole new classes of people as rights holders – slaves, foreigners, property-less persons, women – in ways previously unheard of, and in ways that literally speaking, were misuses of the concept. (Consider "barbarian rights." The very concept of *barbarian* seems to preclude one of them being one of "us," i.e. Greeks, i.e. rights holders.) This malleable rhetoric of rights also in part *created* "rights holders." Persuading someone that he or she has a right to something, for example, or persuading a whole class or group that their rights have been violated, dramatically changes his, her, or their behavior, and ultimately reconstructs his, her or their belief systems and experiences. Even now the creative and rhetorical possibilities of the concept of rights have not been exhausted. It is possible to read the sweeping and inclusive notion of rights in the United Nations Declaration of Human Rights in this light, for instance, rather than dismissing it as conceptually confused, as do legalistic thinkers.[23]

Moreover, the process of co-evolving values and practices at originary stages is seldom a smooth process of progressively filling in and instantiating earlier outlines. Instead, we see a variety of fairly incompatible outlines coupled with a wide range of proto-practices, even social experiments of various sorts, all contributing to a kind of cultural working-through of a new set of possibilities. The process *seems* smooth in retrospect only because the values and practices that ultimately win out rewrite the history of the others so that the less successful practices and experiments are obscured – much as successful scientific paradigms, according to Kuhn, rewrite their own pasts so that in retrospect their evolution seems much smoother, more necessary and more univocal than they actually were. Great moments in the canonical history of rights, for example, include the Declaration of Independence and the Declaration of the Rights of Man, capitalism's institutionalization of rights to property and wealth, and now the persistent defense of a non-positivistic notion of rights for international export. *Not* included are the utopian socialists' many experimental communities, which often explicitly embraced (what *became*) non-standard, even anti-capitalistic notions of rights, such sustained and massive struggles as the labor movement's organization around working persons' rights, and the various modern attempts by most social democracies to institutionalize rights to health care.

A long period of experimentation and uncertainty, then, ought to be expected and even welcomed in the originary stages of any new ethics. Again, as I suggested above, even the most familiar aspects of personhood co-evolved with a particular, complex and even wildly improbable set of ideas and practices. Protestantism contributed not just a theology, and not just Calvin's peculiar and (if Weber is right) peculiarly world-historical "inner-world asceticism," but also such seemingly simple projects as an accessible Bible in the vernacular.

Imagine the extraordinary impact of being able to read the holy text oneself after centuries of only the most mediated access. Imagine the extraordinary self-preoccupation created by having to choose for the first time between rival versions of the same revelation, with not only one's eternal soul in the balance, but often one's earthly life as well. Only against such a background of practice did it become possible to begin to experience oneself as an individual, separate from others, beholden to inner voices and "one's own values," "giving direction to one's life" oneself, as Taylor puts it, and bearing the responsibility for one's choices.

Since we now look at the evolution of the values of persons mostly from the far side, it is easy to miss the fundamental contingency of those values and their dependence upon practices, institutions and experiences that were for their time genuinely uncertain and exploratory. Today we are too used to that easy division of labor that leaves ethics only the systematic tasks of "expressing" a set of values that is already established, and abandons the originary questions to the social sciences. As a result, ethics is incapacitated when it comes to dealing with values that are *now* entering the originary stage. Even when it is out of its depth, we continue to imagine that systematic ethics, such as the ethics of the person, is the only kind of ethics there is. We continue to regard the contingency, open-endedness and uncertainty of "new" values as an objection to them, ruling them out of ethical court entirely, or else as a kind of embarrassment to be quickly papered over with an ethical theory.

This discussion has direct application to environmental ethics. First and fundamentally, if environmental ethics is indeed at an originary stage, we can have only the barest sense of what ethics for a culture truly beyond anthropocentrism would actually look like. The Renaissance and the Reformation did not simply actualize some pre-existing or easily anticipated notion of persons, but rather played a part in the larger *co-evolution* of respect for persons. What would emerge could only be imagined in advance in the dimmest of ways, or not imagined at all. Similarly, we are only now embarking on an attempt to move beyond anthropocentrism, and we simply cannot predict in advance where even another century of moral change will take us.

Indeed, when anthropocentrism is finally cut down to size, there is no reason to think that what we will have or need in its place will be something called *non*-anthropocentrism at all – as if that characterization would even begin to be useful in a culture in which anthropocentrism had actually been transcended. Indeed it may not even be any kind of "centrism" whatsoever, i.e. some form of hierarchically structured ethics. It is already clear that hierarchy is by no means the only option.[24]

Second and correlatively, at this stage, exploration and metaphor are crucial to environmental ethics. Only later can we harden originary notions into precise analytic categories. Any attempt to appropriate the moral force of rights language for (much of) the transhuman world, for example, ought to be expected from the start to be *im*precise, literally confused. (Consider "animal rights." The very concept of *animal* seems to preclude one of them being one of "us," i.e. persons, i.e. rights holders.) It need not be meant as a description of prevailing practice; rather, it should be read as an attempt to *change* the prevailing practice. Christopher Stone's book *Should Trees Have Standing? Toward Legal Rights for Natural Objects*, for example, makes a revisionist proposal about legal arrangements; it does not offer an analysis of the existing concept of rights.[25]

Something similar should be understood when we are invited to conceive not only animals or trees as rights holders, but also the land as a community and the planet as a person. All such arguments should be understood to be rhetorical, in a non-pejorative, pragmatic sense: they are suggestive and open-ended sorts of challenges, even proposals for Deweyan kinds of social reconstruction, rather than attempts to demonstrate particular conclusions on the basis of premises that are supposed to already be accepted.[26] The force of these arguments lies in the way they open up the possibility of new connections, not in the way they settle or "close" any questions. Their work is more creative than summative, more prospective than retrospective. Their chief function is to provoke, to loosen up the language, and correspondingly our thinking, to fire the imagination: to *open* questions, not to settle them.

The founders of environmental ethics were explorers along these lines. Here I want, in particular, to reclaim Aldo Leopold from the theorists. Bryan Norton reminds us, for example, that Leopold's widely cited appeal to the "integrity, stability, and beauty of the biotic community" occurs in the midst of a discussion of purely economic constructions of the land. It is best read, Norton says, as a kind of counterbalance and challenge to the excesses of pure commercialism, rather than as a criterion for moral action all by itself. Similarly, John Rodman has argued that Leopold's work should be read as an environmental ethic *in process*, complicating the anthropocentric picture more or less from within, rather than as a kind of proto system, simplifying and unifying an entirely new picture, that can be progressively refined in the way that utilitarian and deontological theories have been refined over the last century.[27] Leopold insists, after all, that

> the land ethic [is] a *product of social evolution*. . . . Only the most superficial student of history supposes that Moses "wrote" the Decalogue; it evolved in the mind [and surely also in the practices!] of the thinking community, and Moses wrote a tentative summary of it. . . . I say "tentative" because evolution never stops.[28]

It might be better to regard Leopold not as purveying a general ethical theory at all, but rather as simply *opening* some questions, unsettling some assumptions, and prying the window open just far enough to lead, in time, to much wilder and certainly more diverse suggestions or "criteria."

Third and more generally, as I put it above, the process of evolving values and practices at originary stages is seldom a smooth process of progressively filling in and instantiating earlier outlines. At the originary stage we should instead expect a variety of fairly incompatible outlines coupled with a wide range of proto-practices, even social experiments of various sorts, all contributing to a kind of cultural working-through of a new set of possibilities. In environmental ethics, we arrive at exactly the opposite view from that of J. Baird Callicott, for example, who insists that we attempt to formulate, right now, a complete, unified, even "closed" (his term) theory of environmental ethics. Callicott even argues that contemporary environmental ethics should not tolerate more than one basic type of value, insisting on a "univocal" environmental ethic.[29] In fact, however, as I argued above, originary stages are the worst possible times at which to demand that we all speak with one voice. Once a set of values is culturally consolidated, it may well be possible, perhaps even necessary, to reduce them to some kind of consistency. But environmental values are unlikely to be in such a position for a very long time. The necessary period of ferment, cultural experimentation, and thus, *multi*-vocality is only *beginning*. Although Callicott is right, we might say, about the demands of systematic ethical theory at later cultural stages, he is wrong – indeed wildly wrong – about what stage environmental values have actually reached.

V ENABLING ENVIRONMENTAL PRACTICE

Space for some analogues to the familiar theories does remain in the alternative environmental ethics envisioned here. I have argued that although they are unreliable guides to the ethical future, they might well be viewed as another kind of ethical experiment or proposal rather like, for example, the work of the utopian socialists. However unrealistic, they may, nonetheless, play a historical and transitional role, highlighting new possibilities, inspiring reconstructive experiments, even perhaps eventually provoking environmental ethics' equivalent of a Marx.

It should be clear, though, that the kind of constructive activity suggested by the argument offered here goes far beyond the familiar theories as well. Rather than systematizing environmental values, the overall project at this stage should be to begin *co-evolving* those values with practices and institutions that make them even *un*systematically possible. It is this point that I now want to develop by offering one specific example of such a co-evolutionary practice. It is by no means the only example. Indeed, the best thing that could be hoped, in my view, is the emergence of many others. But it is *one* example, and it may be a good example to help clarify how such approaches might look, and thus to clear the way for more.

A central part of the challenge is to create the social, psychological and phenomenological preconditions – the conceptual, experiential or even quite literal "space" – for new or stronger environmental values to evolve. Because such creation will "enable" these values, I call such a practical project *enabling environmental practice*.

Consider the attempt to create actual, physical spaces for the emergence of transhuman experience, *places* within which some return to the experience of and immersion in natural settings is possible. Suppose that certain places are set aside as quiet zones, places where automobile engines, lawnmowers and low-flying aeroplanes are not allowed, and yet places where people will live. On one level, the aim is modest: simply to make it possible to hear the birds, the winds and the silence once again. If bright outside lights were also banned, one could see the stars at night and feel the slow pulsations of the light over the seasons. A little creative zoning, in short, could make space for increasingly divergent styles of living on the land – for example, experiments in recycling and energy self-sufficiency, Midgleyan mixed communities of humans and other species, serious "reinhabitation" (though perhaps with more emphasis on place and community than upon the individual reinhabiters), the "ecosteries" that have been proposed on the model of monasteries, and other possibilities not yet even imagined.[30]

Such a project is not utopian. If we unplugged a few outdoor lights and rerouted some roads, we could easily have a first approximation in some parts of the country right now. In gardening, for example, we already experience some semblance of mixed communities. Such practices as beekeeping, moreover, already provide a model for a symbiotic relation with the "biotic community." It is not hard to work out policies to protect and extend such practices.

Enabling environmental practice is, of course, a *practice*. Being a practice, however, does not mean that it is not also philosophical. Theory and practice interpenetrate here. In the abstract, for example, the concept of "natural settings," just invoked, has been acrimoniously debated, and the best-known positions are unfortunately more or less the extremes. Social ecologists insist that no environment is ever purely natural, that human beings have already remade the entire world, and that the challenge is really to get the process under socially progressive and politically inclusive control. Some deep ecologists, by contrast, argue that only wilderness is the "real world."[31] Both views have something to offer. Nevertheless, it may be that only from within the context of a new practice, even so simple a practice as the attempt to create "quiet places," will we finally achieve the necessary distance to take what we can from the purely philosophical debate, and also to go beyond it towards a better set of questions and answers.

Both views, for example, unjustly discount "encounter." On the one hand, non-anthropocentrism should not become anti-anthropocentrism: the aim should not be to push humans out of the picture entirely, but rather to open up the possibility of reciprocity *between* humans and the rest of nature. Nevertheless, reciprocity does require a space that is not wholly permeated by humans either. What we need to explore are possible realms of *interaction*. Neither the wilderness nor the city (as we know it) are "the real world," if we must talk in such terms. We might take as the most "real" places the places where humans and other creatures, honored in their wildness and potential reciprocity, can come together, perhaps warily, but at least openly.

The work of Wendell Berry is paradigmatic of this kind of philosophical engagement. Berry writes, for example, of "the phenomenon of edge or margin, that we know to be one of the powerful attractions of a diversified landscape, both to wildlife and to humans." These margins are places where domesticity and wildness meet. Mowing his small hayfield with a team of horses, Berry encounters a hawk who lands close to him, watching carefully but without fear. The hawk comes, he writes,

> because of the conjunction of the small pasture and its wooded borders, of open hunting ground and the security of trees. . . . The human eye itself seems drawn to such margins, hungering for the difference made in the countryside by a hedgy fencerow, a stream, or a grove of trees. These margins are biologically rich, the meeting of two kinds of habitat.[32]

The hawk would not have come, he says, if the field had been larger, or if there had been no trees, or if he had been plowing with a tractor. Interaction is a fragile thing, and we need to pay careful attention to its preconditions. As Berry shows, attending to interaction is a deeply philosophical and phenomenological project as well as a practical one – but, nonetheless, it always revolves around and refers back to practice. Without actually maintaining a farm, he would know very little of what he knows, and the hawk would not – *could* not – have come to him.

Margins are, of course, only one example. They can't be the whole story. Many creatures avoid them. It is for this reason that the spotted owl's survival depends on large tracts of old-growth forest. Nonetheless, they are still part of the story – a part given particularly short shrift, it seems, by all sides in the current debate.

It is not possible in a short article to develop the kind of philosophy of "practice" that would be necessary to work out these points fully. However, I can at least note two opposite pitfalls in speaking of practice. First, it is not as if we come to this practice already knowing what values we will find or exemplify there. Too often the notion of practice in contemporary philosophy has degenerated into "application," i.e. of prior principles or theories. At best, it might provide an opportunity for feedback from practice to principle or theory. I mean something more radical here. Practice is the opening of the "space" for interaction, for the reemergence of a larger world. It is a kind of exploration. We do not know in advance what we will find. Berry had to *learn*, for example, about margins. Gary Snyder and others propose Buddhist terms to describe the necessary attitude, a kind of mindfulness, attentiveness. Tom Birch calls it the "primary sense" of the notion of "consideration."[33]

On the other hand, this sort of open-ended practice does not mean reducing our own activity to zero, as in some form of quietism. I do not mean that we simply "open, and it will come." There is not likely to be any single and simple set of values that somehow emerges once we merely get out of the way. Berry's view is that a more open-ended and respectful relation to nature requires constant and creative *activity* – in his case, constant presence in nature, constant inter-action with his own animals, maintenance of a place that maximizes margins. Others will, of course, choose other ways. The crucial thing is that humans must neither monopolize the picture entirely nor absent ourselves from it completely, but rather try to live in inter-action, to create a space for genuine encounter as part of our ongoing reconstruction of our own lives and practices. What will come of such encounters, what will emerge from such sustained interactions, we cannot yet say.

No doubt it will be argued that Berry is necessarily an exception, that small unmechanized farms are utterly anachronistic, and that any real maintenance of margins or space for encounter is unrealistic in mass society. Perhaps. But these automatically accepted common-places are also open to argumentation and experiment. Christopher Alexander and his colleagues, in *A Pattern Language* and elsewhere, for example, make clear how profoundly even the simplest architec-tural features of houses, streets, and cities structure our experience of nature – and that they can be consciously redesigned to change those experiences. Windows on two sides of a room make it possible for natural light to suffice for daytime illumination. If buildings are built on those parts of the land that are in the worst condition, not the best, we thereby leave the most healthy and beautiful parts alone, while improving the worst parts. On a variety of grounds, Alexander and his colleagues argue for the presence of both still and moving water throughout the city, for extensive common land – "accessible green," sacred sites, and burial grounds within the city – and so on. If we build mindfully, they argue, maintaining and even expanding margins is not only possible, but easy, even with high human population densities.[34]

VI CONCLUSION

In the last section, I offered only the barest sketch of enabling environmental practice: a few examples, not even a general typology. To attempt a more systematic typology of its possible forms at this point seems to me premature, partly because ethics has hitherto paid so little attention to the cultural constitution of values that we have no such typology, and partly because the originary stage of environmental values is barely underway.

Moreover, enabling environmental practice is itself only one example of the broader range of philosophical activities invited by what I call the co-evolutionary view of values. I have not denied that even theories of rights, for instance, have a place in environmental ethics. However, it is not the only "place" there is, and rights themselves, at least when invoked beyond the sphere of persons, must be understood (so I argue) in a much more metaphorical and exploratory sense than usual. This point has also been made by many others, of course, but usually with the intention of ruling rights talk out of environmental ethics altogether. A pluralistic project is far more tolerant and inclusive. Indeed, it is surely an advantage of the sort of umbrella conception of environmental ethics I am suggesting here that nearly all of the current approaches may find a place in it.

Because enabling environmental practice is closest to my own heart, I have to struggle with my own temptation to make it the whole story. It is not. Given the prevailing attitudes, however, we need to continue to insist that it is *part* of the story. Of course, we might still have to argue at length about whether and to what degree enabling environmental practice is "philosophical" or "ethical." My own view, along pragmatic lines, is that it is both, deeply and essentially. Indeed, for Dewey the sustained practice of social reconstruction – experimental, improvisatory, and pluralistic – is the most central ethical practice of all. But that is an argument for another time. It is, nevertheless, one of the most central tasks that now calls to us.

ACKNOWLEDGMENTS

"Before Environmental Ethics" is republished from *Environmental Ethics* Vol. 14, No. 4 (Winter 1992). I would like to thank Holmes Rolston, III, Jennifer Church, Jim Cheney, Tom Regan, Tom Birch and two anonymous referees for many helpful comments on an earlier version of this paper.

NOTES

1 Some landmarks of this body of work come into view in the later discussion. For a general overview of work on ethical ideas in particular

from this perspective, see Maria Ossowska, *Social Determinants of Moral Ideas* (Philadelphia: University of Pennsylvania, 1970).

2 I distinguish *anthropocentrism* as a philosophical position, issuing in an ethic, from the practices and institutions in which that ethic is embodied, which I call "anthropocen*trized.*"

3 See Anthony Weston, "Non-Anthropocentrism in a Thoroughly Anthropocentrized World," *The Trumpeter* 8, No. 3 (1991): 108–112.

4 Paul Taylor, *Respect for Nature* (Princeton: Princeton University Press, 1986), p. 3.

5 Mary Midgley, *Animals and Why They Matter* (Athens: University of Georgia Press, 1983), p. 118. See also Arne Naess, "Self-Realization in Mixed Communities of Humans, Bears, Sheep and Wolves," *Inquiry* 22 (1979): 231–241.

6 A tradition beginning with Max Weber, *The Protestant Ethic and the Spirit of Capitalism,* trans. Talcott Parsons (New York: Scribner's, 1958) and *Economy and Society: An Outline of Interpretive Sociology,* ed. G. Roth and C. Wittich (Berkeley: University of California Press, 1978), and carried into the present in different ways by, e.g., Morris Berman, *The Reenchantment of the World* (Ithaca, NY: Cornell University Press, 1981) and Albert Borgmann, *Technology and the Character of Contemporary Life* (Chicago: University of Chicago Press, 1984).

7 For an argument in defense of this point, see Anthony Weston "Beyond Intrinsic Value: Pragmatism in Environmental Ethics," *Environmental Ethics* 7 (1985): 321–389 (reprinted in this volume).

8 Jim Cheney, "The Neo-Stoicism of Radical Environmentalism," *Environmental Ethics* 11 (1989): 293–325.

9 John Rawls, "Kantian Constructivism in Moral Theory," *Journal of Philosophy* 77 (1980): 318; and "Justice as Fairness: Political, not Metaphysical," *Philosophy and Public Affairs* 14 (1985): 228.

10 John Arras, "The Revival of Casuistry in Bioethics," *Journal of Medicine and Philosophy* 16 (1991): 44.

11 Michael Walzer, *Spheres of Justice* (New York: Basic Books, 1983).

12 Alasdair MacIntyre, *After Virtue* (Notre Dame, IN: University of Notre Dame Press, 1981).

13 Charles Taylor, *Sources of the Self* (Cambridge, MA: Harvard University Press, 1989).

14 Sabina Lovibond, *Realism and Imagination in Ethics* (Minneapolis: University of Minnesota Press, 1983).

15 Weber, *The Protestant Ethic and the Spirit of Capitalism* and *Economy and Society.*

16 For classic examples of selves in other keys, see Louis Dumont, *Homo Hierarchichus* (Chicago: University of Chicago Press, 1980) and Colin Turnbull, *The Forest People* (New York: Simon and Schuster, 1961).

17 Unavoidable here is the Kantian objection that ethical values actually offer "reasons" rather than anything in the merely "causal" universe. My dogmatic response is that, despite its patina of logical necessity, this insistence on seceding from the phenomenal world actually derives from the same misconception of "causal" stories criticized in the text. Let me add, however, that, in my view, the idea that one can somehow

understand and systematize ethical values in ignorance of their origins and social dynamics also partakes of the spectacular overconfidence in philosophical reason implicitly criticized in this paper as a whole. For some support on this point, see Kai Nielsen, "On Transforming the Teaching of Moral Philosophy," *APA Newsletter on Teaching Philosophy*, November 1987, pp. 3–7.

18 Witold Rybezynski, *Home: A Short History of an Idea* (New York: Viking, 1986).

19 I don't mean to deny that rapid change (both cultural and biological) occasionally does occur, perhaps precipitated by unpredictable but radical events. Drastic global warming or a Chernobyl-type accident outside of Washington DC might well precipitate a drastic change in our environmental practices. Still, even in moments of crisis we can only respond using the tools that we then have. From deep within our anthropocentrized world it remains hard to see how we can respond without resorting either to some kind of "enlightened" anthropocentrism or to a reflex rejection of it, still on anthropocentrism's own terms. Thus, when I speak of "fundamental" change, I mean change in the entire system of values, beliefs, practices and social institutions – not just in immediate practices forced upon us by various emergencies.

20 For this way of putting the matter, I am indebted to Rom Harré.

21 In general, the possibility of invoking dissonant strands in a complex culture is part of the reason that radical social criticism is possible in the first place. Cf. Lovibond, *Realism and Imagination in Ethics*; Walzer, *Interpretation and Social Criticism* (Cambridge, MA: Harvard University Press, 1987); and Anthony Weston, *Toward Better Problems: New Perspectives on Abortion, Animal Rights, the Environment, and Justice* (Philadelphia, PA: Temple University Press, 1992), pp. 167–174.

22 Richard Bernstein, *Beyond Objectivism and Relativism* (Philadelphia, PA: University of Pennsylvania Press, 1983).

23 While Hugo Bedau (in "International Human Rights," in Tom Regan and Donald VanDeveer, eds, *And Justice Toward All: New Essays in Philosophy and Public Policy* [Totowa, NJ: Rowman and Littlefield, 1982]) calls the declaration "the triumphant product of several centuries of political, legal, and moral inquiry into . . . 'the dignity and worth of the human person'" (p. 298), he goes on to assert that "It is . . . doubtful whether the General Assembly that proclaimed the UN Declaration understood what a human right is," since in the document rights are often stated loosely and in many different modalities. Ideals, purposes, and aspirations are run together with rights. At the same time, moreover, the declaration allows considerations of general welfare to limit rights, which seems to undercut their function as protectors of individuals against such rationales (p. 302n). In opposition to Bedau's position, however, I am suggesting that the General Assembly understood what rights are very well. Rights language is a broad-based moral language with multiple purposes and constituencies: in some contexts a counterweight to the typically self-serving utilitarian rhetoric of the powers that be; in others, a provocation to think seriously about even such often-mocked ideas as a right to a paid vacation, etc.

24 See, for example, Bernard Williams, *Ethics and the Limits of Philosophy* (Cambridge, MA: Harvard University Press, 1985); Walzer, *Spheres of Justice*; and Karen Warren, "The Power and Promise of Ecofeminism," *Environmental Ethics* 12 (1990): 125–146.

25 Christopher Stone, *Should Trees Have Standing? Toward Legal Rights for Natural Objects* (Los Altos, CA: William Kaufmann, 1974). G. E. Varner, in "Do Species Have Standing?" *Environmental Ethics* 9 (1987): 57–72, points out that the creation of new legal rights – as, for example, in the Endangered Species Act – helps expand what W. D. Lamont calls our "stock of ethical ideas – the mental capital, so to speak, with which [one] begins the business of living." There is no reason that the law must merely reflect "growth" that has already occurred, as opposed to motivating some growth itself.

26 See Chaim Perelman, *The Realm of Rhetoric* (Notre Dame, IN: University of Notre Dame Press, 1982) and C. Perelman and L. Olbrechts-Tyteca, *The New Rhetoric* (Notre Dame, IN: University of Notre Dame Press, 1969) for an account of rhetoric that resists the usual Platonic disparagement.

27 Bryan G. Norton, "Conservation and Preservation: A Conceptual Rehabilitation," *Environmental Ethics* 8 (1986): 195–220; John Rodman: "Four Forms of Ecological Consciousness Reconsidered," in Donald Scherer and Thomas Attig, eds, *Ethics and Environment* (Englewood Cliffs, NJ: Prentice-Hall, 1983): 89–92. Remember also that Leopold insists that ethics are "products of social evolution" and that "nothing so important as an ethic is ever 'written'" – which again suggests that we ought to rethink the usual reading of Leopold as an environmental-ethical theorist with a grand criterion for ethical action.

28 Aldo Leopold, *A Sand County Almanac* (New York: Oxford University Press, 1949), p. 225.

29 J. Baird Callicott, "The Case against Moral Pluralism," *Environmental Ethics* 12 (1990): 99–124.

30 On "ecosteries," see Alan Drengson, "The Ecostery Foundation of North America: Statement of Philosophy," *The Trumpeter* 7, No. 1 (1990): 12–16. On "re-inhabitation," a good starting point is Peter Berg, "What is Bioregionalism?" *The Trumpeter* 8, No. 1 (1991): 6–12.

31 See, for instance, Dave Foreman, "Reinhabitation, Biocentrism, and Self-Defense," *Earth First!*, 1 August 1987; Murray Bookchin, "Which Way for the US Greens?" *New Politics* 2 (Winter 1989): 71–83; and Bill Devall, "Deep Ecology and its Critics," *Earth First!*, 22 December 1987.

32 Wendell Berry, "Getting Along with Nature," in *Home Economics* (San Francisco: North Point Press, 1987), p. 13.

33 Gary Snyder, "Good, Wild, Sacred," in *The Practice of the Wild* (San Francisco: North Point Press, 1990); Tom Birch, "Universal Consideration," paper presented at the International Society for Environmental Ethics, American Philosophical Association, 27 December 1990; Jim Cheney, "Eco-Feminism and Deep Ecology," *Environmental Ethics* 9 (1987): 115–145. Snyder also speaks of "grace" as the primary "practice of the wild"; Doug Peacock, *The Grizzly Years*

(New York: Holt, Henry and Co., 1990), insists upon "interspecific tact"; Berry writes of an "etiquette" of nature; and Birch of "generosity of spirit" and "considerateness." All of these terms have their home in a discourse of manners and personal bearing, rather than moral discourse as usually conceived by ethical philosophers. We are not speaking of some universal categorical obligation, but rather of something much closer to us, bound up with who we are and how we immediately bear ourselves in the world – though not necessarily any more "optional" for all that.

34 Christopher Alexander, et al., *A Pattern Language* (New York: Oxford University Press, 1977). On windows, see secs. 239, 159 and 107; on "site repair," sec. 104; on water in the city, secs. 25, 64 and 71; on "accessible green," secs. 51 and 60; and on "holy ground," secs. 24, 66 and 70.

8

COMPATIBILISM IN POLITICAL ECOLOGY

Andrew Light

In two earlier papers I explored the distinction in ecological and environmental political theory between what I called ontologists[1] and materialists.[2] In this paper I will review that distinction and some of the representative theorists who hold these positions, before formulating and evaluating a strategy for negotiating these competing claims in their conceptions of environmental political theory. I will propose a form of environmetal pragmatism as the best framework within which to make competing political theories compatible in practice. In exploring the potential of environmental pragmatism as a meta-philosophical guidepost for a radical political ecology, I will here draw from the sometimes maligned neo-pragmatism of Richard Rorty which he never specifically directs towards environmental concerns. There are, of course, other sources from which to construct an environmental pragmatism which may have broader appeal. If this framework emerging out of Rorty's work can be made intuitively plausible then a more inclusive position should not be too difficult to formulate.

The main thrust of my argument is that the urgency of the environmental crisis forces a need for a new form of metatheoretical compatibilism on the part of both environmental materialists and ontologists. One would hope that this would be a fairly uncontroversial position to take as both of these schools of thought claim that environmental concerns are the pre-political conditions around which a political theory must be formed. A radical pragmatist stance can give us the compatibilism necessitated by dire environmental conditions, without necessitating the surrender of the philosophical commitments of these two forms of political ecology. But again, the form of pragmatism advocated here is one importantly predicated as *environmental*, as a reminder that environmental concerns play a

161

foundational role in directing how radical ecologists decide to form bonds of agreement towards ecological renewal in practice.

MATERIALISTS AND ONTOLOGISTS

For both environmental materialists and ontologists, environmental problems require much deeper analysis than classical liberal policy-making approaches would suggest. Materialists, like Murray Bookchin's school of social ecologists and Herbert Marcuse's version of environmental critical social theory, see the crisis of environmental degradation, and human suffering as a result of that degradation, as presupposed by the material conditions of a capitalist (or state capitalist) economy. Included in their primary concerns are the technological processes that are part of these economies as well as the political systems which sustain them globally.[3] Material conditions (such as who owns and controls the technological processes that are used to sustain economic growth, expand markets and consume natural resources) are for such thinkers the starting points for unpacking the complex web of environmental problems. From such an analysis, Bookchin, Marcuse and other environmental materialists are led to the conclusion that the solutions to these problems should begin with an analysis of the possible range of alternatives that a political or economic system can sustain with an eye towards pushing the boundaries of these systems and changing the material conditions of society as a whole. Changes in individual consciousness, while sometimes necessary for sustaining new social conditions will, on this general theory, follow material changes. By demonstrating that environmental problems emanate from material conditions of society, environmental materialists identify systemic facets of environmental problems that go deeper than the evaluations found in most liberal policy analysis.[4]

The basic premises of environmental materialism can be seen at work in Bookchin's opposition to any political system which assumes a view of nature under the rubric of demands for unlimited economic growth. From this position, he can attack both market-driven capitalist economies and statist planned political economies for their view of nature as an object only to be used to garner more resources. He argues that any economy structured around the maxim "Grow or Die" will necessarily pit itself against the natural world and inevitably lead to ecological ruin.[5] In the press for economic growth, Bookchin says, the state and its people replace the organic with the inorganic, soil with concrete, living forest with barren earth and the diversity of life forms with simplified ecosystems. Such a move, he sees, as a "turning back of the evolutionary clock . . . to a world incapable of supporting complex life-forms of any kind, including the human species."[6]

More specifically in liberal society, as Bookchin reads it, material power relationships are structured around a social status quo which serves the interests of property owners. Therefore, the interests of that part of the public which is property-less, and importantly the property itself, will always be sacrificed to the power interests of the landlord, the developer or the seeker of resources.[7] The best that liberal environmentalists can do, because they are committed to the preservation of the free-market system (or more accurately, they are unwilling to oppose it through revolutionary means), is to garner a few compromises and trade-offs with business interests within the market. These commitments render the liberal approach ineffectual as an authentic challenge to the pro-property-owning interests of the market system. Bookchin concludes from this that the willingness of liberal environmentalists to accept such piecemeal advances must rest on a presupposition that certain fundamental institutions in the liberal, capitalist state are here to stay: "All of these 'compromises' and 'trade-offs' rest on the paralyzing belief that a market society, privately owned property, and the present-day bureaucratic nation-state cannot be changed in any sense."[8] This focus on the material impediments to environmental reform is at the root of all theories that I identity as environmental materialist.

Of course, there are individual differences between the representatives of these theories as I am grouping them together (differences, however, that I do not think break the back of the general distinction).[9] One unique difference of Bookchin's materialism, important to understand for a full appreciation of his critique of ontologists, can be seen in his argument concerning the relationship of the social order to the development of technology.

Adapting a thesis made popular in the nineteenth century by Peter Kropotkin, Bookchin claims that anthropological data suggest that "participation, mutual aid, solidarity, and empathy were the social virtues early human groups emphasized within their communities."[10] After making this point, it is a short step to Bookchin's next argument that socialization itself stems from nature and that "every social evolution, in fact, is virtually an extension of natural evolution into a distinctly human realm. . . . Social ecology clearly expresses the fact that society is not a sudden 'eruption' in the world."[11]

163

For Bookchin, even though every socialization is a product of natural evolution, "we are substantially less than human today in view of our still unknown potential to be creative, caring and rational."[12] Therefore he claims that human society now inhibits rather than realizes its evolutionary potential. It is from this very strong view that we can see the potential for a striking amount of disagreement with other political ecologies. Social evolution is a foundational idea for Bookchin that distinguishes his claims from others and ultimately in his view, how a political ecology should be practiced.

His rationale for this argument concerning evolutionary theory is that since humans are having an adverse effect on the environment they must not be living up to their potential. The suppressed premise in this argument is that human potential involves a form of life mutually exclusive with environmental destruction. Behind this argument is Bookchin's belief that at some period in history humanity was separated from its roots in nature, or that the natural progress of social evolution was derailed. If humans had continued to live naturally, without resorting to current relationships of inequitable domination (racism, sexism, anti-humanism, etc.), then they could have fully developed the potential for technical insight, culture and self-reflective thought that is indicative of a post-scarcity society.[13]

I will return to Bookchin's specific form of environmental materialism soon. But first, contrast to environmental materialism, environmental ontologists, who attempt to place humans and nature in some sort of philosophically informed symbiotic context. This context is not ground in some argument for a uniquely *human* evolutionary social history, but is instead based on the inseparable ontological roots of humans *and* non-human nature. Such theorists argue that humans should be identified with nature not as a separable organism or set of organisms, but as an integrated part of a larger life/world system.

Like some eighteenth-century contractarian political theories, environmental ontology draws on a conception of nature to form definitions of the self and hence to inform some common notions that may serve as a basis for political and social organization.[14] These articulations of the relation of humans to nature in turn influence how one reflects on the natural world and places it in discussions of politics, economics and other areas of human interaction.

Environmental ontologists focus their critique of mainstream environmentalism on the need for more analysis of and changes in individual human consciousness with respect to the relationship between humans and the non-human natural world. For ontologists, the locus of significant solutions to environmental problems is also to be found primarily in a redescription of human ontology which then serves as the basis for environmental practice and policy. The focus for political reform, then, is in a reform of the self as expressed in individual identity – in contrast to the materialist focus on the social group or the institution as the primary mechanism of environmental renewal. One example, and perhaps the clearest example of an ontological theory can be seen in the development of deep ecology, originally by Arne Naess, and later by his American and Australian disciples.

As is now well known, the term "deep ecology" was coined by Naess in his 1973 article "The Shallow and the Deep, Long-range Ecology Movements."[15] In this article Naess attempted to explain the spiritual, idealistic approach to nature which he saw implicitly behind the writings of people such as Rachel Carson. It was his argument that these more sensitive approaches to nature were the result of a more open perspective towards non-human life. Naess, however, like Bookchin, did not develop his theory in a vacuum. It is not surprising that as an ontologist he does not primarily draw from the tradition of a distinct political movement (as Bookchin does with revolutionary anarchism), but instead on the legacy of the natural philosophy developed as a part of Norwegian mountain culture.[16]

The theory expands from basic principles of ecology to structure a normative and ontological dimension which, when fully developed, employs distinctions that can be attributed to metaphysical rational-ism, mysticism and sometimes Eastern philosophy. The "deep" in deep ecology stems from a kind of multi-traditional vitalism which comes in what Theodore Roszak calls, "an awakening of wholes greater than the sum of their parts."[17]

Deep ecology, if we formulate it through Naess' original state-ments on the theory, certainly posits a primarily ontological and not a materialist criticism of human interaction with the non-human natural world. Naess takes great pains to distinguish his theory from a theory of environmental ethics. He asks that we change the way in which we (individuals) "perceive and construct our world."[18] David Rothenberg cites the Australian deep ecologist Warwick Fox as having given a formulation of this suggestion. Fox argues that:

> The appropriate framework of discourse for describing and presenting deep ecology is not one that is fundamentally to do with the value of the non-human world, but rather one that is

fundamentally to do with the nature and possibilities of the self, or, we might say, the question of who we are, can become, and *should* become in the larger scheme of things.[19]

Naess argues that the experience of the world provides the belief and identity basis for deep ecology, an identity which makes it more than simply an ethical system. Says Naess, "If deep ecology is deep it must relate to our foundational beliefs, not just to ethics. Ethics follow from how we experience the world. If you articulate experience then it can be a philosophy *or* a religion."[20]

I emphasize the "or" in this last reference to highlight again a source of tension between materialists and ontologists, particularly the two exemplary representatives here. This failure to distinguish between the sorts of understanding and reasoning at work in religious *belief,* and that at work in philosophical *explanation,* is at the root of Bookchin's worries of deep ecology as a theory *and* as a practice. Given the general source of his worries, this critique of deep ecology applies equally to all forms of environmental ontology. In fact, Bookchin is very quick to lump all forms of ecospiritualism into one foul mass, usually characterized by its most outrageous representatives and their most counter-intuitive claims.[21]

The focus of Bookchin's claim is that due to its spiritual dimensions (and I would suggest environmental ontological commitments) deep ecology is notable for its "absence of reference to social theory" which makes it incompatible with his social ecology (and hence with the priorities for reform of his environmental materialism). A clear struggle therefore emerges for Bookchin:

> In America, the rapidly forming Green movement is beset by a macho cowboy tendency [the reference is to deep ecologists of the Dave Foreman/Earth First! variety] that has adopted Malthusianism with its racist implications as a dogma, an anti-humanism that among some of the wilderness oriented "campfire" boys has become a brutalized form of misanthropy, and a "*spiritualist*" tendency that tends to extol irrationalism and view ecology more as a religion than a form of health naturalism. It has become primarily the task of American eco-anarchists to develop a sustained resistance to these primitivistic, misanthropic, and *quasi-religious* tendencies.[22]

The philosophical and practical terrain that gets divided here is momentous on Bookchin's scheme, and can reveal important differences between the deep ecology and social ecology camps that

166

cannot simply be reduced to sloganeering. Bookchin attacks, for example, the notion that deep ecologists do not draw distinctions between human and non-human animals. True to this charge, Warwick Fox comments that implicit in the deep ecology movement "Is the idea that we can make no firm ontological divide in the field of existence: That there is no bifurcation in reality between the human and the non-human realms . . . to the extent that we perceive boundaries, we fall short of deep ecological consciousness."[23]

Bookchin focuses on the failure of deep ecologists to make this distinction suggesting that in doing so they ignore important, peculiarly human, social dimensions of environmental problems. To reduce humans to only one species among many is to diminish the importance of human social distinctions, such as class distinctions, that are not found in non-human species.[24] When deep ecologists do not distinguish between human social classes, by lumping human and non-human animals together, the result is a broad distribution of the responsibility for environmental problems among humans. If no social distinctions are recognized between classes then all humans are equally to blame for the destructive impact of humans on the environment. Without an explicit acknowledgment of the specific sources of environmental deterioration, it becomes possible to "comfortably forget that much of the poverty and hunger that afflicts the world has its origin in the corporate exploitation of human beings and nature – in agribusiness and social oppression."[25]

While particular charges like this one can serve as the foundation for an interesting dialogue on important social and political concerns, Bookchin's overall attitude towards deep ecology, characterized by his lumping and splitting of the opposition in the "cowboy" quote, is exemplary of the anti-compatibilism that makes environmental political theory difficult to translate into practice. This is particularly problematic when practice may require large-scale coalitions across a broad array of philosophical commitments. But the blame cannot be laid at one doorstep: Bookchin and Naess are both guilty of a kind of essentializing of the theoretical ground of their positions (the former in Kropotkinian evolutionary theory and the latter in religious ecospiritualism) which would make any kind of communication between the two difficult. But it is not only at the theoretical level that divisions are drawn. It is clear that Bookchin really does see the struggle in practice of ecoanarchists to be directed as much against deep ecologists as it is against liberal environmentalists and growth-oriented polluters of the earth.

167

But once we move beyond the particular example of Bookchin versus Naess, the materialist/ontological distinction gets more complicated and difficult to sustain as a basis for hard and fast theoretical divisions. For one thing, there is a clear ontological dimension to some materialists' work, like that seen in Marcuse's implicit political ecology in *One-Dimensional Man*. There, Marcuse argues that the transformation of technology from its repressive forms in advanced industrial capitalism, which is necessary for environmental renewal, occurs in a framework that includes an ontological dimension. It is only through such a combined materialist-ontological framework that an articulation of true and false needs (necessary to determine among other things the limits to growth Bookchin finds necessary) can be made.[26] A careful reading of Bookchin with Marcuse shows that Bookchin shares many of the same views in his development of the role of false needs in creating one-dimensional growth-oriented societies, as well as in his conception of the relationship between the social structures of a post-scarcity society to the attitudes of individuals in that society.[27]

Even with a materialist base, as it were, there is an acknowledgment in Marcuse of the role of individual ontological transformations as important in the struggle of humans and nature against debilitating forms of technological rationalism and instrumentalism. (This view comes out in his explicit political ecology articulated in *Counter-revolution and Revolt* and in later essays.[28]) This should not be so surprising. On the other side of the environmental spectrum, as I am describing it, even Naess' version of deep ecology includes a dimension that is clearly materialist. In his opus *Økologi, samfunn, og livsstil* (translated as *Ecology, Community and Lifestyle*[29]), he has several chapters discussing the social and political order, the types of economic organization that are best suited to a "no growth" society and references to the types of policy changes that are needed to enact a deep ecology program.[30] While I still want to categorize Naess as an "environmental ontologist," because of his strong commitment to the priority of ontological changes in the transformation to and preservation of any lasting shift in the human relationship with the non-human natural world, we should acknowledge that he also makes a strong contribution to the materialist dimension of political ecology both in theory and practice.

Despite this overlap the distinction between materialists and ontologists in environmental thought in general and political theory in particular is still useful as an heuristic mechanism that aids in analysis rather than as a hard and fast division that points to necessary contentions of deep disagreement. The distinction will help us to predict the *propensity* for certain serious points of contention, such as the earlier example of the status of non-human animals as agents, but it need not be the case that such disagreements are necessary for the distinction to work. The distinction identifies where one's priorities fall on general political questions of environmental reform. But even assuming that split, we can imagine materialists and ontologists agreeing on a particular question of reform. Given, for example, the choice of whether a particular site should be preserved as a wilderness area, both groups might agree that the area should be set aside even if they differed on the role of wilderness areas in the ultimate transformation to a new social order (or to the correct individual attitudes towards nature). From the theorists presented so far, it is noteworthy that Bookchin has even reached out to members of the deep ecology camp to at least talk about what they can potentially agree upon.[31]

Attempts at mixing these positions demonstrates the usefulness of developing both types of theories, materialist and ontological, in the pursuit of answers to questions pertaining to both. That is, in the development of a materialist conception of political ecology, certain ontological lessons will be learned along the way which can ultimately be useful to a strict environmental ontologist. The same is true in the other direction.

VARIETIES OF ENVIRONMENTAL PRAGMATISM

Unfortunately, as has already been indicated, the respective theorists who embrace these arguments do not usually see the merits of trying to combine the benefits of these two broadly construed positions. While I think that both of these approaches to environmental problems are useful and have recognizable advantages over mainstream policy approaches to environmental issues, their utility is sometimes limited by the importance their proponents give to the formulation of the foundations of such positions. In turn, both materialists and ontologists get caught up in arguing with each other over the differences between the theoretical implications of each others' philosophies which may have wearied the public to their mutual concerns.[32] Some may claim that in the case of the social ecology–deep ecology debate that this is a result of the strong personalities involved. Other explanations, however, are needed. From the sketch I have drawn here it seems obvious that in both the case of Bookchin and Naess there is a strong foundationalism at work which, though commendable as an attempt at theoretical rigor, may *in practice* hold up attempts at broad organizational unity.[33]

Explanations for what account for the different intuitions behind the environmental thought of theorists like Naess and Bookchin are wide ranging. Assuming the foundations of Bookchin's evolutionary theory in the biology of Peter Kropotkin one can argue along with Steven Jay Gould that there is something of a geographical origin behind the idea of "mutual aid" as a necessity in social evolution (supervening on natural evolution). Gould argues that Kropotkin's thesis emerged quite intuitively as the sort of assumptions one might make about human nature and social structures from observations of cooperative animal behavior on the Russian tundra where Kropotkin spent his early years.[34] A similar geographical explanation is found in Naess' admission of the central role of the peculiarly Northern Norwegian experience of the wild in the development of his thought, and of the experience of mountaineering in general in his assumptions concerning the ability of all humans to be connected with nature in a central way.[35] But importantly, even if we could come up with some definitive explanation of the differences between the intuitions of materialists and ontologists, at some point as political activists we must negotiate a settlement between these two firmly established camps in order to get things done. Sigmund Kvaløy, the most materialist of the theoretical forerunners of deep ecology, has said something similar in his formulation of "ecophilosophy." He argues that we theorists "should strive to be as wide in scope as the attack on the life struggle of the ecosystem and human society is today."[36]

Even though I count myself as a straightforward environmental materialist and am committed to its theoretical development, I believe that some theoretical questions (and sometimes some strategic materialist principles) while valuable often get in the way of an attempt to formulate a broad-based radical plan to solve environmental problems. In order to achieve that goal some sort of mutual toleration of competing theories is demanded by the overwhelming need for action on the environmental front. But such compatibilism, even though it may appear to violate the foundational claims in the work of some representatives of the two forms of environmental thought just described, need not lead inevitably to relativist environment philosophy. I propose, then, a principle of tolerance in the form of a pragmatic position which would require radical environmentalists to leave some questions which divide them to private dispute. At the same time this pragmatism would require theorists and practitioners to communicate a straightforward public position that endorses the overriding ethical and political environmental considerations on which they agree and the practices which best meet the needs of their mutually desired goals.

There have been several attempts to express the connections between classical American philosophy and environmental ethics,[37] and a general pragmatist account of some debates in environmental ethics,[38] but no comprehensive attempt as of yet specifically to outline the structure of a politically pragmatist and compatibilist account of ecological theory and practice. Part of the reason for this gap is explained by the failure of those interested in pragmatism and environmental thought to distinguish between two formulations of the relationship between environmentalism and pragmatism. The first form is found in the strategy of using philosophical pragmatism directly to set up substantive positions against other theories in environmental philosophy. Much of the pragmatism in contemporary environmental ethics (of both sorts just described), which gets directed into a move to create an ethical pluralism as a response to what are now mainstream positions of biocentrism or anthropocentrism in the field, occupies this ground. Such moves do not necessarily contribute to the politically pragmatist principles that I am looking for here and wind up usually adding to the number of intractable sides in environmental ethics debates.[39] Unfortunately, many theorists seem to take this to be the sole contribution that pragmatism can make to environmental thought.

In contrast, a second application of pragmatism to environmentalism emerges out of contemporary pragmatist inspired sources which have evolved as answers to theoretical log jams in other areas. This second position would use the theoretical predisposition of a generalized philosophical pragmatism to weed through debates in environmental ethics and environmental political and social theory but would not attempt to articulate a full and complete ethical or political theory on its own. The roots of this sort of theory are in the use of philosophical neo-pragmatism (practiced today by figures like Rorty, Bernstein and Fish) to try to call an end to certain debates in philosophy and critical theory based at least in part on their contributions to practical concerns.

I take the first project to be a direct *philosophical* use of pragmatism, and the second project to be a more directly *metaphilosophical* use of pragmatism. I do not have the space here to give a full critique of the first project. I hope however that it will suffice to say that in principle nothing prohibits a classical pragmatist philosophical or political view from being just as intractable as a rigid materialist or ontological view. Although I realize this is not necessarily the case, it is none the less not my goal here even potentially to contribute a new *side* to that terrain.

So now I will focus on the latter form of pragmatism. It is different in kind from the straightforward philosophical view: it is used more like a yardstick (perhaps a heavy one wielded overhand) to say to people, "All right, enough with some of these debates, they are going nowhere and a lot of this stuff is not very interesting as a philosophical or political problem." Because it is still an *environmental* metaphilosophical pragmatism though, the reason appended to this argument is the serious nature and immediacy of the area to which we have drawn our philosophical attention. Like Kvaløy's version of eco-philosophy, this strategy is in a sense not freely chosen, but appears as a "necessity – as a response required by the total [environmental] system crisis we are experiencing in the world."[40]

Metaphilosophically inclined environmental pragmatists would argue that we need to give up on some of the debates in political ecology for no other reason than the fact that there is much that we do agree on (whether we are materialists or ontologists) that has not yet been effectively put into policy or communicated to the public. From this metatheoretical perspective, environmental pragmatists are not wedded to any particular theoretical framework from which to evaluate specific problems, but can choose the avenue which best protects the long-term health and stability of the environment, regardless of its theoretical origin.

But this position needs to be fleshed out. How does one "give up" on certain debates and continue others while still retaining the integrity of one's political-philosophical position? In what remains of this paper I want to sketch out one way of constructing a version of metaphilosophical environmental pragmatism, specifically as a guidepost for political ecology, which tries to endorse this theoretical–practical balance. Along the way I will give a concrete example of the sorts of choices which result from embracing such a pragmatism.

METAPHILOSOPHICAL COMPATIBILISM

For environmental pragmatists, as is true with all political ecologists, concern for the environment becomes a pre-political condition that any future politics must have the ability to address. That is, these theories suggest that political thought in general, and specific political decisions in particular, can only be determined within the context of their effects on the state of the environment. The subject of ecology is inherently part of the public sphere for such theories, and the health of the public sphere is the object of any political theory. Such a connection between environmental issues and political organization has in some ways characterized some green politics already, particularly in Europe. Christopher Manes has pointed out that, "With the looming threats of radiation, acid rain, and toxic waste in their own backyard, European environmentalists did not have the *luxury of separating ecology from politics.*"[41] Of course there are many ways to embrace a specifically ecological political theory. One could construct such a theory from within the bounds of an authoritarian system out of the same motivation emanating from the environmental crisis as we have seen in the egalitarian visions of Bookchin and Naess.[42] In this paper I will not seriously consider the authoritarian option for political ecology, so all of the theories considered here can more accurately be described as examples of democratic political ecology (DPE), where "democratic" does not denote a specific form of government but it does indicate a rejection of all anti-egalitarian types. A specifically democratic political ecology, then, is one that also takes as a presupposition the necessity of maintaining some form of participatory political order within the public sphere. The rules of the formation of political ecology, as it were, must emerge within some kind of participatory context and the implementation of the policies emerging from political ecology must accord with reasonable egalitarian principles. So what marks the DPE theorist who embraces my form of environmental pragmatism?

The environmental pragmatist takes the commitment to stopping environmental destruction and increasing the likelihood of sustaining and expanding existing organic environmental systems a bit further than the non-metaphilosophically pragmatic inclined political ecologist or DPE theorist. Here the ecological-political claim is extended from its role as a pre-political condition for formulating a theoretical political scheme to serve the function of a regulative ideal emanating from the lessons of practice, and ranging over the use and function of theory in political ecology. Difficulties in political ecological practice lead us therefore to new considerations of the status of our theories. Again, I find an ally to this approach in Sigmund Kvaløy, who, not surprisingly, Peter Reed and David Rothenberg describe as closest to Marx, among the Norwegian environmental philosophers, in linking practice to theory through action (thus appropriately challenging their separation from the beginning).[43] Kvaløy interestingly enough finds a plethora of good action and a dearth of good theories coming out of the first decade of environmental activism of which he played a major role in Europe. He maintains that out of this process, "We need . . . to sit down and do ecophilosophy again, to philosophize under direct influence of the reaped experience."[44]

In accordance with this goal, metaphilosophical environmental pragmatism is a good principle to embrace. One strategy within this framework is for the pragmatist not to be concerned in public with the outcome of some debates among environmentalists and ecologists within a certain range of issues. For example, some questions would be shelved for now, such as: What is the *intrinsic* value of species that fuels a duty to insure their diversity? What is *the* ontological relationship between humans and non-human animals? Does technology *necessarily* exercise a pejorative influence on humans or is it a natural development of human cognitive evolution? In all these cases the environmental pragmatist searches for answers to these questions in private while publicly pursuing the best possible solutions to practical environmental questions. The pragmatist also recognizes that the current state of the world, either economically, politically or ontologically, is the result of a contingent history: existing individual or social relationships with the non-human natural world are not assumed to be the only ones that must be worked within in looking for solutions to environmental problems. There are no universal statements on the state of nature available for the environmental pragmatist (or rather, the pragmatist does not permit such statements in public) – descriptions of the relationship between humans and nature are descriptions of states of affairs that could have been otherwise. And because they could have been otherwise pragmatists do not essentialize their conceptions of the identity of nature or our duty to it which would limit their ability to re-conceptualize nature in such a way that it can be most expeditiously protected.

What sort of choices are the result of this view? One example (of interest to all kinds of political ecologists): my metaphilosophical environmental pragmatist would have no public theoretical problem with supporting and undertaking projects like ecological restorations which attempt to reconstruct previously damaged lands. Such projects, like the prairie restorations at the University of Wisconsin's Arboretum, which require some hard choices about what state of the prairie to restore to (pre-Columbian or not, etc.), would none the less be fully supported. Restoration makes sense because on the whole it results in many advantages over mere preservation of ecosystems that have been substantially damaged by humans.[45] Therefore an environmental pragmatist at best would find it only mildly intellectually interesting, and at worst morally irresponsible, to contribute to the raging debate in *Environmental Ethics* over the question of whether ecological restoration inherently can result in a full restoration of the value of the bit of nature being restored.[46]

What is the point of such arguments now, ask the pragmatists? And what answer do we offer the hard-working volunteer restorationist (of which there are thousands) who asks whether environmental philosophers support what he or she is doing? "Well, yes of course we support what you do," says the non-pragmatist engaged in this debate, "but the problem is that no matter what you think you're doing, you can never really *restore* nature (in its full normative sense), because nature has certain definitive value, and the best arguments concerning that value suggest that *a priori* restoration can never really restore nature." Cold comfort it seems for the environmental activist who is engaged in a truly radical practice. This point is particularly cogent when such a restorationist is compared with reform environmentalists who sometimes seems content either to ignore restoration all together, or to leave it in the hands of those who will do it for profit and never really care about the result for the land itself.[47] For purposes of interacting with the world of environmental practice, pragmatists let the demands of the environmental crisis regulate their interaction with the *practice* of restoration. (Assume for the moment that there is no essential normative issue about the moral status of the restored land as a part of the larger assumption, for now, that there is no necessary value of nature that need concern us here.)

Privately though, the pragmatist may hold definite views concerning the moral or political status of restored nature and may even wish to publish these views in obscure academic journals.[48] But the pragmatist will understand that publicly to voice such worries leads to too much confusion in the practice of a project that almost everyone can agree is a good idea: restoring damaged lands to something approximating their former native species diversity to the point where the land can once again regenerate itself. I qualify this description of the

restored landscape to indicate what I hope is obvious – being a pragmatist on questions like this does not mean that one lends full support
to every act of restoration. Sometimes the environmental pragmatist
will want publicly to be *very* vocal about what should count as a
good restoration (he or she may, as I do, want to exclude corporate-
sponsored restorations as good restorations). In that case, it is the
duty of the ecophilosopher, following Kvaløy, to become deeply
involved in the debate on this issue with practitioners, rather than
only privately discussing the problem. The philosopher or political
theorist certainly may have much to add to a full analysis of the
relevant questions concerning the practice.

This version of the public/private distinction employed in this
description of environmental pragmatism is drawn from Richard
Rorty's *Contingency, Irony, and Solidarity*,[49] but in a very qualified
sense that I will explain soon. In general, Rorty argues that modern
philosophy's focus on the pursuit of metaphysical and transcendental
truth has not been useful in the pursuit of freedom and instead has
distracted from attempts to increase the autonomy of individuals or
communities. Rorty has said in other discussions that we should
"look for regulative ideals . . . stick to freedom and forget about truth
and rationality."[50] Instead of pursuing foundations for knowledge
in all cases, Rorty argues that liberal culture needs "an improved self-
description." This description amounts to an openness to redescribe
our beliefs in terms that do not provide us with a reason as to why we
might be wrong about things through any sort of *normative* force,
but rather as suggestions as to how things might have been otherwise.
Reworking truth in such a way justifies redescriptions of our present
state which can then serve as a litmus test for our conceptions of
whether our beliefs are still applicable to our current situation. It is
something like a standard that, in never being met, reminds us that
we might be wrong.

But such redescriptions are limited for Rorty: in public the agents
of redescription (who he calls "liberal ironists") are concerned with
problems of humiliation and suffering, and how bonds of solidarity
can be forged (by redescribing others in terms that place them as agents
towards whom we can easily feel solidarity) between persons to limit
such suffering. In private, liberal ironists can redescribe others in whatever terms they please, terms according to Rorty, "which have nothing
to do with my attitude toward your actual or possible suffering." He continues, "My private purposes, and the part of my final vocabulary which
is not relevant to my public actions, are none of your business."[51]

Environmental pragmatists employ a similar strategy with the exception that their public redescriptions are concerned with the suffering of all of nature, rather than only with the suffering of humans. (Some may find this contentious to speak of nature as "suffering" but whether nature *really does* suffer is again, not important for now for the metaphilosophical pragmatist.) Since they are compelled to speak in public of redescriptions rather than truths, environmental pragmatists will not fall into the trap of articulating essentialist positions that serve as obstacles to the development of a healthy pluralism concerning our duties to and descriptions of nature. Redescriptions, remember, are only guidelines for how things could be different rather than rigid positions on how things must be.

Pragmatists, on my account, may disagree about the framework in which to explore these questions. Some may start with inclinations towards communitarian political ideas and socialist economic frameworks while others may have leanings towards market solutions and various shades of liberalism. Some will be materialists, others ontologists, and others something else. But what my pragmatists agree on is that the truth of these approaches are not always fundamental for purposes of environmental practice and the appropriateness of any one theory in a particular case is contingent on historical, cultural, social and resource conditions. What is constant is the environmental pragmatist's acknowledgment that environmental needs are the foreground and the background of such contingencies and the controlling interest in deciding which framework is most appropriate in deciding how to solve environmental problems.[52] Importantly, the environmental pragmatist because she is not committed to a dogmatic defense of her starting frameworks (which may direct her research or activist interests at first), and because she is not looking for a totalizing definition of the true state of the natural world when so many practical considerations are at hand, is open to acknowledging that at some time her framework may be inappropriate for obtaining the goals of preservation and protection of the environment. On some issues the ontologist must defer to the materialist, for example, when questions of social structures are on the table. Materialists generally have stronger historical traditions considering such questions.

Environmental pragmatists are not concerned with articulating some positive conception of what nature is for the purposes of *practice*, but are instead concerned with negative conceptions of what restricts or restrains things in the natural world, that is, what makes nature (including humans as part of nature) suffer.[53] Such negative conceptions of the "freedom" of nature to "be for itself" are not *necessarily* connected to any strong monistic ideas of nature, but are instead merely an articulation of a sense of pain in the world.

Fortunately, pragmatists in general and environmental pragmatists in particular also understand that inquiry into social problems need not come to an end in order for action to take place. As Rorty suggested some time ago, "reference to action . . . can take place at any step in the eternally incomplete series of interpretations."[54] When we feel the urgency to act, in order to provide aid to nature, we will find temporary stopping places in our ongoing conversations on how best to act for nature and how best to interpret the needs of nature.[55] For pragmatists as I conceive them there may be no other public solutions to some problems other than "pragmatic" ones as we have given up on the idea of a totalizing discourse *for now* about conceptions of nature. But pragmatic solutions are not something we settle on, they are things we strive for while privately pursuing, if we choose, our individual redescriptions of nature in positive, totalizing or hegemonic terms.

This picture, however, is not complete. While it is true that Rorty shares with Marx a desire to detranscendentalize philosophy and connect it with practical critique,[56] his *politics* still is decidedly not radical (and probably reformist), and his philosophical neo-pragmatism is too weak as a whole to sustain the sort of environmental political theory I have in mind. So if in the first part of this paper I am celebrating the environmentalism and the radicalism of materialist and ontological political ecologists, how can I then embrace a version of Rorty's notorious liberalism? The answer is that it is not Rorty's political theory that I am endorsing but only his theory's organizational principles.

Rorty's liberal neo-pragmatism, by itself, is too restrictive for the purposes of political ecologists. At the end of the day, his position is clear: Liberal Democracies seem to be the best form of government created in history so far (even with their more obvious faults), so there is no compelling reason to break from their way of doing things.[57] But following Bookchin, Naess, Marcuse, the early Bahro, Gorz, Jim O'Connor and others, I think there are a lot of reasons why liberal democracies, embedded in capitalist growth-oriented structures, are insufficient by themselves for achieving the goals of long-term environmental sustainability. Classical liberalism, *per se*, is insufficient as a principle within which to compose a political theory which acknowledges the priority of ecology in political structures, planning and decision-making that emerges from a DPE.

I hope I have demonstrated however, that the compatibilism which is generated as a necessary part of Rorty's neo-pragmatist philosophy for constructing bonds of solidarity between different people (and specifically the public/private split as part of that strategy), can be very useful for structuring a metaphilosophical position within which radical environmental materialists and ontologists can work together. At least Naess realizes the importance of some sort of political tolerance within the deep ecology movement: "those working on any goal of global dimensions should have certain principles in common. But these principles should not imperil deep differences in ultimate metaphysical or religious views."[58] Why not use part of Rorty's neo-pragmatism then to extend Naess' call for such a principle beyond the borders of deep ecology?

Finally, I want to point out that this pragmatist strategy works not only within the bounds of DPE but also seems to be a good way to get at a workable DPE that all political ecologists can live with. Why is this the case? A metaphilosophical pragmatist, of whatever origin, when looking at a political-ecological situation, takes politics in general to be a constantly shifting contested terrain. This view follows from their non-essentialist approach to practical political questions. But again, the *environmental* part of this pragmatism forces such assessments within the context of constructing the best approach that accords with established ecological principles. Looking out at the wealth of political theories, the environmental pragmatist (here concerned with constructing a DPE) is going to use the theory that best embraces the fact of the contested terrain of politics towards the goal of laying the foundation of egalitarian practices which are in the best interests of the long-term health of the environment. Nothing in principle restricts the pragmatist from such broad considerations and choices, while at the same time, it could be shown that the pragmatist is obligated to undergo such a search.

This idea reminds us that neither materialists nor ontologists are completely constitutive of the terrain of democratic political ecology. DPE exists on a different theoretical plane, and in a different relationship to materialists and ontologists than they do to each other. The relationship is similar to that of radical political economics as a whole to both Althusserian influenced social structures of accumulation theorists, and classical labor theory of value Marxists: the last two are both approaches to radical political economics and neither exclusively amounts to what counts as a radical economic theory. Environmental pragmatism, as I imagine it, stands in a relationship to DPE, and materialists and ontologists, as providing something like rules for the game of how the last two should work in practice as forms of DPE.

ACKNOWLEDGMENTS

Many of the key ideas in this paper were first published in "Materialists, Ontologists and Environmental Pragmatists," *Social Theory and Practice*, Summer 1995. The present paper is however an expanded version of that piece.

NOTES

1 A. Light, "The Role of Technology in Environmental Questions: Martin Buber and Deep Ecology as Answers to Technological Consciousness," in *Research in Philosophy and Technology*, Vol. 12, 1992.

2 A. Light, "Rereading Bookchin and Marcuse as Environmental Materialists," in *Capitalism, Nature, Socialism*, 4:1, 1993.

3 I mean by materialism here, a thin interpretation of the term. Rather than the thick version of materialism advocated by Marx and most Marxists, which entails a strong form of ontology, I mean to suggest by the term simply an analysis of social problems which takes into account and looks primarily at the physical structures and institutions which make up a society.

4 See, for example, Bookchin's critique of environmental liberals in his *Rethinking Society: Pathways to a Green Future* (Boston: South End Press, 1990), and Marcuse's general arguments concerning the chilling effect of liberal society on activism in his *One-Dimensional Man* (Boston: Beacon Press, 1972).

5 *Toward an Ecological Society* (Montreal: Black Rose Books, 1980), p. 15.

6 *Rethinking Society*, op. cit., p. 20.

7 Ibid., p. 15.

8 Ibid.

9 For an exchange on the utility of the distinction and a defense of that utility, see Bookchin's "Response to Andrew Light's 'Bookchin and Marcuse as Environmental Materialists'" and my, "Which Side Are You On? A Rejoinder to Murray Bookchin," *Capitalism, Nature, Socialism*, 4:2, 1993.

10 *Rethinking Society*, op. cit., p. 23. Why it is that such behavior is not simply behavioral or self-protective is not known; for Bookchin such behavior seems to be automatically normative, as if anthropologists are now finding buried virtues while they unearth primitive villages. For a more detailed critique see "Which Side Are You On?" op. cit.

11 *Remaking Society*, op. cit., p. 23.

12 Ibid., p. 35.
13 Ibid., p. 27. This of course begs the question, what forms of technical insight are consistent with the parameters of social evolution?
14 I am thinking here of the importance of property in defining human freedom in Locke's *Second Treatise of Government*. For a discussion of topics similar to this see Carolyn Merchant's *The Death of Nature* (New York: Harper and Row, 1980).
15 *Inquiry* 16, 1973: 95–100.
16 See the introduction to *Wisdom in the Open Air: The Norwegian Roots of Deep Ecology*, eds Peter Reed and David Rothenberg (Minneapolis: University of Minnesota Press, 1993).
17 Cit. in B. Devall and G. Session, *Deep Ecology: Living as if Nature Mattered* (Salt Lake City: Peregrine Smith Books, 1985), p. 65.
18 Cit. in Rothenberg's introduction to A. Naess' *Ecology, Community and Lifestyle*, trans. David Rothenberg (Cambridge: Cambridge University Press, 1989), p. 19.
19 Cit. in Rothenberg from Fox, "Approaching deep ecology: a response to Richard Sylvan's critique of deep ecology" Hobart: University of Tasmania Environmental Studies Occasional Paper 20, 1986.
20 Ibid., p. 20., emphasis added.
21 See the first chapter of *Remaking Society* which lumps indiscriminately goddess worship with deep ecology. While some deep ecologists may invite this comparison, certainly not all warrant it – see for example David Rothenberg's remarks on the differences between various branches of theoretical deep ecology in his introduction to *Ecology, Community and Lifestyle*, op. cit.
22 "New Social Movements: The Anarchic Dimension," in *For Anarchism: History, Theory, and Practice*, ed. David Goodway (London: Routledge, 1989), p. 273, emphasis added.
23 Cit. in Devall and Sessions, *Deep Ecology*, op. cit., p. 16.
24 *Remaking Society*, op. cit., p. 12.
25 Ibid., p. 10.
26 For a more thorough discussion of this materialist/ontological tension in Marcuse's work see Andrew Feenberg's "The Bias of Technology", in *Marcuse: Critical Theory and the Promise of Utopia*, ed. Robert Pippin, A. Feenberg and C. Webel (Massachusetts: Bergin and Garvey Publishers, 1988).
27 See my "Rereading Bookchin and Marcuse", op cit.
28 See Chapter 2 of *Counterrevolution* (Boston: Beacon Press, 1972) and his "Ecology and the Critique of Modern Society", in *Capitalism, Nature, Socialism* 3:3, 1992.
29 Op. cit.
30 Also see Naess' "The Politics of the Deep Ecology Movement", reprinted in *Wisdom in the Open Air*, op. cit.
31 See *Defending the Earth: A Dialogue Between Murray Bookchin & Dave Foreman*, ed. Steve Chase (Boston: South End Press, 1991). Even in this fairly conciliatory exchange the priority of Bookchin's environmental materialism shines through. See my analysis of the book in "Which Side Are You On?", op. cit.

ANDREW LIGHT

32 National green gatherings have been known to notoriously end in debilitating arguments between deep and social ecologists. Disputes have also often spilled over into the popular press in Europe and America in magazines like *The Nation* and *Z.* as well as some national newspapers.

33 One cannot stress enough the need for pragmatic conciliation at times (within the bounds of maintaining moral commitment) in order to get things done, as has been evidenced in the alliances formed around the Redwood Summer activities in Northern California. For a fascinating account of the environmental politics there see the interview with Judi Barry in issue 17 of *Capitalism, Nature, Socialism* 5:1, 1994.

34 See Gould's essay, "Kropotkin was no Crackpot", in *Bully for Brontosaurs* (New York: Norton, 1991).

35 See David Rothenberg's interviews with Arne Naess in *Is it Painful to Think?* (Minneapolis: University of Minnesota Press, 1993). Of course Naess supplemented his experience with rigorous philosophy, drawing from Spinoza, Ghandi and Dogen. Of the latter, however, Dean Curtin has argued recently that the appropriation of Dogen's Buddhism as support for Norwegian deep ecology is based on a misunderstanding of Dogen's thought. See Curtin's "Dogen, Deep Ecology, and the Ecological Self", in *Environmental Ethics*, Summer 1994.

36 Kvaløy, "Complexity and Time: Breaking the Pyramid's Reign", in *Wisdom in the Open Air*, op. cit., p. 119.

37 See for example William Chaloupka's "John Dewey's Social Aesthetics as a Precedent for Environmental Thought", *Environmental Ethics*, Fall 1987; Bob Taylor's "John Dewey and Environmental Thought", *Environmental Ethics*, Summer 1990; and Robert Fuller's "American Pragmatism Reconsidered: William James' Ecological Ethic", *Environmental Ethics*, Summer 1992.

38 See, for example, Anthony Weston's "Beyond Intrinsic Value: Pragmatism in Environmental Ethics", *Environmental Ethics*, Winter 1985, as well as his "Before Environmental Ethics", *Environmental Ethics*, Winter 1992 (both reprinted in this volume); and Kelly Parker's "The Value of a Habitat", *Environmental Ethics*, Winter 1990.

39 At first glance the Katz–Weston debate (reprinted in this volume) appears to be such a debate. It may however represent an occasion for reconciliation between pragmatists and non-pragmatists. See the final chapter of this volume.

40 Kvaløy, "Complexity and Time", op. cit., p. 119.

41 My emphasis, from C. Manes' *Green Rage* (Boston: Little, Brown, 1990), p. 124.

42 See, for example, William Ophuls' political ecology in his *Ecology and the Politics of Scarcity* (San Francisco: W. H. Freeman and Co., 1977).

43 Reed and Rothenberg, "Sigmund Kvaløy", in *Wisdom in the Open Air*, p. 114.

44 Kvaløy, "Complexity and Time", op. cit., p. 116.

45 For an overview of this example of ecological restoration and others, see *Beyond Preservation*, eds Anthony Baldwin, Judith DeLuce and Carl Pletsch (Minneapolis: University of Minnesota Press, 1994).

182

46 See Robert Elliot's "Extinction, Restoration, Naturalness," *Environmental Ethics*, Summer 1994; Alastair Gunn's "The Restoration of Species and Natural Environments," *Environmental Ethics*, Winter 1991; as well as Eric Katz's "The Big Lie: Human Restoration of Nature," in *Research in Philosophy and Technology*, Vol. 12, 1992, reprinted in part as "Restoration and Redesign: The Ethical Significance of Human Intervention in Nature," in *Restoration and Management Notes* 9:2, 1991. Elliot and obviously Katz are the perpetrators here.

47 For a good critique of some of Robert Elliot's arguments on restoration and an argument for the literal terrain up for grabs in restoration as a practice, see Eric Higgs' "A Quantity of Engaging Work to be Done: Ecological Restoration and Morality in a Technological Culture," *Restoration and Management Notes* 9:2, 1991.

48 It should be noted that Katz did originally publish "The Big Lie" outside of the restoration literature, and his essay was only reprinted in *Restoration and Management Notes* on the invitation of the journal's editor. The responsibility for where and when to publish theoretical contributions to practical questions ought also to rest on journal editors and not only authors. I'll be the first to admit that following through on this suggestion is going to be difficult, but there is nonetheless an important question to consider whether the Katz debate that is still continuing in the pages of *Restoration and Management Notes* is uncontroversially healthy for the practice. See Steven Rassler's letter "Naturalness and Anthropocentricity," and Donald Scherer's "Between Theory and Practice: Some Thoughts on Motivations Behind Restoration," both in *Restoration and Management Notes* 12:2, Winter 1994.

49 Cambridge: Cambridge University Press, 1989.

50 "Truth and Freedom: A Reply to Thomas McCarthy," *Critical Inquiry* 16.3, 1990, p. 634.

51 *Contingency, Irony and Solidarity*, op. cit., p. 91.

52 Such a non-theoretical approach to encountering nature has been an integral part of some radical political ecology. Christopher Manes describes John Seed's (one of the early leaders of the Australian environmental movement) conception of nature as "an intense, tangible reality, not a theoretical issue involving resource depletion and land management." *Green Rage*, op. cit., p. 120.

53 Importantly, environmental pragmatism should not be confused with the "authentic" environmentalism (or environmentalism based solely on ecological concerns) of the Herbert Gruhl wing of the German Green Party. Environmental pragmatism is simply a less publicly theoretical version of radical ecology than that characterized by non-pragmatic materialists or ontologists. Pragmatists can begin with the assumption that there is some connection between ecological and social justice issues if they feel compelled to and find commonalities between these issues based on redescriptions of them in terms of overlapping notions of jeopardy. Pragmatists must realize though that such combined concerns should not be theoretically straight-jacketed

through overdetermining conceptions of the identity positions of subjects in these justice issues, that is, pragmatists cannot embrace some notion of "natural" rights (or "natural" law) for humans or the non-human natural world.

54 R. Rorty, "Pragmatism, Categories and Language," *Philosophical Review* 70 (April, 1961), p. 219.
55 I am indebted to Meredith Garmon for pointing out this last passage in Rorty's work. For a much more complete discussion of Rorty's position with respect to action see Garmon's *Pragmatist Critiques of Jurisprudence*, unpublished diss. (University of Virginia, May, 1992).
56 The strongest version of this argument is the argument that Rorty's is a call not so much for a new philosophy (*à la* Marx) but a claim that any theory which is disconnected from practical critique does not count as philosophy. See Garmon, ibid.
57 See Rorty, "The Priority of Democracy to Philosophy," reprinted in *Reading Rorty*, ed. Alan Malachowski (Oxford: Blackwell, 1990).
58 Naess, *Is it Painful to Think?*, op. cit., p. 136.

Part 3

PRAGMATIST APPROACHES TO ENVIRONMENTAL PROBLEMS

9

PRAGMATISM AND POLICY
The case of water

Paul B. Thompson

One criticism of pragmatism in ethics is that it never results in policy prescriptions, or in a positive program of action. If pragmatic environmental ethics is to inform or influence environmental policy, it must ultimately say something about how human society should comport itself with respect to the broader environment. Bryan Norton has given a convincing account of how Leopold's concept of extending the moral community connects up with the general climate of pragmatist thought,[1] but that essay stops short of a policy prescription. His book *Toward Unity Among Environmentalists* places Leopold's thought in the context of environmental policy disputes that began with John Muir and Gifford Pinchot,[2] but pragmatism is at best an implicit theme in *Toward Unity*. The question remains: how does pragmatic environmental philosophy translate into public policy?

Among the classical pragmatists, Josiah Royce, John Dewey and James Hayden Tufts each wrote many essays on important public policy problems of their day.[3] While to my knowledge none of these policy essays refers directly to environmental issues, reading the pragmatists in light of these essays provides an entrée to pragmatic environmental ethics. James Campbell has replied to the suggestion that pragmatism is not prescriptive with a review of Dewey's prescriptive essays, linking them to Dewey's writings on philosophical method.[4] Campbell argues that pragmatism is prescriptive by offering a method for arriving at prescription, rather than by offering substantive solutions that conform to a pre-established pattern.

While a pragmatist might endorse a policy framed in the language of rights or utility, the philosophical justification for this endorsement will be procedural, and hence not an endorsement of rights or utility theory. This means that pragmatists do not understand the

187

connection between ethics and policy as one of applying the correct moral theory to a specific situation. Pragmatic environmental ethics, then, will relate to policy in a way that is quite different from that of traditional applied ethics. Water policy is a paradigm example of environmental policy disputes. This paper will examine what applied philosophy, and environmental pragmatism, respectively, offer in the way of a recommendation for water policy.

TWO CASES IN WATER POLICY

The conflict of interests over water use is one of the most frequently repeated dilemmas in environmental policy. The particular groups disputing water policy differ from region to region and from time to time, but there is a constancy in the perspectives that are likely to be represented. I will illustrate this by considering two cases. The Chatham River Case was designed as a model for teaching undergraduates about water policy disputes. It fictionalizes a real case in North Carolina, but is intended to typify tensions between those who advocate different uses for water. Chatham River was not written with ethical analysis in mind, and the case includes technical data that will not be summarized here. For present purposes, the fact that Chatham River was written for generalized technical, rather than ethical, applications makes it all the better as a representative case. The second case, offered more as confirmation of the Chatham's typicality than as an alternative, is the ongoing dispute over water in the Edwards Aquifer, which lies beneath Central Texas.

The fictional town of Springdale borders the fictional Chatham River in the non-fictional state of North Carolina, USA. The case was developed for teaching purposes and published in Wilson and Morren's 1990 book on systems analysis for agriculture and natural resources. It generalizes problems in North Carolina water policy studied by Robert Morse, and the textbook treatment applies not only hydrological and economic data, but real decisions from North Carolina riparian law.[5] The basics of the case can be summarized in a few sentences, however. Like many rural communities, Springdale is desperately seeking economic growth. The Springdale Town Council is convinced that economic growth can be assured if only water can be made available for light industry and for residential use. Herein lies the rub, for Springdale's wells are producing at their limit. Springdale, however, borders the Chatham River, so the Town Council's proposal is to divert water for residential and light industrial use.

188

The Town Council's plan is opposed by two groups. First are the existing riparian rights holders, who are small farmers. These farmers have used Chatham River waters for over a century. Their use has expanded and contracted over the years, and at various times and at various places, has included watering of stock and irrigation, as well as household uses. The State Water Board study of stream flow in the Chatham River indicates that in years of reduced flow, the Town Council's plan would not allow farmers to maintain current levels of use, much less to permit the expansion of current use that they associate with their property rights as irrigators. As such, farmers are contemplating legal action to protect these rights, and are resentful of the Town Council's assumption that Chatham River waters could simply be taken without their permission.

The second group is a loose-knit coalition of environmentalists and sportsmen called the Friends of the Chatham. The Chatham River supplies sustenance to fish and wildlife up and down its length, including (of course) an endangered species of salamander unique to Chatham River habitats. The Town Council's plan will threaten the Chatham ecosystem on which this wildlife depends during years of reduced flow. Harm to the Chatham ecosystem is offensive to Friends of the Chatham for two reasons. One is that recreational users of the Chatham ecosystem come from around the State, and have a strong attachment to the fishing, camping and nature observation that the ecosystem provides. The second is that members in Friends of the Chatham who do not use the Chatham River for recreation nevertheless believe that harm to the wildlife and, indeed, the eco- system itself is a moral affront. The Town Council, they hold, has offended nature simply by proposing a plan that displays so little apparent regard for its interests. To follow through on the plan is entirely unacceptable.

Now the Chatham River case is emendable in many different ways, some stressing legal or economic analysis, and others offering technological solutions. My proposal is to examine the moral arguments that each group might offer to justify their position, or to align their interests in accordance with an image of the public good. The most plausible version of each group's moral argument appeals to distinct philosophical principles. When these arguments are made explicit, the policy dispute replicates conflicts that are typical of philosophical debates between moral theorists.

The riparians, it is clear, must base their claim on a strong inter- pretation of property rights. The libertarian position in ethics defines justice as performance of key "negative duties" of *not* acting in ways that interfere with the activities of others. Libertarian non-interference rights include life, liberty and property. These rights are thought to constrain acts that harm others (clearly a case of interference). The limitation of rights to only those that require negative duties (e.g., *not* acting so as to harm another) is thought to preserve the maximum amount of freedom that can be enjoyed simultaneously by all. Property rights are thought to give property owners authority to utilize their property according to their wishes, subject only to the constraint that such use not interfere with or harm others. In particular, the fact that a property owner uses property in ways that are wasteful or even contrary to self-interest is not thought to provide a rationale for interference. Put another way, the mere fact that someone else (or government) can make a better use of your property than you can is no rationale for taking your property without your permission. The farmers think they own the water, and that this ownership bestows a right to any use up to the point at which such use interferes with the rights of others. Efficiency of use is irrelevant.

This point is especially relevant to the case of the Town Council, for Springdale's appropriation of Chatham River water is justified primarily because its use for economic growth will benefit far more people and to a far greater extent than will the water's continued use by riparians. This is a classically utilitarian argument. Property rights are treated not as protecting essential moral claims, but as instruments for producing benefits or reducing costs. When the farmers' control of water produces less total utility than control by the town, the town is justified in taking the water and using it for something else, be it light industry or residential supply. The Town Council thinks its position valid and correct because its plan produces the best consequences, measured in terms of economic benefits. Protecting riparian rights is inefficient.

The Friends of the Chatham are unlikely to be impressed by the arguments of either the Town Council or their riparian critics. There are at least two main arguments available to the Friends. One argument is anthropocentric; it would stress the rights of recreational users, future generations and those who use the Chatham River very indirectly – as an ecosystem preserving habitat, as a conservation reserve, and as a buffer against pollution and other environmental insults. The second argument extends rights directly to the flora and fauna of the Chatham River, and perhaps to the river itself. Non-human use, then, is protected along with the anthropocentric uses noted in already. Since both arguments appeal to rights of use that are separate from standard property rights, they claim opportunity rights on behalf of uses thought consistent with the public good. In this respect, both anthropocentric and extensionist arguments appeal to the standard pattern of egalitarian moral philosophy. Like libertarian rights of non-interference, opportunity rights override or "trump" the efficiency arguments of the utilitarian Town Council. In proposing that non-property owners should have the opportunity to enjoy and use Chatham River ecosystems, however, the Friends reject the libertarian limitation to negative duties, and propose a set of norms that could compromise the irrigators' use of Chatham River water just as surely as it challenges the economic growth plans of the Town Council.

Now any further analysis of even this fictionalized case will involve a multitude of puzzles that will be familiar to students of environmental ethics. The two arguments available to the Friends, for example, are emblematic of the dispute between anthropocentric and non-anthropocentric approaches to environmental ethics, a dispute that pervaded environmental ethics for its first two decades. However, the details of these arguments are not what is of interest here. The point has been to show how each of the interested parties can give *prima facie* accounts that justify their interest in the Chatham River as the legitimate one. Furthermore, this analysis shows why the policy stand-off can become protracted. The availability of incompatible moral justifications for each position can form the basis for a brand of self-righteousness on the part of each interest that bodes ill for a consensus solution of the problem. This is not a circumstance that will surprise philosophers, for a well-defined stand-off among libertarian, utilitarian and egalitarian philosophies has existed now for at least 200 years.

A single analysis of one fictionalized case does not make a generalization, but the elements described above are evident in cases that are not fictional. The Edwards Aquifer in Texas extends in a crescent from the southern hill country near Uvalde to the springs that feed the Guadalupe, Comal and Colorado rivers along I–35 between New Braunfels and Austin. In between is the City of San Antonio, a deceptively large metroplex, and one of the fastest-growing urban areas in the United States. In the west, where hill country rains recharge the aquifer, dryland farmers have dropped wells to water stock for two centuries, and have more recently taken to irrigating selected fields. At the north-eastern tip of the crescent, recreational users fish and float on scenic rivers and bathe in the springs, and wildlife (including an endangered salamander) depend on the habitat created by the springs. San Antonio, for its part, wants to tap the water for economic growth, and their use is projected to grow from a measured 270,000 acre feet per year in 1980 to 760,000 acre feet per year by 2040.[6] The triangular structure of libertarian property owners, utilitarian developers, and egalitarian environmentalists is, thus, repeated in the Edwards case. Other disputes over land and forest policy would, I think, support the generalization of this structure to many issues in environmental policy.

The support that the Edwards Aquifer case lends to the validation of the Chatham River model must be qualified by several differences that are crucial to pragmatist environmental ethics. First the arguments offered by each group of disputants are more complicated. Irrigators, for example, were quick to note the economic value of agriculture to hill country communities, as if to rebut the utilitarian logic of the City of San Antonio. This strategy was gradually (though never totally) withdrawn as more and more farmers came to appreciate how it required one to favor the economic health of relatively less populated hill country towns over that of San Antonio. For their part, advocates of San Antonio countered the ranchers' argument by noting that economic growth would benefit the city's large Hispanic underclass; policies favoring middle-class Anglo irrigated farmers over a relatively poor ethnic minority group were not justifiable. In this respect, San Antonio's argument adopted egalitarian rhetoric, if not its fundamental commitment to opportunity rights. Environmentalists also attempted both utilitarian and libertarian arguments. For example, economic valuation models have been particularly well developed for making the environmentalist case in Texas, permitting the measurement of the economic value of recreational use in the New Braunfels region. What is more, riparians on the rivers downstream from the springs have claimed their own property rights in water, rights that are most easily protected when spring flow remains at historical levels.

The appearance of water-owning agricultural producers among the environmentalists is indicative of the second point. Interest groups are not defined along philosophical lines in the Edwards case. Down river riparians need the same water policy sought by environmentalists, though for very different reasons. Even some of the hill country ranchers have thrown in their lot with San Antonio in the belief that they will eventually make more money selling water to Sea

World, the amusement park that has become emblematic of urban water use, than they do growing row crops. The complexity of interests and of arguments makes the Edwards case more fluid (no pun intended) than the Chatham case study. Water, in short, is too slippery; its flow does not conform to tidy lines of philosophical theory. While one might imagine a classic three-way philosophical debate deciding the Chatham case in favor of the group with the most articulate, rational and persuasive philosophical view, the untidy realism of the Edwards case makes one doubt that even a philosophically decisive blow to any one of the three moral perspectives would have great effect. This is an observation that will become important as we consider how environmental pragmatism bears on water policy. A pragmatic approach, not the vindication of any particular theoretical doctrine, will be necessary to solve real-life environmental problems.

THE TWO CASES: ETHICS APPLIED

Given disputes such as the Edwards or the Chatham, what do philosophers have to say? One response is certainly to develop each of the representative arguments that might be offered by the disputants with rigor and detail. As Varner, Gilbertz and Peterson argue in this volume, moral theories can be understood as tools that sharpen and clarify positions and that more clearly delineate the terms of the debate.[7] Indeed more detailed and explicit versions of libertarian, utilitarian and egalitarian arguments can readily be found in the literature on water and natural resource policy. Three such papers have been collected side by side in Charles Blatz's anthology, *Ethics and Agriculture*. The libertarian in this collection is Donald Scherer. Scherer analyzes the problem as arising from an absence of reciprocity that characterizes what he calls "upstream/downstream" environments: "What happens upstream is perceived to cause effects downstream, but not vice-versa. Consequently, those who live upstream do not fear harm from downstream."[8] The lack of reciprocity leads "upstream" users to ignore harm to others, as the biological facts of their situation make them invulnerable to similar abuses from the use of others. While simple self-interest is sufficient to establish shared norms of non-interference when reciprocity holds, it fails to produce an adequate protection of rights in upstream/downstream environments.

Scherer talks about the practice of an upstream user as one of "externalizing" costs. His use of this term is not that of a resource

193

economist's, however. Scherer's externalities are, in fact, impositions or interference in the liberty or property rights of weak parties by relatively stronger ones. He concludes by arguing that the function of morality in such situations is to protect the weak from interference by the strong. Scherer's argument is libertarian in that it denies the validity of efficiency criteria, insisting on protection of non-interference rights without regard to cost. This view stands in direct opposition to the Town Council/City of San Antonio viewpoint. Scherer thinks that his view supports environmentalists and, indeed, libertarian arguments support strong environmental laws up to the point that practices such as pollution or waste interfere in the exercise of well-established rights. In either of the cases at hand, however, an argument based on property rights favors ranchers and farmers over environmentalists in virtue of the fact that the environmentalists are attempting to initiate new claims on Chatham or Edwards water. Although they might like to use the waters for recreation, or simply to know they are there as habitat, it is hard to see how environmentalists are materially harmed by the riparians' use, and easy to see how interference in the irrigators' legally sanctioned pattern of historical use constitutes a taking of something they formerly owned.

The utilitarian analysis is provided by resource economists Terry Anderson and Donald Leal. Anderson and Leal present the case for resolving water disputes by establishing water markets. Their argument depends upon two key elements: the doctrine of allocative efficiency as the norm for effective resolution of conflicts and an analysis of water disputes as forms of market failure. Stated non-technically, allocative efficiency holds that goods are justly distributed when they have been allocated to their highest valued use. The doctrine can be seen as a special case of the utilitarian maxim, act to produce the greatest good for the greatest number, applied to the question of distributive justice. "Most valued use" is understood so that each user of a good is thought to have subjective preferences that determine its value, and economists assume that if each user is free to exchange their use of a good (e.g., the water) for something else, these subjective values are revealed in economic behavior. Since people will only make voluntary trades when every party to the contract expects to benefit, so-called free markets – free, that is, from coercion by government – produce an allocation that cannot be improved upon without making at least one party worse off.

The assumptions of neo-classical economics would be rejected by many utilitarians, but the point here is to examine an illustrative application of the utilitarian viewpoint, not to critique Anderson and Leal's particular version of utilitarian norms. Like Scherer, they see the problem as one of externalities, but they interpret externalities not as simple harms, but as instances of market failure. Market failure means that resources are not allocated to their most valued use by ordinary exchange, although the failure is not a result of coercion. Markets fail when people who could cooperate (often by making a trade) are unable to do so, usually because the structure of the market or natural traits of the good in question prevents cooperation. The right of free capture, for example, is the historical basis for water policy in much of western North America. This right permits land owners to use as much water as they can capture from rivers or underground sources, but gives them no privilege of ownership over water that remains in streambeds or aquifers, unused.[9] Anderson and Leal point out that this situation creates a disincentive for conservation, since landowners have no way to profit from conservation, or even to recoup costs due to uses that reduce (much less minimize) consumption.

If landowners could sell the water that they don't use, such incentives would be supplied and conservation would increase. Thus the Anderson and Leal solution is to establish property rights in water that could be exchanged for cash or other goods. This would establish a market for water, and would provide incentives for conservation. It would also bring water use practices in line with the pattern demanded by allocative efficiency, establishing a utilitarian justification for whatever use eventuated from the exchange of property rights.[10] Since municipalities can charge far more for water than agricultural users are willing to pay, they can easily purchase water from riparians and divert it to urban uses. Thus the Town Council/City of San Antonio would be quite happy with Anderson and Leal's position, though the utilitarian logic on which it is based is truly impartial to geographically based interests. Farmers, town planners and environmentalists are all treated equally; it is the greater value of water to town planners that determines the result.

It is probably worth taking some pains to clarify the difference between the libertarian and the utilitarian positions just described. Both use the concept of externalities, and both rely on property rights in water for the preferred policy instrument. Scherer's argument is libertarian because the word "externality" is used as a synonym for "interference," the chief sin in the libertarian theory of justice. The reason to stabilize property rights in water is because secure property rights are the way to protect "downstream" users (e.g., relatively weaker parties) from abuse by those who have morally irrelevant advantages over them. Scherer is arguing that both upstream and downstream users have the same natural right to water, but that current public policies may unjustly allow upstream users to violate rights of others. Anderson and Leal, on the other hand, understand externalities as costs or losses that are external to the market-clearing mechanisms of ordinary exchange. They are not reflected in the price that users of water must pay, or in the return that they might achieve if they were able to capture value that flows to others. The distortion of prices creates incentives for inefficient water use. Those who would be willing to pay for conservation have no way to buy it, while those who are in a position to conserve have no incentive to do so. Property rights are the utilitarian's solution, but there is nothing morally important about rights. They are merely the policy instrument that maximizes utility in this particular case. The convergence of libertarian and utilitarian arguments on the importance of property rights suggests a larger (and in this context tangential) theme in environmental ethics.

Helen Ingram, Lawrence A. Scaff and Leslie Silko present a highly nuanced account of equal opportunity in their discussion of water policy from the Blatz volume. They argue that water is a social good, meaning that it should be removed from ordinary market transactions and allocated according to a democratic political process that "entails respect for legal or constitutional rules and procedures arrived at in an 'impartial' and 'fair' manner" and requires the sharing of power by "members of a community having equal access to the law, to social goods, and to political positions."[11] Ingram, Scaff and Silko go on to enumerate five principles that characterize the policies they support:

- distributive advantages and costs should be shared equally by all members of the relevant community;
- users' rights to employ water to pursue whatever values they consider legitimate should be respected, provided use does not degrade the resource or harm others;
- members of society having claims consistent with other stated values should always be accommodated in resource allocation and in the decision process;
- equity assumes the obligation to obey promises agreed to in good faith in the course of negotiation and compromise;

- present use of water resources should be accompanied by responsibility for future generations.[12]

Does this amount to an egalitarian view? There are reasons to think that it does. First, these principles clearly undercut the efficiency arguments of the utilitarian view. Ingram, Scaff and Silko also interpret their libertarian sounding second principle to be constrained by each of the other four, so that exercise of property rights is permitted only when equal sharing of cost and opportunity by *all* members of the community has been assured.[13] Furthermore, the commitment to opportunity rights for future generations and the third principle's guarantee of a seat at the negotiating table for environmentalists, along with developers and traditional owners of property rights, are clearly egalitarian elements that might be unnecessary on libertarian or utilitarian grounds. However, the resulting argument is less sharply egalitarian than one would expect from a member of the faith. It leaves much of the substantive policy on water use to be settled by the political process. While one could strengthen the environmental dimensions of this argument by stipulating more carefully what is entailed by the constraint on degradation, some environmentalists will clearly want an argument that eschews the term "resource" entirely. Nevertheless, some version of the argument endorsed by Ingram, Scaff and Silko will be part of the entrée into policy-making that environmental action groups like the Friends of the Chatham or the defenders of Texas springs must claim.

More explicit examples of the egalitarian view can be drawn from the broader literature on environmental ethics. Certainly Kristen Shrader-Frechette is one of the most sophisticated spokespersons. Shrader-Frechette's work on a variety of environmental issues applies Rawls' difference principle to environmental policy. Shrader-Frechette follows Rawls in arguing for a criterion that would prevent any group or individual from gaining the upper hand, or benefiting disproportionately, from a public policy decision. As codified in the *Difference Principle*, the egalitarian ethic requires that at each juncture society should select policies that will most benefit those in the worst off group. Such a policy criterion should tend to create relatively equal distributions of wealth and power in the long run, and permits unequal distributions in those cases where doing so is most beneficial to the poor and weak. The key to Shrader-Frechette's adaptation of the Rawlsian position in environmental affairs often turns upon the disproportionate power of different agents seeking to influence environmental policy, and upon the disproportionate risks that the relatively weak must ultimately bear. Shrader-Frechette stresses the relative wealth of actors who would propose the uses of land or water that bring about commercial gain. Both farmers and cities, the perspectives represented by libertarian and utilitarian viewpoints, are included among those who stand to benefit financially from the uses of land and water that they propose.[14]

In brief, Shrader-Frechette argues quite plausibly that a policy decision favoring the economically and politically dispossessed would also favor the environment. While there are cases (especially in the developing world) where this argument may not be applicable, there is little doubt that her strongly egalitarian view would produce policies more favored by environmentalists in the United States. The alternative, again, would be to apply egalitarian forms of reasoning, but to make an explicit extension of political rights to non-human entities. The two Nashes (Roderick and James) writing on environmental issues have advocated this approach, for example.[15] While such an innovation is sure to cause philosophical discomfort, the view serves as an adequate example of how the extensionist side of the egalitarian view might be emphasized in a manner that would justify the policy positions taken by the Friends of the Chatham and by environmental organizations active in Central Texas.

APPLIED ETHICS: WHAT HAVE WE WROUGHT?

Environmental ethicists will certainly find points to question in the above characterization of the three viewpoints. No doubt each has been constructed as something of a straw man representing a particular foundational view in ethics.[16] However, the point here is to provide examples of how philosophical analysis is or can be used to illuminate and bolster the unsystematic opinions that are already held by disputants in policy disputes. In this respect, we can make three observations based on the preceding two sections. First, the fit between policy prescriptions and recognized positions in moral theory supports the judgment that we are truly making *applications* of theoretical views when we diagnose water policy disputes in this way. Second, applied philosophy largely *consists* in classifying political views according to their particular theoretical cast. Third, the practical effect of such application and classification, if any, is likely to be negative. These observations lead to a decidedly negative assessment of applied philosophy's role in environmental policy, and it is worth expositing each of them in detail.

First, the libertarian, utilitarian and egalitarian analyses of water policy that are suggested by Chatham River and Edwards Aquifer and actually made by Scherer, Anderson and Leal, Ingram, Scaff and Silko, and Shrader-Frechette are each examples of applied ethics in the following sense. Key philosophical assumptions or axioms are either espoused by or attributed to disputants, much as religious tenets might be adopted by the faithful. The specific policy perspectives that are advocated represent the logical implications of these ethical tenets, combined with contingent factual statements about the case at hand. Ethics is, in this view, truly applied theory. The assumptions or axioms of theory are stipulated almost exclusively in *a priori* terms; key tenets take the form of a universal statement. When combined with true empirical statements that assign names or definite descriptions to the free variables, or which describe physical laws that predict the course of events, the metaphor of application captures the role of ethics quite well. Ethics is identified with moral theory; application consists in performing the instantiation of variables correctly and in deducing the implications of instantiated terms with logical consistency. No moral content is added in the process of application, nor is there anything comparable to falsification in science, where the moral content of the theory might be modified (or at least rejected) by the attempt to apply it.

Second, once the attitudes and positions of the respective parties to these water disputes have been grounded in recognizable philosophical positions, the applied philosopher considers the job to be done. Theory has been applied; a prescription has been generated. What more is there to say? If there are continuing disputes among libertarian, utilitarian or egalitarian viewpoints, surely these are disputes that must be addressed in the pure philosophical terms that are the familiar truck of political theory. Whichever theory carries the day philosophically determines the correct policy; so the applied ethicist need do nothing more than ground the preferred policy in some philosophical foundation. Since application cannot modify theory, these foundation disputes are safe from interference by those who muddy themselves in water policy. The fact that philosophical disputes in political theory do not appear to be going anywhere fast is not seen as either relevant to an analysis of the policy conflict, nor philosophically interesting as a dimension of applied ethics.

Finally, the practical effect of philosophical analysis is to provide disputants with robust accounts of their earlier opinions, namely to freeze the actual political debate into stasis. Philosophers love such stasis, for it provides the "dilemma" that is so useful for classroom pedagogy. With respect to water policy, however, the result is an icy political stalemate. Clearly nothing in these applications would lead one to modify ethical tenets; only a foundational argument can do that. It is questionable whether we are really better off than we were when disputants had vague, intuitive understandings of their own rationale. Arguably, clarity only hardens a dispute that might be resolved had disputants not been so clear about the burdens of proof they expected their opponents to provide. If this is what applied philosophy yields, we might be better off with lawyers!

THE PRAGMATIST RECONSTRUCTION: NECESSITY

What does pragmatist environmental philosophy have to offer in this situation? First, a pragmatist will offer a different problem identification from the foundational philosopher. For a foundational applied philosopher, the problem is "How do I apply my theory to water policy?" For the pragmatist, the problem is "Different interests are in contention over water use. How can the contention be resolved in a manner that is consistent with our political ideals?" While either of these questions might be appropriate in a particular context, when water policy is really at stake, the second formulation has a greater claim on the pragmatic philosopher's attention. The reason for this is central to all of pragmatist philosophy: the second formulation has pragmatic necessity.

In several essays, William James argues that we are epistemologically warranted in believing (or accepting as true) propositions which are neither demonstrated nor demonstrably false, even when little of the evidence that might be brought to bear upon them supports their truth. We are warranted in assenting to such beliefs just in the circumstance that either believing or not believing makes some material difference in our conduct or our psychological well-being. We cannot, James notes, remain uncommitted on matters such as what is right and wrong in the conduct of personal affairs; we must do one thing or another.[17] Such circumstances are forced choices, that is what *pragmatic necessity* means.

James does not mince words, either. In *Pragmatism* he writes, "The world stands really malleable, waiting to receive its final touches at our hands. Like the kingdom of heaven, it suffers human violence willingly. Man *engenders* truths upon it."[18] How uncanny James' view is here, and how unlike most philosophy! Does James simply conflate the ontological condition for truth and epistemic conditions of justification? John J. McDermott cites the above quote from James within the context of claiming that the pragmatic approach to ethics is pervaded by the transiency of pragmatist metaphysics. He writes, "The pragmatic approach is clearly temporalistic and finite, thereby placing enormous emphasis on the moral gravity of all our activities."[19]

James' primary achievement in his theory of truth is to notice that philosophical thinking goes awry when it is given over to problems of philosophical technique. Following McDermott, I suggest that we read James as reaching a forced choice of his own in this essay. He must choose between two strategies for developing a theory of truth. One option treats beliefs and claims (or their propositional content) as models or representations of the world, and construes philosophical problems in terms of the sign–signified relation. If successful, this strategy can ground the propositional content of claims and beliefs in a foundational relation between word and object. The other option treats the truth of beliefs and claims[20] as conditioned by the signifier, who for James is always a human being, as well as the sign and its reference, and situates truth itself within the situatedness of a human being's use of language.

The moral gravity of this choice is revealed in two radically different visions of how philosophy relates to practice. The foundational view permits a complete separation between the truth of a claim and the conditions under which it should be accepted or believed in practice by human beings. The truth conditions of a belief's propositional content must be independent of the conditions under which a human being could see that a given claim is true. Foundational philosophy attempts to specify both sets of conditions and usually concludes that justified true belief occurs when one correctly believes that the truth conditions of the claim obtain analytically or in reference to the world.[21]

The doctrine of pragmatic necessity states James' grounds for rejecting the foundational approach. As a psychologist, James had come to doubt that belief had the hierarchical structure implied by the foundational view. In practice, people often adopt and discard beliefs with little attempt to examine their empirical or logical adequacy, using instead a rough estimate of their serviceability for the task at hand. To be sure, people often discard beliefs on logical and empirical grounds, but perhaps this is because in certain instances these grounds are as relevant to the serviceability of a belief as to its place in a logical hierarchy. In any case, the crucial point for James is that in divorcing truth from the problems that lead people to adopt and discard beliefs, one was adopting a conception of truth, and in turn a conception of philosophy, that could become irrelevant to the practical problems of ordinary people. Although the foundational theory of truth may be very useful for those engaged in the common project of constructing a hierarchy of knowledge, it problematizes truth in way that is inconsistent with the way that truth becomes problematic in practice.

THE PRAGMATIST RECONSTRUCTION: DECONSTRUCTION

Arguably, too much of Dewey's philosophical writing consists in tearing down philosophical abstractions that prevent people from responding to their real problems with the creativity and vitality imagined by James. Reading Dewey today I am always reminded of the Monty Python sketch where two English working-class women carp aloud about what the youth of today are coming to. "It's all that Cartesian dualism gone mucking up their heads!" one says. Like Python's worried elders of the working class, Dewey sees Platonic fascination with forms, Cartesian dualism and British empiricism conspiring in an ill-conceived quest for certainty that ultimately prevents philosophical thinking from being brought to bear upon the problems of importance.[22] As Richard Rorty argues,[23] one must deconstruct or somehow get out of these traps if one is to think philosophically about what matters, especially for public policy.

It is worth distinguishing deconstruction, understood as a method of refuting an interpretation or view of the world by exposing implicit and locally contingent[24] elements of its logic, from de-construction*ism*, or the view that the omnipresent potential for deconstruction entails a kind of nihilism. Dewey may have been the first philosopher to perfect the method of deconstruction, but he was no deconstructionist in the current sense. Reconstruction must always follow deconstruction, and the function of reconstruction is to build shared ideas and a sense of community. For the public policy pragmatists, Royce, Tufts and Dewey, it is only through community that we have a chance of meeting the crises of our time.

Community is needed because community is the method of science, and the basis for a pragmatic theory of truth. Non-pragmatic (e.g., foundational) applied ethics is part of a philosophical project that takes the value or use of truth and knowledge as inessential to the theory of truth or knowledge, then excludes even the reflexive question of why the philosophical project is worthwhile. Philosophers who adopt this project eliminate those with largely practical concerns from their community. They simply forgo their advice and commentary. As Larry Hickman has argued, Dewey understood science as a communal project, and saw the technological means and offshoots of pure scientific inquiry as checks and balances that reign in the theorizing component, preventing in practice the abuses that critics have associated with relativist philosophy.[25] The theoretical position is that truth is *for* and *by* a community of inquiry, hence relative. In practice, however, communities that include practitioners – bridge builders, farmers, policy-makers – have a reliable mechanism of self-criticism: the ideas must work. Thought and theory must help solve practical problems, and when it does not, we must revise in a fashion that is methodologically equivalent to falsification. Why not just call it falsification, and say that when ideas don't work, there is something about them that is not true?

Pragmatic deconstruction is a form of moral pedagogy that goes before reconstruction, and hence before any attempt at prescription. It presumes with post-structural deconstruction that any symbolic apparatus – be it a text or a set of practices – consists of connections that are contingent, rather than necessary. Like the post-structuralists (but more so), Dewey's philosophy is mindful of the power that becomes concentrated when contingent connections are unquestioned or treated as necessary. Like the post-structuralists, Dewey is especially wary of constructions that allege metaphysical foundations. Unlike them, however, Dewey recognizes the socially and morally beneficial, even essential function that such constructions perform. He hopes to reconstruct common visions of life and purpose on more honestly social (and hence quite temporary) foundations.

AND BACK TO WATER

Libertarianism, utilitarianism and egalitarianism provide model forms of reasoning and argument that might be adopted by each of the three model interest groups illustrated in the two cases discussed.

A foundational approach to ethics requires that one of these three theories (or perhaps a fourth alternative not yet brought forward) is true (or possesses some trait functionally equivalent to truth). The foundational philosopher beckons disputants into the academy, where they may watch their champions do battle on behalf of each viewpoint, and when the victor has been declared, we will know who was morally right, and who was morally wrong in advocating a particular use for Chatham River or Edwards Aquifer waters. Then, presumably, we go back to North Carolina or Texas and just do the right thing.

No one really thinks that this will happen, of course, and I have argued that the more likely consequence of the three-way analysis is that each group of disputants will go away with a firm rationale for why their view is morally right, and a dispute that might have been settled legally, economically or politically becomes a moral one. The real-life complexities of the Edwards Aquifer case demonstrate the potential of political means. The Texans at odds over Edwards water did not derive their positions on policy issues from ethical theories that they had already adopted as personal beliefs. Policy preceded theory. It is incompatible use, not ethical viewpoints, that caused the political controversy. If economically feasible technical solutions become available, the dispute will recede. In San Antonio, the proposed technical solution is to build reservoirs. The dispute continues because it is not clear who will pay for them.

Furthermore, the disputants use their best (meaning most politically potent) argument, without regard to its philosophical origins. If the developers' utilitarian case can be defeated on it its own terms (perhaps because they have underestimated either the agricultural or the environmental value of the water), riparians and/or environmentalists will not deprive themselves of this argument in the name of philosophical purity. Similarly, if a developer or an environmentalist can defend a clear property right, they are likely to do so. The factors that determine both technical solutions and the availability of arguments are contingent; they will vary from case to case. It is true that property rights, utility and fairness (to other humans or to other creatures) represent principles that strike to the heart of three interest groups likely to be at odds with one another in many disputes over the environment, but the particular arguments that will win the day vary on a case by case basis.

Given this circumstance, Gary Varner's suggestion that conventional moral theories are tools must be met with the question "Tools for what?" There is some danger that clarifying the moral dimension of water debates will lead disputants to insist upon an interpretation of problems that conforms to an underlying philosophical position, just as applied philosophy suggests that it should. It is at least arguable that the bitterness and inflexibility of the abortion debate today is due to both sides not only having adopted philosophically incompatible positions, but also having buttressed these positions with moral philosophy that demonizes the other side. Environmental debates over habitat for endangered species are showing signs of a similar pattern, as those who countenance any standing for economic considerations are characterized as greedy and corrupt, while those who advocate for nature are portrayed as uncaring and disrespectful of human needs.

Yet if the problem of moralistic policy gridlock is created in part by adopting the approach of applied ethics, it could also be very useful to show how each of the three viewpoints influences policy debates over water. It would be useful, for example, to show disputants how their opponents have used reasoning that they themselves might have used in a different situation to arrive at the opposing point of view. It would be useful to deconstruct the images of the irrigator as a demented rustic, of the Town Council as out-of-touch technocrats and of the Friends of the Chatham as ignorant (and interloping) malcontents. It would be useful for each group to see themselves as part of the same community, at odds on a given issue, perhaps, but drawing from common moral traditions and headed towards a common future. Such a community might find political solutions that reciprocate each interest, even while they may demand compromise on the case at hand.

Pragmatic necessity implies that any analysis of water problems that does not facilitate the formation of broader community and action to address problems is philosophically flawed. An analysis that freezes disputants into fixed positions has just failed to get at what is important about water, and about environmental problems in general. Pragmatic deconstruction is the pedagogical tool for ending moral gridlock, and beginning the reconstruction of community. Nevertheless, it is not appropriate to propose solutions here. For one thing, proposing a policy mechanism may be inappropriate in advance of the reconstruction needed to form a community capable of addressing key problems. For another, there will be few universal solutions to generalized problems. What works in the Chatham may not work in Texas, for much depends not only on the biological features of an environment, but on the cultural and political characteristics of the contending parties, as well.

Analysis through conventional applied philosophy produces multiple viewpoints and this can be useful if the viewpoints enter the policy debate as a tool in the pedagogy of community. If political action is to address the problem of water, multiple ethical viewpoints must be integrated into a community of interest and hope. If their impact is to moralize a debate and preclude political action, they are worse than useless, for if political action does not address our environmental issues, how are they to be addressed at all? James and Dewey, like Bryan Norton and other contributors to this volume, challenge environmental ethicists to embrace the political. To do so without abandoning their beliefs is to embrace the task of constructing a broader and more inclusive community, one that includes ranchers, farmers and, yes, even developers! But that, of course, only brings us back to those opening paragraphs of "The Land Ethic," and it is there that this Odyssey must, for a time, come to an end.

NOTES

1 Bryan Norton, "The Constancy of Leopold's Land Ethic." Reprinted in this volume.

2 Bryan Norton, *Toward Unity Among Environmentalists* (Oxford: Oxford University Press, 1991).

3 See Royce's essays on war, on insurance and on Western settlement collected in Josiah Royce, *Basic Writings* 2 Volumes, John J. McDermott, ed. (Chicago: University of Chicago Press, 1969). See Tufts essays on foreign affairs and educational policy in *The Selected Writings of James Hayden Tufts*, James A. Campbell, ed. (Carbondale, Il: Southern Illinois University Press, 1992).

4 James Campbell, *The Community Reconstructs* (Urbana: University of Illinois Press, 1992), pp. 39–42.

5 Katherine Wilson and George Morren, *Systems Approaches for Improving Agriculture and Natural Resource Management* (New York: Macmillan Publishing Co., 1990). I heartily recommend this case for classroom applications.

6 Paul B. Thompson, Robert Matthews and Eileen Van Ravanswaay, *Ethics, Public Policy and Agriculture* (New York: Macmillan Publishing Co., 1994) pp. 143–153.

7 Gary Varner, Susan Gilbertz and Tarla Peterson, "The Role of Ethics Education in Environmental Conflict Management" (In this volume pp. 226–282).

8 Donald Scherer, "Towards an Upstream-Downstream Morality for Our Upstream-Downstream World," in *Ethics and Agriculture*, Charles Blatz, ed. (Moscow, ID: University of Idaho Press, 1991) p. 418.

9 Significantly, only Texas still maintains a pure interpretation of free capture. Water policies vary dramatically, but many states that allocate a limited right of capture do not allow the alienation of water from the site of capture, and hence preclude the sale of the water allocation.

10 Terry L. Anderson and Donald R. Leal. "Going with the Flow: Expanding the Role of Water Markets," in *Ethics and Agriculture* Charles Blatz, ed. (Moscow, Id: University of Idaho Press, 1991) pp. 384–393.

11 Helen M. Ingram, Lawrence A. Scaff and Leslie Silko, "Replacing Confusion with Equity: Alternatives for Water Policy in the Colorado River Basin," in *Ethics and Agriculture* (Moscow, Id: Idaho University Press, 1991), p 400.

12 Ibid. pp. 402–406.

13 Ibid. p. 404.

14 Kristin Schrader-Frechette, "Agriculture, Coal and Procedural Justice", In *Policy for Land: Law and Ethics*, Lynton Keith Caldwell and Kristin Schrader-Frechette (Lanham, MD: Rowman and Littlefield, 1993), pp. 111–131. I'm taking some liberties here, since Schrader-Frechette herself describes the position as "procedural" rather than egalitarian. However, the position of both Schrader-Frechette here and Rawls in *A Theory of Justice* is procedural in the sense that both argue that a fair procedure for arriving at constitutional principles results in an egalitarian constitution. As such, both are more firmly committed to the substantive claims of the egalitarian positions than are pragmatists, who take a procedural view of policy that might combine elements of each of these positions as readily as it might settle on any one.

15 Roderick Frazier Nash, *The Rights of Nature: a History of Environmental Ethics* (Madison, Wi: University of Wisconsin Press, 1990); James A. Nash, *Loving Nature: Ecological Integrity and Christian Responsibility* (Nashville, TN: Abingdon Press, 1991).

16 I do *not* mean to imply that any of the persons discussed are foundationalists. It is doubtful that any of the water policy authors have worked out a position on the foundationalist/anti-foundationalist issue, at all. Shrader-Frechette, for her part, is pretty consistent in taking an egalitarian line on issues, but one can find very pragmatic sounding passages in her work, as well. The *arguments* discussed here apply classically libertarian, egalitarian and utilitarian forms of reasoning, but if we were all more careful in distinguishing the argument and the person (including, ironically, ourselves) we might find out that more of us are pragmatists than we ever suspected.

17 William James, *The Will to Believe and Other Essays* (Cambridge, Ma: Harvard University Press, 1979), Original 1897.

18 William James, *Pragmatism* (Cambridge, Ma: Harvard University Press, 1976), p. 123.

19 John J. McDermott, "Pragmatic Sensibility: The Morality of Experience," in *New Directions in Ethics: The Challenge of Applied Ethics*, Josephe P. DeMarco and Richard M. Fox, eds. (New York: Routledge and Kegan Paul, 1986), p. 124.

20 Notice that James does not relativize truth entirely. The will to believe

provides no epistemic warrant for believing something that can be shown to be false.

21 Gettier problems aside.

22 John Dewey, *The Quest for Certainty: A Study of the Relation of Knowledge and Action* (New York: Minton, Balch and Company, 1929).

23 Richard Rorty, "Human Rights, Rationality and Sentimentality," in *On Human Rights: The Oxford Amnesty Lectures 1993*, Stephen Shute and Susan Hurley, eds (New York: Basic Books, 1993) pp. 111–134. I recommend Rorty's view of how deconstruction figures in pragmatist ethics, a theme that is, I think, underappreciated by those who study the classical pragmatists without reading the articles on public policy.

24 Local to a particular time, place, or community of discourse, that is.

25 Larry Hickman, *John Dewey's Pragmatic Technology* (Bloomington, In: University of Indiana Press, 1991).

10

TOWARDS A PRAGMATIC APPROACH TO DEFINITION

"Wetlands" and the politics of meaning

Edward Schiappa

This paper is part of an ongoing project in which I argue that definition should be treated less as a traditional philosophical or scientific question of "is" and more as an ethical and political question of "ought."[1] My approach to definitional disputes is heavily informed by the pragmatist writings of Richard Rorty. Against the traditional theory of "real" definitions – which suggests we must "find" the true essences of things – I argue that disputes over new definitions are a matter of deciding what sort of world we wish to "make." My argument is that *all* definitions are "political" in two respects: first, definitions always function to serve particular interests; second, the only definitions of consequence are those that have been *empowered* through persuasion or coercion. I examine the debate over the legal definition of "wetlands" as a case study because it illustrates how definitions are political in the two respects just mentioned.

DEFINITIONS AND THE POLITICS OF MEANING

To claim that definitions are political because they must be empowered through persuasion or coercion is not to say, simply, that "Might makes meaning," but rather that for a particular definition to be *shared*, people must be moved to adapt their linguistic and non-linguistic responses according to the understanding instantiated in the definition.[2] Such responses "may be shaped through the application of various forms of power from logical or moral suasion, through bribery, to coercion."[3]

Definitions devised by scientists usually are not described as "political." Scientific definitions are usually described as being more "objective" (that is, more *real*) than non-scientific definitions, and informed by "rational" and "neutral" criteria rather than based on value-laden political factors. I believe that such distinctions are misleading and unproductive. As has been illustrated repeatedly in the past several decades of the history and philosophy of science, "facts," "data" and "observations" are theory-dependent such that what is considered "objectively real" can vary from theory to theory and change over time. Thomas Kuhn argues that *no* scientific description (or definition), however strongly believed today, should be considered invulnerable to change in the future.[4] In lieu of a more thorough defense, I will for the moment simply assert that there is no compelling theoretical or practical reason to treat definitions by scientists as more "objective" or "real" than definitions by non-scientists.[5] Though we can distinguish the two groups by reference to different social practices, little is gained and much is lost by attempting to do so on an ontological or epistemological basis.[6] Kuhn provides an anecdote that demonstrates how definitions serve particular interests. Two scientists were asked whether a single atom of helium is or is not a molecule: "Both answered without hesitation, but their answers were not the same. For the chemist the atom of helium was a molecule because it behaved like one with respect to the kinetic theory of gases. For the physicist, on the other hand, the helium atom was not a molecule because it displayed no molecular spectrum."[7] In this example there were two different theory-driven answers to the question "What is a molecule?" What *counts* as a molecule differs according to the current needs and interests of chemistry and physics. It is pointless to ask which answer is "really" correct since the implicit definitions involved are theory-dependent and, equally importantly, what may be the most appropriate conceptualization for one group of specialists may not be so for another. The question of what *ought* to count as X for a particular language community is a normative and prescriptive question; what we consider X to be "really" is the *result* of our answer, not its *cause*.

Definitions proffered by scientists may serve different interests than those put forth by non-scientists, but they serve interests none the less. Typically "scientific" interests can be described as those "internal" to the language community to which a scientist belongs. How well a definition serves the shared purposes of the community might be discussed in terms of coherence with other concepts, clarity, amenability to quantification, or other predictive and explanatory interests. Intentionally or otherwise, "external" interests often are served by scientific definitions as is made evident in the case of defining "wetlands."[8]

BACKGROUND OF THE WETLANDS CONTROVERSY

Although words such as bog, marsh and swamp have been in use for centuries, the collective term *wetland* "came into broad usage only during the late 1960's and early 1970's."[9] Generally speaking, the term is used to denote areas "sufficiently saturated by water that only specially adapted plants can grow there. Saturation with water prevents oxygen from working its way into the soil and therefore creates conditions of no oxygen."[10] Only *hydrophytes*, vegetation that has adapted to such anaerobic conditions, can survive in wetlands. Furthermore, because the soil in such areas is periodically or permanently saturated with water, it has higher than average moisture content and is classified as *hydric soil.* The degree or type of water saturation of an area is known as its *hydrology.* These three factors – hydrology, hydric soil and hydrophytes – are the traditional defining characteristics of wetlands.

Wetlands are "open systems." That is, wetlands interact with other ecological systems, such as groundwater tables and rivers, in a way that enhances the overall environment and, in particular, water quality. When water flows in and out of the wetland area, "sediments and other pollutants tend to remain, and the nutrients are converted into plants."[11] Wetlands produce vegetation that photosynthesize at much higher rates than non-wetlands, which creates material vital to the aquatic food chain. A wide variety of plant and animal life flourish in wetlands.

So-called "drier" wetlands are areas that are saturated for relatively short periods of time, but still serve some of the important ecological functions of wetlands. Ironically, "drier" wetlands are among the most valuable of wetland areas from an environmental perspective: "Many have a powerful intuition that the wetter the wetland the more valuable it is. This intuition is false."[12] Among the valuable functions of "drier" wetlands are the following:

1 They are particularly effective natural flood controls: "Their relative dryness gives them greater capacity to absorb floodwater.

Their strong vegetation slows down floodwaters and limits their destructive force."[13]

2 They are especially useful filtration systems because they "trap and absorb pollutants *before* runoff can mix with deeper waters. Scientific studies have confirmed that many drier wetlands provide the most effective treatment of water quality."[14]

3 Certain animals can live in shallow wetland areas when the "wetter" wetlands become too deep: "Loss of these areas leaves these animals nowhere to go in periods of high water."[15]

4 During dry periods, a good deal of plant and tree growth occurs. Then, during seasonal or temporary periods of saturation and flooding, certain plant material is carried into deeper waters where it becomes an important food supply for various fish species. In sum, certain wetlands appear to be "dry" much of the time. Nonetheless, the saturation they receive is sufficient to facilitate valuable ecological functions that distinguish them from non-wetlands.

Specific definitions of "wetlands" differed somewhat from state to state when the term first became popularized in the late 1960s. It was not until the mid-1970s that efforts were made to produce a standardized definition that could be used nationwide.[16] Virtually from the beginning, those most interested in defining "wetlands" were interested in identifying and preserving the ecological functions of wetlands. In strictly academic settings, conflicting definitions can coexist without serious problems (for example, in rival textbooks). It is assumed that there is sufficient overlap in the competing definitions that no harm results from a lack of strict uniformity. Besides, normally no one in academic settings has the authority to declare *one* specific definition to be that which everyone in a given discipline must follow. Public laws, on the other hand, are aimed at precisely this sort of denotative conformity. Section 404(f) of the 1977 Clean Water Act was designed to halt widespread wetland destruction. The subsequent definitions put into service by the relevant federal agencies after that point in time were backed by the power of federal law. In 1979, a standard ecological definition was published by the US Fish and Wildlife Service:

[W]etlands are lands where saturation with water is the dominant factor determining the nature of soil development and the types of plant and animal communities living in the soil and on its surface. The single feature that most wetlands share is soil or substrate that is at least periodically saturated with or covered by water. The water creates severe physiological problems for all plants and animals except those that are adapted for life in water or in saturated soil.[17]

Such a definition uses all three factors mentioned previously: hydrology (wetness), hydric soil and hydrophytes. While the temporary or permanent presence of water is what makes a given area a wetland ecology, the total amount of water *on* and *in* the soil varies tremendously over the seasons and is very difficult to document directly. Accordingly, the 1979 definition, like most of those to follow, defines "wetlands" as areas that have any one of three features: wetland vegetation, soil or hydrology:

> For the purposes of this classification, wetlands must have one or more of the following three attributes: (1) at least periodically, the land supports predominantly hydrophytes, (2) the substrate is predominantly undrained hydric soil, and (3) the substrate is nonsoil and is saturated with water or covered by shallow water at some time during the growing season of each year.[18]

Because the amount of water necessary to produce wetlands is highly variable and difficult to measure, most efforts to define "wetlands" throughout the 1980s focused on hydric soil and hydrophytes. Hydric soils have been defined as those that receive sufficient saturation to produce anaerobic conditions – conditions that sharply limit the types of vegetation and animal life that can live in or on the soil. Hydrophytes are those plants that have adapted to such anaerobic conditions. A specific list of hydrophytes was drafted by the US Fish and Wildlife Service in 1977 and has been reviewed and updated many times since.[19]

Based on these early ecological definitions, it has been estimated that wetlands in the contiguous states are being destroyed by natural and human causes at a rate of nearly 300,000 acres annually. If present trends continue, an additional 4,250,000 acres of wetlands will be lost between 1990 and 2000.[20] Given that approximately 56 per cent of wetlands have already been lost over the past two centuries,[21] the potential cumulative impact in terms of loss of flood control, habitat protection and water quality is enormous.[22]

During the 1980s, four different Federal agencies had jurisdiction relevant to the regulation of wetlands: the US Fish and Wildlife

Service (FWS), the Environmental Protection Agency (EPA), the Army Corps of Engineers (CE) and the Agriculture Department's Soil Conservation Service (SCS). All four had the legislative or administrative power to define "wetlands" according to their respective needs and interests. As noted by Max Peterson of the International Association of Fish and Wildlife Agencies: "At one time Fish and Wildlife Service had a *habitat* classification. Soil Conservation Service had a *soils* classification, and other agencies had a definition based on *water presence*."[23] Each of these regulatory bodies had the statutory or administrative power to designate specific areas as "wetlands" and to affect peoples' behavior accordingly. For example, according to the so-called "Swampbuster" provision of the 1985 Food Security Act, farmers wishing to sell wetland acreage to commercial developers first must obtain a federal permit. If the acreage is classified as "wetlands," according to federal definitions, the permit can be denied.

Just how disparate the different federal regulatory agencies' definitions of "wetlands" were, prior to 1989, is a matter of some dispute. Though some contend that the various agencies used "very similar approaches,"[24] others complain that the lack of standardized methods "resulted in inconsistent determinations of wetland boundaries."[25] To ensure a reasonable degree of uniformity, the four responsible federal agencies began a series of meetings beginning in early 1988 to produce a standardized manual for delineating wetlands. In January 1989 the Federal Interagency Committee for Wetland Delineation published the *Federal Manual for Identifying and Delineating Jurisdictional Wetlands* (hereafter *Manual*). According to Francis C. Golet, a Professor of Natural Resources Science who has been involved in wetlands research for over twenty years, "the 1989 Manual represents the culmination of nearly 17 years of efforts by wetland scientists, soils experts, and land managers from throughout the country. It also represents a consensus of the four leading wetland management/regulatory agencies."[26]

Like most federal regulations, the 1989 *Manual* received both praise and criticism from those most directly affected. Critics charge that the *Manual* significantly broadened the definition of "wetlands" such that millions of acres previously *not* considered "wetlands" would now be so designated.[27] Those who defend the *Manual* point out that it did *not* "initiate a significant revision to prior existing standards. Like the other manuals, it most heavily emphasized evidence of soil types and vegetation, and used the limited available evidence of hydrology (of wetness) primarily as a means of verifying the evidence provided by soils and vegetation."[28] Defenders of the *Manual* agree that there have been problems implementing the relevant federal regulations, but argue that the definition of "wetlands" itself is consistent with years of experience and needs no revision.[29] The implementation of federal regulation concerning wetlands, like all public policy procedures, requires an ongoing process of negotiation and mutual adjustment between regulators and those regulated. If not for the campaign rhetoric of George Bush during the 1988 presidential election, hammering out the details concerning the appropriate regulatory definition of "wetlands" very well might have remained a matter of interest solely to specialists. As a result of campaign promises made in the fall of 1988 – promises repeated after Bush took office – how to define "wetlands" became a national controversy.

REDEFINING "WETLANDS" DURING THE BUSH ADMINISTRATION

As part of a bid to be known as "the environmental president," Bush promised in the 1988 presidential election that he would commit his administration to the goal of "no net loss" of wetlands. In October of 1988, as part of a candidate forum in the magazine *Sports Afield*, Bush stated: "My position on wetlands is straightforward: All existing wetlands, no matter how small, should be preserved."[30] Following his election, in a speech before members of Ducks Unlimited in June of 1989, Bush proclaimed that "any vision of a kinder, gentler America – any nation concerned about its quality of life, now and forever, must be concerned about conservation."[31] Noting that "our wetlands are being lost at a rate of nearly half a million acres a year," Bush reaffirmed his commitment to "no net loss":

> You may remember my pledge, that our national goal would be no net loss of wetlands. And together, we are going to deliver on the promise of renewal, and I plan to keep that pledge. . . . Wherever wetlands must give way to farming or development, they will be replaced or expanded elsewhere. It's time to stand the history of wetlands destruction on its head. From this year forward, anyone who tries to drain the swamp is going to be up to his ears in alligators.[32]

Bush described the protection of the environment as "a moral issue. For it is wrong to pass on to future generations a world tainted by present thoughtlessness."[33] Encouraging his audience to judge their actions in light of the verdict of future generations, Bush asked those present to imagine what might be said forty years from now:

It could be they'll report the loss of many million acres more, the extinction of species, the disappearance of wilderness and wildlife. Or they could report something else. They could report that sometime around 1989 things began to change and that we began to hold on to our parks and refuges and that we protected our species and that in that year the seeds of a new policy about our valuable wetlands were sown, a policy summed up in three simple words: "No net loss." And I prefer the second vision of America's environmental future.[34]

The efforts to codify the different federal definitions of "wetlands" began in early 1988, well before Bush was elected president. Nevertheless, by making "no net loss" a centerpiece of his administration's environmental policy, Bush energized governmental efforts to protect wetlands. As Congressperson Gerry E. Studds noted, "this is the first instance I know of where campaign rhetoric rises to the level of statutory law. No loss of wetlands originated in a campaign speech; to my knowledge, it is not the law."[35] In his 1990 budget statement Bush reiterated the goal of "no net loss." The US Department of Interior and the US Fish and Wildlife Service published a "wetlands action plan" in 1990 that was titled *Wetlands: Meeting the President's Challenge*; the publication prominently quoted the above cited passages from Bush's speech before Ducks Unlimited. Congressional hearings were held in part to explore ways in which to meet the president's goal of "no net loss" of wetlands. The responsible federal agencies already were committed to enforcing existing statutory regulations requiring the protection of the nation's wetlands; their efforts to produce a unified *Manual* for delineating "wetlands" was part of that ongoing effort. Bush's policy of "no net loss" created heightened awareness of such federal efforts so that when wetland protection came into conflict with other policy objectives, public controversy was virtually inevitable.

The goal of "no net loss" probably was deemed a politically viable promise because it sounds flexible and absolutist at the same time; thereby placating environmentalists who demand commitment and pro-growth developers who want flexibility. Thus environmentalists tend to emphasize the *no loss* part of the promise, while developers emphasize no *net* loss. In practice, however, the effort to placate both constituencies proved impossible. Congressional hearings held in 1989, 1990 and 1991 document that pressure was mounting from both directions. On the one hand, Bush's call for "no net loss" generated considerable enthusiasm for protecting wetlands. Almost all policy-makers endorsed the goal; the only question was how to implement it. When is it appropriate, for example, to drain a valuable wetland in one location with the expectation that another wetland will be created elsewhere? Pressure mounted on the different regulatory agencies to articulate clearly how they would meet the President's challenge for "no net loss." On the other hand, the coincidental pledge for "no net loss" and the publication of the official *Manual* to delineate wetlands galvanized opponents of federal environmental regulation. Opposition to the protection of wetlands comes most often from farmers wanting to sell their land and from developers wishing to buy, drain and develop wetlands. They argued that the regulatory agencies had run amuck, applying the label "wetlands" much too broadly, and that regulators were not taking local economic needs into adequate consideration.

The Bush administration found itself in a dilemma: Either Bush could lessen his commitment to "no net loss," thereby breaking a highly visible and useful campaign promise, or he could stand by the promise and risk alienating pro-business, pro-development constituents. Bush's "solution" was simple and, had it worked, politically ingenious. In January of 1990, White House Press Secretary Fitzwater announced that "At the President's direction, the Domestic Policy Council, which has created a task force on wetlands, is in the process of examining how best to implement the President's goal of no net loss."[36] "How best to implement the President's goal of no net loss," turned out to be a proposed *redefinition* of "wetlands." By sharply narrowing the scope of the regulatory agencies' definition of wetlands, Bush would be able to claim that he kept his promise.

In August of 1991, the four agencies charged with protecting wetlands published in the *Federal Register* a document entitled "'Federal Manual for Identifying and Delineating Jurisdictional Wetlands'; Proposed Revisions."[37] Although bearing the name of the relevant regulatory agencies, the document was produced under the direction of the Vice President's task force on wetlands and was intended to be codified as a Presidential Executive Order with the force of law. The "Proposed Revisions" were presented and explained as a clarification and refinement of the 1989 *Manual*, but in effect the revisions represented a major departure from the *Manual's* procedures for delineating wetlands. The practical result of the proposed redefinition, if implemented, would be to decrease dramatically the amount of acreage that could be designated protected wetlands. The most modest estimate is that "as much as a third of the 38.4 million hectares (95 million acres) of wetlands in the lower 48 states will be considered wetlands no more and thus will be vulnerable to development."[38] The Environmental Defense Fund's extensive study of the effects of the proposed changes to the *Manual* suggests that an even larger percentage – 50 per cent or roughly 50 million acres – of land previously designated "wetlands" would be excluded by the proposed redefinition.[39] Their estimate corresponds to that by the National Wetlands Technical Council, a group of "independent wetlands scientists."[40]

There are two primary differences between the 1989 *Manual* (which represents the traditional practices of delineating wetlands) and the Bush administration's proposed redefinition. First, where the 1989 *Manual* allowed an area to be designated a wetland if any one of several criteria were met clearly, the 1991 redefinition require that all three criteria (hydrology, hydric soil, hydrophytic vegetation) be met and proved independently. Second, the specific standards by which each criteria is judged was made much more stringent. For example, the 1989 *Manual* required seven consecutive days of inundation or saturation "at or near the surface," while the 1991 redefinition more than doubled the length of time necessary (15 to 21 days) and specified water *at the surface*, not just near it.

The codification of definitions of "wetlands" in the 1989 *Manual* was implemented by the relevant federal agencies without additional authorization by the White House or Congress, and without inviting public comment. The *Manual* was considered a "technical guidance document which is not required by law to go through Administrative Procedure Act rulemaking procedures."[41] In other words, the relevant federal agencies were *empowered* to enforce the *Manual's* definition of "wetlands" without additional authority, since the power to regulate wetlands was given already under current federal law. Opponents of wetland regulation responded in two ways. First, a rider was successfully attached to the Energy and Water Development and Appropriations Act of 1992 that cut off funding for further delineation of "wetlands" using the 1989 *Manual*. The action temporarily *disempowered* the federal agencies from requiring conformity to the *Manual's* definition. Second, opponents criticized the agencies for creating and enforcing the "new" definition without inviting public comment. The Bush administration *could have* enforced the 1991 proposed redefinition by Executive Order or through the same input-free process that the 1989 *Manual* was adopted. However, having criticized the federal agencies for having acted without inviting public comment, the administration felt compelled to extend such an invitation regarding the proposed redefinition.[42] The response was overwhelming: over 10,000 documents were sent to the EPA, requiring the agency to hire an outside consulting firm to collate the input provided.

Though not without supporters, the proposed redefinition was met mostly with intense opposition and condemnation. *Sierra* magazine claimed that the administration's "evisceration of existing wetlands policies demonstrates – more conclusively than any previous actions in this arena – the abandonment" of the "no net loss" pledge.[43] The Bush administration's proposed redefinition "broke his most specific campaign pledge."[44] The redefinition was seen as a cynical ploy: "A teensy redefinition of what constitutes a wetland, and presto – the administration jeopardizes 30 million acres of them, an area about the size of New York state."[45] The Associated Press reported that "Government wetlands experts have concluded that the Bush administration's proposed redefinition of the term is unworkable, unscientific and would leave 'many obvious wetlands' unprotected."[46] By late November of 1991, criticism from inside and outside of the administration had grown so intense that a spokesperson for the President's Competitiveness Council admitted that the proposed redefinition would have to be revised "to honor President Bush's 1988 campaign pledge."[47]

COMPETING INTERESTS, COMPETING DEFINITIONS

Before discussing the specific arguments lodged against the administration, I want to draw attention to a rhetorical strategy that emerged in the criticisms that is particularly relevant. The proposed redefinition was branded "political" and contrasted to the current "scientific" definition. Representative Lindsay Thomas complained that policymakers had no business defining "wetlands": "The problem is not how to define wetlands. That is a science."[48] Similarly, scientist Francis C. Golet suggested that "the definition of wetland is wholly a scientific issue." Although political input is unavoidable, "in matters of science, such as the definition of wetland, scientific arguments must prevail."[49] The image of an objective, bias-free science was invoked frequently to help justify continuing the 1989 *Manual*'s definition as opposed to the administration's proposed redefinition. James T. B. Tripp, general counsel to the Environmental Defense Fund, ended his testimony to Congress as follows:

> The proposed revisions to the manual represent this callous approach to science taken to an extreme – as nonscientists believed they could draw up a manual that would be usable and would accurately cross off some unarticulated category of wetlands that did not perform important functions. The need for more dispassionate, unbiased science has rarely been greater on any environmental issue. I urge this Committee to make an important priority the assurance that accurate science guide public policy on this issue.[50]

In light of the fact that few, if any, scientists engaged with environmental studies were willing to support the administration's proposed redefinition, it is tempting to interpret the dispute in the simple terms of *Politics* (subjective, emotion-driven and biased) versus *Science* (objective, rational and unbiased). As an editorial in *The New York Times* put it: "Mr. Bush's scientists have one definition of what a wetland is, his political advisers another."[51] Consistent with contemporary neopragmatism, however, I believe that we are better off without such a dichotomy.[52]

Pragmatically, the dispute over defining "wetlands" is understood as a matter of *competing interests*. When science and politics are treated as wholly distinct, the tendency is to slip into the rhetoric of "real definitions." "A wetland is a wetland is wetland," as one environmentalist put it.[53] Real definitions, for those who believe in them, ought to be provided by those who are most expert with respect to the slice of Reality being defined. Reality is to be defined by the experts. "The natural sciences, given the frequent presence of a scholarly consensus on the means and ends of inquiry, often approximate the rational ideal of definition," notes political scientist Peter C. Sederberg.[54] If we accept this stereotype, then it is an easy step to granting definitional hegemony to a technocratic elite. One witness to a congressional hearing on wetland protection claimed that "the 1991 revisions represent a knee jerk reaction to political pressure rather than reason."[55] Another witness suggested that the definition of wetlands "probably needs to be turned over to the National Academy of Science because an independent objective standard needs to be used."[56] The problem with such a solution is that it turns "experts" into a technocratic ruling elite. "These technocrats," Sederberg observes, "would be the functional equivalent of the philosopher kings."[57] Rather than relegating the task of defining reality to philosopher kings, I believe a more productive and ethical route is to describe definitional disputes as a matter of competing interests that require arguments that some interests are better than others in particular cases. Accordingly, the questions to ask are: Whose interests are being served by a particular definition, and do we want to identify with those interests?

In the case of "wetlands," the competing interests are fairly easy to identify and to contrast. In the simplest of terms, the 1989 *Manual*'s definitions represent the interests of ecologists; in contrast, critics argue that the proposed redefinition of wetlands is "devised by developers, timber companies, and highway departments."[58] Just how fair and accurate such a simple contrast may or may not be is illustrated by looking at the more specific interests identified in arguments for and against the proposed redefinition. The most thorough critique was provided by the Environmental Defense Fund (hereafter EDF). Together with the World Wildlife Fund, the EDF published *How Wet Is a Wetland? The Impacts of the Proposed Revisions to the Federal Wetlands Delineation Manual* in January of 1992. According to the EDF, forty scientists and specialists were involved in preparing the 175 page report. EDF claims that "an estimated 50 percent of America's remaining wetlands" would be excluded by the proposed redefinition.[59] The long-term result would be "severe environmental and economic impacts." The report identifies five specific areas of harm: flooding, water quality, biological diversity, waterfowl and fisheries. In each area, EDF directly challenges the administration's belief that only "wet" wetlands deserve protection. As noted earlier, "drier" wetlands sometimes protect the most important ecological interests served by wetlands. EDF notes that the proposed redefinition's criteria for determining hydrology "has virtually no relevance to flood control." In fact, "compared to the more permanently flooded wetlands, the wetlands excluded by the proposed manual actually have greater capacity to detain floodwaters because they are less likely to be filled with water before the flood event."[60] A specific example demonstrates the costs of the new definition:

> In the eastern portion of DuPage County, Illinois, the loss of wetlands has led to frequent severe flooding that caused $120

221

million in damages in 1987 and will cost up to $50,000 per damaged residence to remedy. The proposed manual would exclude 86 percent of the similar kinds of wetlands in the western portion of the county – an area that today retains many wetlands and suffers from little flooding.[61]

The reports proceeds to document similar harms to water quality, biological diversity, waterfowl and fisheries. In each area of harm the report specifies the sort of damage that would occur if the new definition were utilized. The problems described are precisely those identified in documents such as the Department of the Interior's *Wetlands: Meeting the President's Challenge* and discussed by Bush himself in addresses such as the one given to Ducks Unlimited in 1989. The difference is that EDF documents, in detail, how much of the damage from loss of wetlands comes from the loss of so-called "drier" wetlands that the proposed redefinition was designed to exclude.

The arguments set forth by EDF are openly pragmatic. There is little or no effort to invoke the sort of circular rhetoric typically associated with "real definitions." Although Chapter 1 is titled "What are wetlands?" the answer is pragmatic and functional. EDF notes that "because wetlands are diverse, few generalizations about them are always true."[62] Instead of looking for unchanging qualities or an essential nature of wetlands, EDF identifies the valuable ecological functions that various sorts of wetlands serve. The current definition, fueled from the beginning by ecological interests, ought to be preserved because the consequences of the proposed redefinition are undesirable. EDF rejects what they call "the misconception that only areas that are wet at the surface for extended periods are 'real' or 'valuable' wetlands."[63] In the process of defending the claim that "wetter wetlands are not better wetlands," EDF does not adopt the position that there are real versus apparent wetlands, but instead focuses on the many valuable functions such lands perform and notes that surface hydrology – the primary defining characteristic of the proposed revisions – has little to do with such functions.

Interestingly enough, both sides in the wetlands definition controversy are interested in producing a definition that will delineate wetlands accurately, consistently and predictably. Both sides want, in other words, *denotative conformity* with respect to the word "wetlands" for the purpose of enforcing current statutes regarding wetlands.[64] Accuracy, consistency and predictability are often considered "scientific" values.[65] Indeed in the proposed redefinition the Environmental Protection Agency claims that "Of paramount importance to us . . . is to maintain and improve the scientific validity of our delineation methods."[66] In a general sense, then, both sides are interested in their definition being considered "scientific." When critics of the proposed redefinition call it "unscientific," as they often have, to what are they referring? The scientists who have charged the administration with being "unscientific" are not *merely* interested in accuracy, consistency, and predictability. They *also* want to continue to study and protect wetlands for their ecological significance. Being "unscientific" in this context translates as "abandoning what scientists have been doing with respect to wetlands". Accordingly, when a scientist such as Francis Golet charges that the redefinition "disregards more than 15 years of scientific research,"[67] I believe his criticism is best understood as a complaint that the redefinition breaks faith with those responsible for many years for our understanding of the ecological importance of wetlands and abandons the values and interests that current statutes were drafted to protect. The EDF's studies concerning the amount of loss of wetlands protection suggest that the EDF is, in fact, able to utilize the new definition to delineate wetlands accurately, consistently and predictably. The problem with the new definition is not so much that it is "unscientific" in general, but rather in particular because it abandons the values and interests of scientists traditionally associated with the study of wetlands.

The interests pursued by those in favor of the proposed redefinition are fairly straightforward. Organizations such as the Tidewater Builders Association, the Forest Farmers Association, the National Association of Homebuilders, Weyerhaeuser Company, Associated Builders and Contractors and the National Association of Realtors testified before Congress in favor of the administration's proposed redefinition. In fact, the National Association of Realtors noted that they had advocated the policy in early 1989 "that *all three parameters* (which include hydrophytic vegetation, hydric soils and hydrology) be utilized in delineation of a wetland" – precisely the policy proposed by the Bush administration.[68] The arguments offered by such organizations boil down to one basic complaint: the 1989 *Manual* prevents people from developing land in the manner of their choosing. As a result, the right to use one's property profitably is obstructed by federal regulations that these developers feel "go too far." A representative of the Tidewater Builders Association complained that real estate "estimated at $50 billion" potentially met the *Manual*'s criteria for wetlands and thus could not be developed.[69] A county commissioner from Georgia claimed that "economic growth has been drastically curtailed" by the 1989 *Manual*. "Engineers, architects, home builders, developers, contractors and their employees were impacted."[70]

A related and persistent justification for the proposed redefinition is that the 1989 *Manual* drastically expanded the amount of land regulated as wetlands. The argument is controversial; as noted earlier, environmentalists as well as government officials claim that such accusations are ungrounded and are the result of misunderstandings that subsequently have been clarified.[71] Nonetheless, advocates of the proposed redefinition consistently argued that the 1989 *Manual* expanded protection to far too many areas that are not "true" wetlands. Robert W. Slocum of the North Carolina Forestry Association argued that "identifying dry land that has no resemblance to *true* wetland ecosystems as 'wetlands' only confuses the public and the landowners and hinders protection of *true* wetlands."[72] Slocum praised the administration's proposal as offering "a more *realistic* definition" that protects "*true* wetland ecosystems."[73] Similarly, the National Association of Realtors stated that they were "pleased with the consensus definition of protected wetlands reached by the Bush administration, which more accurately and clearly defines a *true* wetland."[74] More often than not, advocates of the proposed redefinition expressed their belief that "true" or "real" wetlands still would be protected.[75] Explicitly or implicitly the 1989 *Manual* was condemned for protecting lands that are not "really" and "truly" wetlands.

I have argued previously at length that such dissociative claims in defense of a definition are circular and unhelpful.[76] To claim that one definition is superior to another because it captures what is "really and truly" a wetland simply avoids the pragmatic question of *what ought to count* as a wetland for the purposes of federal regulation. Typically, advocates of the proposed redefinition rely on a "wetter is better" logic. The Delaware Council of Farm Organizations, for example, argued that "farmers are not, in general, opposed to protecting wetlands; that is land that is *truly wet*."[77] The problem with such arguments is that they fail to clash with the case offered by ecologists concerning the value of so-called "drier" wetlands. Rather than invoking the dichotomies of "true" versus "false" wetlands, or "scientific" versus "political" definitions, a more productive discussion would focus on the relative costs and benefits of protecting the lands included by the 1989 *Manual* and excluded by the proposed redefinition. Such a discussion is precisely what the EDF offers in *How Wet Is a Wetland?* The most important question should be whether the benefits of protecting the disputed lands are considered more important or valuable than maintaining the property rights of those who own and wish to profit by developing them. So far, the values and interests expressed and implied by existing legislation would warrant the conclusion that the answer be "yes."

Even setting aside the question of which interests are more important to protect, the Bush administration's attempt at redefinition was logically inconsistent as well as ethically suspect. Bush's early declarations about wetlands, in his role as "the environmental president," depend on traditional definitions of wetlands. For example, in his statements about the quantity of wetlands being lost each year, he relies on statistics that utilize a definition of "wetlands" codified in the 1989 *Manual.* Yet his later statements clearly back away from standard definitions. While insisting that "I am committed to no net loss of wetlands," Bush also insisted that "I am not committed to decisions that take productive land out of production." He complained that "you've got zealots in various levels of the bureaucracy" that require control "from the top on down."[78] Bush effectively abandoned his identification with the agencies charged with protecting wetlands who had been working towards a consistent definition of "wetlands" for over a dozen years. In so doing, he rejected his previous alignment with the interests those agencies represent. During 1992 Bush repeatedly insisted that he was committed to "no net loss" of wetlands, but backed off the traditional definition: "I think what happens during some periods, some of the bureaucrats in our regulatory agencies started defining the wetland problem in a way that really overdefines it. There was not a legitimate wetland we were trying to preserve."[79] Not surprisingly, Bush relies on the rhetoric of real definition to defend the revised policy. When speaking to an agricultural organization with pro-development sentiments, Bush made his interests clear:

> My direction to Vice President Quayle's Council on Competitiveness was to protect environmentally sensitive wetlands and protect the property rights of landowners. I've asked the board [of the Farm Bureau Federation] to send in specific recommendations during this hearing period. Our new guidelines will

distinguish between *genuine* wetlands which deserve to be protected and *other* kinds of land, *including your farmlands*.[80]

Noting that "the extreme environmentalists are not happy" with his new wetlands policy, Bush claimed that the answer is "to try to balance all of these interests."[81] Yet by dramatically narrowing the standard definition of "wetlands," Bush clearly tipped the balance away from environmental interests. Complaining again that "we were too far over between the Corps [of Engineers] and EPA on the regulatory side," Bush warned that we must "be wary of the extremes." His own definition is simple and direct: "I've got a radical view of wetlands. I think wetlands ought to be wet."[82] By identifying himself with the "wetter is better" criteria espoused in the proposed redefinition, Bush explicitly distanced himself from the environmental interests reflected in the traditional definition upon which his pro-environmental statements depend. It is ironic but fitting that Ducks Unlimited – the very organization before which Bush gave his most important and influential wetlands address – came to oppose the proposed redefinition.[83] That Bush's attempts to balance interests has been unpersuasive is suggested by a steady decline in approval rates of his handling of environmental issues. In March of 1991, at the peak of his popularity as president, Bush had a 52 per cent approval rate for his handling of environmental issues. By June 1992, that rate had fallen to 29 per cent with 58 per cent of those polled expressing disapproval.[84]

CONCLUSION

I conclude with two comments about the wetlands controversy. First, the dispute is a useful case study because it throws into relief how definitions are interest-driven and saturated with questions of power and persuasion. What makes "wetlands" unusual is the amount of media coverage the controversy has received. But the fact that definitions *matter* – that there are pragmatic and political results of our choices of definitions – is not unusual at all. Power to define is power to influence behavior. All proposed definitions are devised for specific purposes that can be evaluated according to the interests that they advance. And the success of any definition depends on how effectively its advocates persuade (or coerce) members of a given community to conform and use the term "properly." In the case of "wetlands," the Bush administration was unable to persuade enough regulators and citizens to support the proposed redefinition, and unwilling to coerce them to do so. Such disputes over the scope of government regulation highlight the political dimension of defining that is, I believe, ubiquitous.

Second, note that none of the interests identified so far need be classified as exclusively "scientific" or exclusively "political." Scientists constitute a specific subset of society that is identifiable, in part, by their common interests and values. But interests and values they indeed have, and accordingly it sheds little light to describe such interests and values as "non-political." A variety of social interests are advanced by achieving a level of denotative conformity with certain words. Thus politicians and scientists *share* the goal of denotative conformity with respect to "wetlands." The dream of escaping politics altogether and letting "experts" define tough concepts for us is a powerful one, but the dream potentially ends in disaster. If one considers the outcome of the wetlands controversy a happy ending, it is because one identifies with the interests of the winner. The results could have been otherwise; *politics* is responsible for what we now call "wetlands" and for what we will treat as "wetlands" forty years hence. Interests are always served by definitions: the only question is *which* interests. Prudence requires that, as a society, we learn to tell the difference between the definitional disputes that are exclusively "scientific," in the sense that the outcome only affects the community of scientists, and those disputes that involve us all – scientists included. Both sorts of conflicts are political; recognizing them as such may prompt us to take greater responsibility for the social and political process of defining reality.

NOTES

1 Edward Schiappa, "Arguing About Definitions," *Argumentation* 7 (1993): 403–417.
2 Peter C. Sederberg, *The Politics of Meaning: Power and Explanation in the Construction of Social Reality* (Tucson: University of Arizona Press, 1984), p.56.
3 Ibid. p. 7.
4 Thomas S. Kuhn, "Dubbing and Redubbing: The Vulnerability of Rigid Designation," *Minnesota Studies in the Philosophy of Science* 14 (1990): 298–318.
5 I refer to definitions as set forth by particular social groups rather than using the usual rubric "scientific definitions" to emphasize that sociology, not metaphysics, distinguishes definitions offered by one group or another. For a more thorough defense of the position asserted here, see my forthcoming book *Defining Reality*, manuscript in progress.

6 Richard Rorty, *Objectivity, Relativism, and Truth: Philosophical Papers Volume 1* (Cambridge: Cambridge University Press, 1991).
7 Thomas S. Kuhn, *The Structure of Scientific Revolutions*, 2nd edn (Chicago: University of Chicago Press, 1970), p. 50.
8 There is more involved in the dispute over federal regulations regarding wetlands than the issue of definition. For example: Does a ban of development on a privately owned wetland constitute an uncompensated "taking" by the federal government? Are the costs and delays involved in getting a development permit reasonable? I set aside such issues in this section; though they are important, they are not directly relevant to the question of how "wetlands" ought to be defined.
9 Francis C. Golet, "A Critical Review of the Proposed Revisions to the 1989 *Federal Manual for Identifying and Delineating Jurisdictional Wetlands.*" In US Congress, *Wetlands Conservation.* Hearings Before the Subcommittee on Fisheries and Wildlife Conservation and the Environment of the Committee on Merchant Marine and Fisheries, House of Representatives 16 October 1991 and 21 November 1991. Serial No. 102–150. (Washington, DC: Government Printing Office, 1992), p. 635.
10 James T. B. Tripp, "Comments of the Environmental Defense Fund on National Wetlands Issues." In US Congress, *Wetlands Conservation*, op. cit., p. 203.
11 Ibid., p. 195.
12 Ibid., p. 201.
13 Ibid.
14 Ibid.
15 Ibid.
16 Golet, op. cit., p. 635.
17 L. M. Cowardin, V. Carter, F. C. Golet and E. T. LaRoe, *Classification of Wetlands and Deepwater Habitats of the United States* (Washington, DC: US Fish and Wildlife Service, 1979), p. 3.
18 Ibid.
19 Golet, op. cit., p. 637.
20 US Department of the Interior, *Wetlands: Meeting the President's Challenge* (Washington, DC: US Fish and Wildlife Service, 1990), p. 13.
21 Thomas E. Dahl, *Wetland Losses in the United States 1780s to 1980s* (Washington, DC: US Department of the Interior, Fish and Wildlife Service, 1990).
22 US Department of the Interior, op. cit., p. 15.
23 In US Congress, *Wetlands Conservation*, op. cit., p. 43 (emphasis added).
24 Tripp, op. cit., p. 199.
25 Environmental Protection Agency, et al. "Federal Manual for Identifying and Delineating Jurisdictional Wetlands'; Proposed Revisions," *Federal Register* 56 (14 August 1991): p. 404449.
26 Golet, op. cit., p. 639.
27 Environmental Protection Agency, et al., op. cit., p. 40450.
28 Tripp, op. cit., p. 199.

29 Environmental Defense Fund, *How Wet Is a Wetland?: The Impact of the Proposed Revisions to the Federal Wetlands Delineation Manual* (NY/Washington, DC: Environmental Defense Fund/World Wildlife Fund, 1992).

30 Quoted in Tom Paugh, "Sports Afield and the Candidates," *Sports Afield* 200 (October 1988): p. 15.

31 George H. W. Bush, "Remarks to Members of Ducks Unlimited at the Sixth International Waterfowl Symposium," *Weekly Compilation of Presidential Documents* 25 (1989): p. 860.

32 Ibid. p. 861.

33 Ibid. p. 862.

34 Ibid.

35 In US Congress, op. cit., p. 31.

36 "Statement by Press Secretary Fitzwater on the Development of Wetlands Conservation Policy," *Weekly Compilation of Presidential Documents* 26 (1990): p. 73.

37 Op. cit.

38 Michael D. Lemonick, "War over the Wetlands," *Time* (August 26, 1991): p. 53.

39 Op. cit., p. x.

40 In US Congress, *Wetlands Conservation*, op. cit., pp. 661–663.

41 Environmental Protection Agency, et al., op. cit., p. 40446.

42 Ibid.; see also Philip J. Hilts, "U.S. Aides Retreat on Wetlands Rule," *The New York Times* (November 23, 1991): p. A10.

43 Carl Pope, "That Question of Balance," *Sierra* (November/December, 1991): p. 22.

44 Ibid., p. 23.

45 Tom Dworetzky, "Promises, Promises: What Did Bush Say? What Did He Do?" *Omni* 14 (1992): p. 9.

46 Associated Press Release, "Papers Chastise Bush's Wetland Proposal," *The Daily Collegian* [Pennsylvania State University] (November 22, 1992): p. 7.

47 Hilts, op. cit., p. 1.

48 In US Congress, *Wetlands Conservation*, op. cit., p. 24.

49 Op. cit., pp. 640, 654.

50 Op. cit., p. 208.

51 Editorial, "Back in the Bog on Wetlands," *The New York Times* (26 November 1991): p. A20.

52 Rorty, op. cit., pp. 46–62.

53 Jean Seligmann, "What on Earth Is a Wetland?" *Newsweek* (26 August 1991): p. 49.

54 Sederberg, op. cit., p. 94.

55 In US Congress, *Wetlands Conservation*, op. cit., p. 244.

56 Ibid., pp. 62–63.

57 Sederberg, op. cit., p. 57.

58 Pope, op. cit., p. 23.

59 Environmental Defense Fund, op. cit., p. x.

60 Ibid.

61 Ibid., pp. x–xi.

62 Ibid., p. 2.
63 Ibid., p. xiii.
64 Cf. Schiappa, op. cit.
65 Thomas S. Kuhn, *The Essential Tension* (Chicago: University of Chicago Press, 1977), pp. 320–339.
66 Environmental Protection Agency, et al., op. cit., p. 40446.
67 Golet, op. cit., p. 639.
68 In US Congress, *Wetlands Conservation*, op. cit., p. 368; emphasis in original.
69 *Ibid.*, p. 60.
70 *Ibid.*, p. 226. These claims are challenged explicitly by members of The Georgia Conservancy (see pp. 398–402).
71 Environmental Defense Fund, op. cit., pp. 13–18.
72 In US Congress, *Wetlands Conservation*, op. cit., p. 109.
73 Ibid., p. 113.
74 Ibid., p. 366.
75 Ibid., pp. 336, 367, 386.
76 Schiappa, op. cit.; see also Edward Schiappa, "Dissociation in the Arguments of Rhetorical Theory," *Journal of the American Forensic Association* 22 (1985): 72–82.
77 In US Congress, *Wetlands Conservation*, op. cit., p. 409.
78 George H. W. Bush, "Remarks and a Question-and-Answer Session With the National Association of Agriculture Journalists," *Weekly Compilation of Presidential Documents* 26 (1990): p. 632.
79 George H. W. Bush, "Remarks and a Question-and-Answer Session with the Agricultural Community in Fresno, California," *Weekly Compilation of Presidential Documents* 28 (1992): p. 971.
80 George H. W. Bush, "Remarks to the American Farm Bureau Federation in Kansas City, Missouri," *Weekly Compilation of Presidential Documents* 28 (1992): p. 83; emphasis added.
81 George H. W. Bush, "Remarks and a Question-and-Answer Session with the Agriculture Communicators Congress," *Weekly Compilation of Presidential Documents* 28 (1992): 1177.
82 Ibid.
83 See US Congress, *Wetlands Conservation*, op. cit., pp. 88–90, 311–327.
84 Lydia Saad, "Bush Stance on Environment Unpopular," *Gallup Poll News Service* (10 June 1992): 1–2.

11

A PLURALISTIC, PRAGMATIC AND EVOLUTIONARY APPROACH TO NATURAL RESOURCE MANAGEMENT

Emery N. Castle

INTRODUCTION

I have become increasingly distressed in recent years by the lack of interdisciplinary communication on problems of natural resource policy. As an economist I have been concerned by the tendency of some non-economists to reject, seemingly out of hand, the possible use of economics on this subject. At the same time I have been disturbed by the unwillingness of many economists to consider the philosophical underpinnings of our discipline. The consequence has often been either the rejection of economics or its rigid application. Either approach results in a too narrow a view of policy-making. Representatives of other disciplines probably have similar concerns. I have concluded that the situation described above will not be improved unless dialogue occurs on the necessary characteristics of a public policy for natural resources. Only in the light of such a dialogue can we understand the norms and values inherent in particular scientific disciplines. If these norms and values are brought into the open we can better evaluate their contribution as scientific constructs, as contrasted with their role as policy norms.

Over the past several decades environmental concerns have made natural resource management one of the major policy issues of our time. The response to this concern has been considerable and varied. Some have called for an environmental ethic that, presumably, would either guide or dominate decision-making in this complex arena (Nash, 1989). More recently sustainable development has been advanced as a promising way of dealing with the matter (World Commission on Environment and Development, 1987).

The purpose of this essay is to state the requirements for and some of the implications of a successful approach to natural resource management. In particular, attention is called to two fundamental conditions that seem to be missing from much of the current literature. One is the need to recognize that the natural environment and social systems are in constant change. Many approaches to management assume that the system being managed should be guided towards a steady state or an equilibrium condition. Such an assumption is counter-factual. Unless change is recognized explicitly, reliable management systems cannot be developed. The other fundamental condition is that preferences among people within any human society vary greatly. For example, democracies are seldom pure in the sense that the majority view is imposed without recognition of minority opinions and rights. Thus natural resource management must give explicit attention to the decision-making process if multiple viewpoints are to be accorded respect. The incorporation of the above considerations in natural resource management requires an approach that is capable of adapting to changing conditions, pluralistic in philosophy and pragmatic in application.

Forest management is used in this essay as an example of a class of resource management issues. This is done to provide concrete reference points for the generalizations which are advanced, even though the principles set forth have broader application than to forest management. In both the more and less developed parts of the world, conflicts rage over proper forest management. The traditional approach is often described as providing for the sustained yield of timber with other benefits being viewed as joint products. Whether or not this is a stereotyped, oversimplified view is of no great consequence here. Forestry has become a recognized profession on the implicit assumption that there was a social consensus on desirable forest products and that professionals could be entrusted with forest management. The notion that there is something that can be called "scientific resource management" which can be viewed as desirable public policy is no longer generally accepted. Why is this the case?

Economists have faulted traditional forest management for its neglect of economics and there is little doubt of the validity of their claim (Clawson, 1984; Bowes and Krutilla, 1989), but it is far from clear that the incorporation of economics in forest management would re-establish "scientific resource management" to its traditional eminence. While economics has a contribution to make in this regard, there are limitations to economics as a management tool as it is often practiced. Neither economic development nor economics should be ignored in natural resource management, but there is need to view and utilize them in a broader context than is often done in current practice.

The world today is very different from the one which existed when forestry first emerged as a professional field. The economic development of the more advanced nations has been associated with a transformation in both the means of production and patterns of consumption. Many other nations have moved forward economically at a dramatic pace. Even in the least prosperous nations, human populations have grown dramatically because of decreases in infant mortality and longer human life spans. This enormous economic and human population growth has had a profound effect on natural resource use. Production of the extractive industries – agriculture, forestry (timber), mining (including fossil fuels) – has increased greatly and has been accompanied by technical change in the means of extraction. Even as the output of extractive goods has grown, the demand for the amenity uses of the natural environment has increased. Some of the increased demand has come from greater numbers; some is the result of improved income. These developments have been associated with growing concern on the part of many about physical and biological systems which control all life on earth. Some of this increased concern has stemmed from an improved capacity to observe, measure and communicate information about natural systems.

Four requirements must be satisfied if natural resource management is to accommodate the kinds of changes sketched above.

1 It must provide for economic and social change. Modern economies and societies change and evolve with the passage of time. Much economic analysis ignores this reality which stems, in part, from its base in equilibrium concepts;
2 It must recognize ecological interdependence as well as the interdependence of humans and the natural environment;
3 The welfare of people in the present must be considered relative to the welfare of the unborn. Natural resource management deals with very long time periods and intergenerational equity is legitimate subject of concern, but there exists enormous income and wealth disparity both within and between nations at any

given time. Specifically, this means that the welfare of the less fortunate today must be taken into account as well as inter-generational equity;

4   The process by which group decisions are made regarding requirements (1), (2) and (3) is a matter of importance and requires explicit attention. Process influences conditions which exist at any given time as well as outcomes over time. Because social adaptation is never complete, the concern with process must be continuous as well.

ECONOMIC CHANGE AND CONSEQUENTIALIST POLICY

Consequences matter – in this case economic consequences. It is "given" for this essay that an objective of natural resource management is to produce outputs that have economic value. Capitalism, which utilizes markets and private sector enterprise, is widely credited (or blamed) for much of the economic change described earlier. This method of organizing economic activity has had an enormous impact on living conditions as well as the natural environment. The production and consumption of vast quantities of consumer goods has been the hallmark of the economic change described.

Utilitarianism is often given credit for providing the philosophical base for capitalism. Such an attribution may not be entirely correct. Adam Smith wrote the *Wealth of Nations* before the utilitarian philosophy was fully developed. Furthermore materialism that has been the result of modern capitalism is not necessarily inherent in utilitarianism. Utilitarianism is permissive in this regard – the desired outcome is the one which yields the highest net balance of satisfactions (Rawls, 1971, p. 22). A utilitarian society which (say) places a high value on individual contemplation will provide for leisure time after the basic necessities of life have been achieved. Nevertheless, it is to utilitarianism that an appeal is often made when the performance of capitalism is evaluated or judged. In practice, it is impossible to know when the greatest net balance of satisfactions has been achieved. To know would require that the utility of one person be compared with the utility of another, but no one knows just how to do this. If interpersonal utility could be compared directly in common units of measurement, utilitarianism would have a powerful policy tool and it could be decided when an economy was working according to the utilitarian ethic.

Enter Pareto optimality and benefit–cost analysis. Vilfredo Pareto was an Italian economist and sociologist who lived from 1848 to 1923. He utilized the notion of ordinal or relative utility. Following this line of thought, a social improvement would occur if, and only if, a policy action would make someone better off without making anyone worse off. Taken literally, this would mean there would have to be unanimous consent before a policy action is taken. In practice, benefit–cost analysis is used to measure whether a given policy action would make it possible to make some better off without making anyone worse off. Economic efficiency has come to be defined in this way and a sub-discipline within economics, welfare economics, has arisen which provides the theoretical structure for benefit–cost analysis. If done properly, a benefit–cost analysis which yielded a positive benefit-cost ratio would meet such a condition. In practice, compensation is seldom paid to those made worse off but, theoretically, a positive benefit–cost ratio would make it possible to do so (Kneese and Schultze, 1985).

Even though welfare economics stemmed from utilitarianism, it is important to distinguish between classical utilitarianism and welfare economics. Maximizing the net balance of satisfaction for a society is not the same as permitting some to be made better off without making anyone worse off. Nevertheless, both are consequentialist and attribute importance to economic performance. It is welfare economics and associated benefit–cost techniques that are used to define economic efficiency and measure the welfare gains or losses from proposed policy actions in much current policy literature.

Numerous concerns have been expressed about the use of benefit–cost analysis in natural resource policy. In the context of this paper three are especially important.

The discount rate problem

Natural resource policy involves very long time periods. Despite a voluminous literature, it is far from clear what discount rate should be used to reflect the uncertainties associated with long time periods and fairness across generations. A scholar of this subject has written "searching for the correct rate of discount is searching for a will of the wisp" (Page, 1988).

Why should no one be made worse off?

In practice, the application of benefit–cost analysis for public policy usually does make some worse off. If a dam is constructed with public funds, usually no attempt is made to compensate everyone who sacrifices to make the dam possible. It would be very difficult to identify everyone in this category even if we wished to do so. But the operative criterion is that it would be possible to do so if they were known. In other words, if the benefit–cost ratio is positive, net national income will be greater and, theoretically, it would be possible to make some better off without making anyone worse off. But this begs the question of why the present distribution of income and wealth should be considered morally superior to other possible distributions. For every possible distribution of income and wealth there will be a different base from which economic efficiency (optimality) can be judged.

The answer usually given is that the present distribution is taken as having normative significance because society could change it if it were dissatisfied with it. This is equivalent to saying that the only distortion that matters is the one being studied, or to put it another way, society may make mistakes in the allocation of resources but does not make mistakes in the distribution of income.

Is the economy an evolving or an equilibrating system?

Not only does benefit–cost analysis attribute normative significance to the status quo situation, it also carries with it the implicit predictive assumption that the economy is an equilibrating one over time. Technically this requires that competitive markets and constant returns to size be prevalent in the economy. Recent theoretical and empirical work raises questions about whether modern economies in fact behave in accordance with such assumptions. Increasing, rather than constant, returns may prevail over long time periods for many industries which make competitive markets impossible (Lucas, 1990; Romer, 1990; Grossman and Helpman, 1990). Furthermore, recent work by Arthur (1990) suggests that economies may travel through time on sub-optimal trajectories which, in turn, may be affected by random events (also see Anderson et al., 1988). If such conditions prevail, the normative base for using benefit–cost analysis is badly damaged, to say the least.

Obviously these concerns are far from trivial and should not be dismissed out of hand. The philosophical burden is indeed great if it is argued that the economic efficiency criterion is to be controlling in the policy arena. Those of an environmental persuasion, or others who challenge the sole criterion approach have considerable ammunition in their arsenal. But it is important to be precise when arguing for or against such techniques as benefit–cost analysis. It is difficult to reject completely consequentialist arguments if one attributes any importance to anthropocentric objectives; everyone is not likely to be convinced that material goods are unimportant. Benefit–cost analysis may be used with varying degrees of intensity; the results may be used as a guide to inform rather than control the decision process. It is from the consequentialist framework of economics that the powerful principle of opportunity cost emerges – any course of action requires sacrifices or opportunities forgone. This concept is utilized later in this paper; it is an essential tool for natural resource management.

THE ENVIRONMENTAL CHALLENGE

During the past two decades especially, the environmental movement has provided a challenge to unconstrained economic materialism. The challenge has often been articulated in a consequentialist context by calling attention to the effect of vanishing species, denuded forests and polluted air on the human condition. Economic indicators which show that progress is occurring have been challenged as being misleading, because it is alleged that the real cost of such developments has been underestimated or ignored.

The environmental challenge has been anything but monolithic. Shortly after the turn of the century, there was a reaction in the US to the utilitarian ethic in natural resource management to the extent that ethic was reflected in unconstrained markets. In this view natural resources were to be managed for human use, but markets could not be trusted to be the sole determinant of natural resource value. As a result, numerous government programs came into being in an effort to improve on market performance. For example, national forests were managed to provide a wider range of outputs than markets would have produced. Multiple use was carried to an extreme; commercial timber was harvested in areas where markets would have found commercial timber production to be uneconomic after the first harvest. But the utilitarian ethic prevailed in shaping such programs and multiple use was organized around uses which had commercial value, whether economic or not.

The recent environmental movement has witnessed an intensive search for an environmental ethic not based on utilitarianism. An enormous recent literature has arisen from the inspiration provided by the earlier writings of Thoreau, Muir and Leopold. Nash (1989) and Rolston (1988) may be consulted for recent summaries.

An appreciation of interdependence in nature underlies environmentalism. This seems to place environmentalism immediately at odds with economic development. Adam Smith argued that specialization and trade were the routes that nations should take to increase their wealth. Science and technology has made it possible to separate things that are joined in nature; natural materials can then be reconstituted into forms that are more suitable for commercial purposes. On the surface this seems to fly in the face of an appreciation of and commitment to interdependence in nature. There are two major sources of concern among environmentalists about irreversible change in nature. One stems from the belief that such change may diminish the human prospect – anthropocentric environmentalism. The other reflects a belief in the intrinsic beauty and moral goodness of the natural world – biocentric environmentalism.

Those of an anthropocentric persuasion have a great deal in common with classical utilitarianism or even economic efficiency proponents. Both are concerned about the long-term human prospect; both may be concerned about humans causing irreversibilities in nature. In a practical context the difference between anthropocentric environmentalism and utilitarianism often comes down to where the burden of proof is to rest. Those who tilt towards economic consequentialism may wish to require proof from those who would prevent economic activities that a proposed natural system intervention will, in fact, diminish the human prospect. Conversely, the anthropocentric environmentalist may wish to require proof that a proposed economic activity will not harm the environment to the detriment of people over time. These two groups occupy center stage in the policy arena currently and specific policy issues are often framed to bring this kind of conflict into the open. Even so, the traditional mainline environmental organization and private enterprise advocates are both consequentialist and anthropocentric. Advocates for these different points of view may have a great deal in common with respect to social class and lifestyle. Mainly environmental organizations often solicit and receive support from corporations who believe it to be in their interest to support "reasonable" environmental activities.

Biocentric environmentalism is very different. Taken literally, biocentrism would accord no more importance to humans than to any other part of nature. If carried to its logical conclusion, only a hunting and gathering society would be permitted. But such an extreme position does not do justice to the complexities of bio-centrism in practice. We do not find the earth in a pristine state. Humans have literally transformed the earth and many irreversible changes have been made. Some may conclude that the time has come to draw a line in the sand even though they would not have opposed intervention in nature at another time or under other conditions. It is not surprising that biocentrists are not of one mind with respect either to strategy or to tactics. Some wish to work within human institutions and, for example, obtain legal protection for things in nature. Others may believe that existing institutions are immoral and that it is appropriate to go outside the law to achieve their objectives.

In the context of this paper it is important to recognize that environmentalism has contributed to the public debate in many important ways. It has required that interdependence in nature be taken into account in public policy. It has held that human welfare is dependent upon the preservation of natural systems. It has required that we look beyond markets for a definition of natural resources, and it has forced us to consider nature from a perspective other than that provided by utilitarianism.

EQUITY WITHIN AND AMONG GENERATIONS

Natural resources are typically managed for multiple objectives. A forest may provide timber, outdoor recreation and biodiversity. The welfare of those who tend the forests or are otherwise dependent upon them may also become an explicit objective of forest manage-ment. When the relative weights given to, for example, timber production and forest preservation in management change, it may be necessary to consider the welfare of those in the present generation who are directly affected by such a change. The concern of many with the community effects resulting from the elevation of preserva-tion values in forest management provides an example. In a larger sense this means that the income and wealth distribution both of an enterprise system as well as of public land management is a legitimate concern of forest land management. But attention cannot be confined to the present generation. Because of the long time periods involved, intergenerational equity has long been an important problem in natural resources policy. As the title of this section suggests, it is necessary that equity within be related to equity across generations.

The writings of the philosopher Rawls (1971) can be used to illustrate the conflicts and complexities of fairness within and among generations. The basic moral principle that guides Rawls is that those who are in a favored position because of nature should be permitted to gain from their good fortune only on terms that permit those who are less fortunate to improve their situation. Rawls believes that the correct starting point to achieve this is with the basic structure of society. This structure is to be arrived at from behind a veil of ignorance which permits those who decide on social structure to do so without reference to personal circumstances, including the generation to which they will belong. Rawls develops two principles:

> First: each person is to have an equal right to the most extensive basic liberty compatible with a similar liberty for others and Second: social and economic inequalities are to be arranged so that they are both (a) to the greatest benefit of the least advantaged and (b) attached to offices and positions open to all under conditions of fair equality of opportunity (p. 60).

He states: "Injustice, then, is simply inequalities that are not to the benefit of all" (p. 62). For the purpose of this paper three characteristics of justice as fairness are important:

1 Because the structure of society is determined by those who have no knowledge of the generation in which they will live, the welfare of future generations will be provided for by the social structure;
2 The criterion for an improvement in social well-being is the improvement in the lot of the less fortunate. This is to be contrasted with both classical utilitarianism and Pareto optimality;
3 Intervention by the state is not only tolerated, but required to insure that the two principles of justice stated above are met. The state is necessary for the realization of justice.

While Rawls was not the first to utilize social contract theory and the "veil of ignorance" device, his treatment brings two issues into the open that are of particular relevance in natural resource management. One is the spotlight it places on intragenerational equity; the other is the special attention it gives to the circumstances of people in judging fairness.

Issues of intergenerational equity are coming to be recognized and related to intergenerational equity in natural resource policy. Those who become dependent on the natural resource extractive industries must frequently live and work in geographically remote areas. When the extractive industry declines, they may have few alternative economic opportunities in the area in which they live. Unless there is explicit recognition of this problem, interests of future generations may be enhanced at the expense of the less fortunate of the present generation. This is not to argue that the occupation, lifestyle or location of anyone should be guaranteed by natural resource policies. It is to say that when primary emphasis is shifted from timber production to the preservation of biodiversity, fairness requires that the circumstances of those who will gain and those who will lose from such a policy shift be compared. On average and in the near-term, those who constitute the environmental movement are more economically advantaged than some who are adversely affected by it (Mitchell, 1979). Furthermore, given the trend of rising real incomes over times, the least fortunate in future generations may be better off than the least fortunate in the present generation.

THE GROWING IMPORTANCE OF PROCESS

None of the philosophic positions advanced above, when taken alone, provides either an adequate or generally accepted framework for resource management. Each, however, provides a viewpoint that incorporates values and objectives to which some people relate. The essence of democracy pertains to the process by which values and objectives are selected and implemented. Natural resource management requires public participation; process issues related to that participation deserve explicit attention. Adapting to nature is a process; it is not something that can be done once and then forgotten.

The environmental literature has not given extensive treatment to process. Some writers (e.g., see Baden and Stroup, 1981) have emphasized the inefficiency and counter-productivity of government in the accomplishment of environmental goals. Classic utilitarianism views government as a means for the accomplishment of individual objectives rather than an end in itself. Sagoff (1988) notes that people may have group or social objectives that are different from those they hold as individuals. According to him, liberalism provides a philosophic justification for using government to achieve a social objective. Justice as fairness accords less importance to process; it places reliance on the basic structure of society for the accomplishment of its objectives.

As this is written, people over much of the world are challenging the authority of large central government. How much of this is occurring because it is believed that such governments are inefficient or ineffective, and how much is because of a desire for individual autonomy is unclear. With the Nature Conservancy and the American Farmland Trust serving as cases in point, it is instructive to consider the political environment that would provide the greatest possible latitude and largest role for voluntary associations. Nozick's (1974) work is instructive in this regard. It seeks to make a case for the minimal state and the greatest possible role for individual autonomy. In the preface to this work Nozick states:

> Our main conclusions about the state are that a minimal state, limited to the narrow functions of protection against force, theft, fraud, enforcement of contracts, and so on is justified, that any more extensive state will violate persons' rights by forcing them to do certain things, and is unjustified; and that the minimal state is inspiring as well as right. Two noteworthy implications are that the state may not use coercive apparatus for getting some citizens to aid others, or in order to prohibit activities to people for their *own* good or protection.

Nozick specifies how assets must be held if such a philosophy is to be implemented. Three considerations are important: (1) the original acquisition of holdings, (2) the transfer of holdings, and (3) the certification of injustice in holdings. An original acquisition of holdings is considered to be just if it was acquired in accordance with the principle of justice – that is, in the absence of stealing, fraud, enslavement and the use of force. Holdings may be transferred in exchange for something of value or by voluntary gift. If, at any time, holdings can be shown to have been acquired by unjust means, it is appropriate to rectify past injustices. Thus private holdings of property with minimal interference in the way those holdings are used and transferred is consistent with the tenets of libertarianism. Any patterned system of distribution or redistribution is to be rejected. Acquisition, transfer and redistribution of assets are to be in accordance with the law of the minimal state. "Taxation of earnings from labor is on a par with forced labor" (Nozick, 1974, p. 169).

Two fundamentals of human behavior are inherent in libertarianism.

One is that each person is best able to judge that which is in their own best interests and there is no social or group good that is independent of the individual. The second is that any cooperation beyond that provided by the minimal state must be voluntary (requires unanimous consent). Deficiencies of individual action will be recognized by the individuals affected and addressed by mutual agreement; to do otherwise is to infringe an individual's rights and entitlement. Individual liberty becomes the measuring rod by which group action is judged. Such a philosophy cannot view government in a neutral way. Government is the problem, not a possible solution, except for those issues addressed by the minimal state.

Libertarianism, then, would provide the greatest latitude for voluntary associations. It is advanced here as a limiting case that is helpful in isolating voluntary activities that, under other social structures, would be performed by the state. It is a "rights"-based philosophy concerned with the structure of a society rather than the output or consequences of a system. Even though it is not generally considered to be process oriented in theory, in practice libertarians are continuously concerned with minimizing the role of government. It provides one vision of the direction that would be selected if individual autonomy is emphasized.

Voluntary associations, of course, may flourish in political systems other than libertarianism. Indeed, as is the case in the US at present, voluntary associations may be given protection in the law as well as receive other forms of encouragement by government. As this is written, they have become one of the most rapidly growing avenues of collective action in the resource and environmental field. The classic work on voluntary group action is to be found in the work of Mancur Olson (1965). This seminal work has done much to shape the public choice literature which has grown from the disciplines of political science and economics. The existence of a field such as public choice emphasizes the continuing importance of process in public policy generally and natural resource management in particular.

WHY A PLURALISTIC, PRAGMATIC AND EVOLUTIONARY APPROACH TO NATURAL RESOURCE MANAGEMENT IS NECESSARY

To this point in the paper a philosophical base has been presented for economic development, environmental preservation and enhancement, intra- and intergenerational equity, and the process by which conflicts are resolved. This survey has necessarily been brief; there is no pretense that all philosophic positions relevant to natural resource management have been identified. However, the presentation has established the framework for the final portion of the paper. In this final section I am specific as to why adequate natural resource management must be pluralistic in philosophy, pragmatic in process and evolutionary or adaptive in nature. Such a management framework places importance on communication among academic disciplines. In many democracies natural resource management is no longer entrusted to technocrats who are asked to manage a resource consistent with pre-established guidelines or specified objectives. In the US all three branches of government are involved at the local, state, regional and federal level. This broad involvement is a result, not just of our form of government, but stems also from different preferences among citizens with respect to desired outcomes. It may also reflect fundamental disagreement on the probable outcomes from alternative management systems. No one philosophical system is sufficiently broad or flexible to encompass this diversity of viewpoints. One of the leading writers on environmental philosophy has argued in a similar vein (Stone, 1988). He has concluded that attempts to discover an all-encompassing environmental ethic are not likely to be fruitful. Pluralism, however, is not a philosophically respectable position in the eyes of many because it does not forbid inconsistencies and fails to guarantee conflict resolution (Callicott, 1990). Of course one of the major functions of political systems is to resolve conflict; it is not necessary that people agree beyond acceptance of a means for conflict resolution. The political system gives explicit recognition and definition to pragmatism in natural resource management. This need not diminish the role of different philosophies in natural resource management; many may have the capacity to provide important insights even though no one philosophic position will necessarily dominate in all or even in particular circumstances.

The evolutionary, as well as the pluralistic, nature of natural resource management also needs to be appreciated. Natural systems are not static even when undisturbed by humans; neither are social (including economic) systems highly predictable. There is a growing body of literature in economics concerned with increasing returns, and the non-exclusiveness of new intellectual discoveries (Romer, 1990; Grossman and Helpman, 1990). The possibility that economies may proceed on sub-optimal rather than optimal trajectories is a subject of investigation within economics as well. These lines of thought have not been accommodated in the resource and environmental economics literature which is generally based on assumptions of competitive markets, constant returns and equilibrating markets. The theory underlying natural resource economics as it is currently practiced is more akin to static physics than evolutionary biology.

The possibility that economies may proceed through time on suboptimal trajectories and may not obtain a long-term equilibrium condition is of great significance for benefit–cost analysis of decisions that affect the distant future. Neither the predictive nor the normative base of the existing economic system may be of great relevance under such circumstances. Economies may evolve in ways that are not anticipated on the basis of present trajectories, which diminishes the predictive capacity of economic models with their base in contemporary conditions. The possibility of increasing returns and random events which affect existing economic systems does damage to the normative content of such models as well. This is not to argue that contemporary resource and environmental economics is of no relevance in natural resource management. It is to say that it should be used in accordance with its limitations.

The evolutionary nature of both natural and social systems does not seem to be adequately recognized in some of the recent rhetoric about sustainable development. What would be the *a priori* distinguishing characteristics of a sustainable economic system? Non-declining even flow, for example, was advanced at one time as such a system for forests. The problem was that the attributes of the forestry system that were neglected in this scheme have turned out to be more limiting than those attributes the system was designed to sustain. An evolutionary approach rejects the notion that techniques based on a synoptic view of natural resources problems provide a sufficient basis for natural resource management. Norgaard (1984), for instance, has long argued for a co-evolutionary approach to natural resource policy-making. Our understanding of Norgaard is that he believes social and biological systems should be interdependent as they adapt and change.

An evolutionary approach requires that the enormous uncertainty inherent in natural and social systems be recognized and accommodated. Part of this accommodation requires that new information be reflected in management decisions. An approach which extends legitimacy to different philosophical systems places a premium on intellectual constructs that are acceptable in more than one such system. Approximately four decades ago Ciriacy-Wantrup (1952) proposed the safe minimum standard (SMS) as a tool of conservation policy. The safe minimum standard is attractive from both an evolutionary and pluralistic point of view. As proposed by Wantrup, the SMS is deceptively simple – a resource should be preserved unless the costs of doing so are "immoderately high." The question is begged as to what constitutes "excessive costs" but it is clear that the preservation criterion is less rigorous than would result from a benefit–cost analysis on a resource-by-resource basis. The concept is openly anthropocentric; it recognizes the opportunity costs of forgone alternatives.

The current concern over the possibility of a permanent climate change from increased carbon in the atmosphere can be used to illustrate the advantages of the approach advocated in this paper. Most scientists agree there is uncertainty as to whether a long-term warming trend is under way. But there appear to be two major responses to this uncertainty from the scientific community. One response emphasizes the economic and social adjustments that can be made to climate change on a geographic basis. This approach considers climate change as comparable with other exogenous forces that have shaped economic and social life; little importance is attributed to natural world irreversibilities. The conclusions stemming from this general response have been reassuring and suggest that policy actions which would reduce carbon accumulations in the atmosphere are not of high priority. It may be said, for example, that most of the adjustment will fall on agriculture. Because agriculture is a relatively small part of the economy, the cost of the irreversible climate change probably is also small. Another response has been to argue for policies that would minimize climate change with little recourse to economic reasoning. Fortunately, these two responses do not exhaust the possibilities. Economic systems should not be regarded as sacred. If it is possible to predict how an economy will adjust to climate change (natural irreversibility), it should also be possible to predict how an economy will adjust to reducing carbon discharges into the atmosphere (preventing a natural irreversibility). It is appropriate to know what the costs will be of such an adjustment. In a technical sense, if the costs, on balance, are positive there will be a reduction in goods and services over some finite time period. If such costs are extremely high a society may rationally conclude to permit carbon discharges to continue. But this, of course, is very different from arguing that economic efficiency considerations should dominate when massive, global irreversibilities are a possibility. Such an argument is enough to give economics a bad name!

SUMMARY AND CONCLUSIONS

The message of this paper can now be summarized.

1   No single environmental ethic or philosophic system exists nor is one likely to be discovered that will guide natural resource and environmental policy. Several philosophic approaches help illuminate and bring into the open the values and choices inherent in such policy. Natural resource policy is necessarily pluralistic.
2   Pluralism is not an acceptable comprehensive philosophic system because it does not forbid inconsistencies. For this reason natural resource policy must be pragmatic; it must provide for the making of choices when relevant philosophies come into conflict. The institution of democracy is such a pragmatic device. In practice, of course, democracy is far more than majority rule, providing, though imperfectly perhaps, for minority rights and viewpoints.
3   Social and natural systems coexist as they pass through time. The forces affecting their survival and performance are unlikely to be either stable or predictable far into the future. Such systems will be required to adjust and adapt if they are to perform near their potential or, perhaps, survive. Thus natural resource and environmental policy needs to be evolutionary.

If these three characteristics of a successful natural resource and environmental policy were generally accepted, policy dialogue and scientific work could become more productive. Insistence on an unattainable consistency in philosophy would be abandoned. There would be explicit recognition of the policy norms implicit in many intellectual disciplines. The future would be recognized as being inherently unpredictable, and policies would be evaluated on the basis of their capacity to take new information into account and thereby provide for adaptation and change.

A pluralistic, pragmatic and evolutionary approach makes it clear that many academic disciplines can contribute to but that none provide an adequate or sufficient intellectual base for natural resource management. Academic disciplines have at their core presuppositions regarding that which is scientifically attainable. Ciriacy-Wantrup (1956) labeled such constructs "scientific fiction"; they may be counterfactual even though useful in scientific work. Perfect vacuums, perfect competition and ecological climax types provide examples. When disciplines based on such constructs are applied to policy issues or management problems, the temptation and the tendency is for them to become policy norms or policy objectives. The consequence is that only those who happen to hold philosophical or value positions consistent with the derived policy norm may find the policy implications acceptable. Because some resist the policy implications that seem to flow from a discipline, they may reject the discipline itself. This is unfortunate. Academic disciplines are anthropocentric tools that may be helpful in achieving particular ends. Taken independently they provide a particular view of part of reality. Taken collectively they illuminate more of reality than when used independently (see, e.g., Hyman and Wernstedt, 1991).

The practical implications of the framework advanced here are numerous, but concrete examples are helpful in making specific the principal message of the chapter. Some have been referred to in the course of the chapter but, in closing, the evaluation of an action to preserve an ecosystem at the expense of timber harvesting is considered. The application of a utilitarian ethic might result in developing a benefit–cost model "with" and "without" the ecosystem. The determination of benefits of preserving an ecosystem far into the future becomes an exceedingly complex undertaking. This would include questioning people in the current generation about their willingness to pay for environmental preservation to benefit future generations.

Environmentalists, both anthropocentric and biocentric, as well as others, may reject such an approach but thereby run the risk of losing any contribution of economics in making such a difficult decision (Norton, 1987; Sagoff, 1988). An alternative might be to use economics to measure the sacrifice, and the incidence of the sacrifice, in the present generation from the preservation of the ecosystem. This is a task well within the capacity of economics to perform. Judgements can then be better made about intra and intergenerational equity as well as the process to be used. These equity considerations need to be considered relative to the uniqueness and long-term importance of the ecosystem which clearly requires contributions from biological scientists. It is encouraging that this kind of research is now under way in a number of institutions.

ACKNOWLEDGMENTS

"A Pluralist, Pragmatic and Evolutionary Approach to Natural Resource Management" is republished from *Forest Ecology and Management* Vol. 56 (1993). Numerous individuals have contributed to my thinking on this subject. This list includes but is not necessarily limited to Stephen Daniels, Brian Greber, Carl Stoltenberg, Richard Fisher, Douglas Brodie, Andrea Clark, Robert Berrens, Peter List, Allen Kneese, Richard Stroup, David Brooks and Kenneth Godwin. Support from the Forest Research Laboratory, College of Forestry, Oregon State University, is acknowledged gratefully.

REFERENCES

Anderson, P. W., Arrow, K. J. and Pines, D (eds), 1988, *The Economy as an Evolving Complex System: Proceedings in the Sante Fe Institute in the Sciences of Complexity,* Addison-Wesley Company, Redding, Ma, 336 pp.

Arthur, W. B., 1990, "Positive Feedbacks in the Economy," *Scientific American,* February 1990, pp. 92–99.

Baden, J. and Stroup, R., 1981, *Bureaucracy Versus the Environment,* University of Michigan Press, Ann Arbor, Fl, 238 pp.

Bowes, M. D. and Krutilla, J. V., 1989, *Multiple-Use Management: The Economics of Public Forestlands. Resources for the Future,* Washington, DC, 357 pp.

Callicott, J. B., 1990, "The Case Against Moral Pluralism," *Environmental Ethics,* 12: 99–124.

Ciriacy-Wantrup, S. V., 1952, *Resource Conservation: Economics and Policies,* University of California Press, Berkeley, Ca, 393 pp.

Ciriacy-Wantrup, S. V., 1956, "Policy Considerations in Farm Management Research in the Decade Ahead," *Journal of Farm Economics,* XXXVIII (5): 1301–1311.

Clawson, M., 1984, *Forests for Whom and for What? Resources for the Future,* Baltimore, Md, 175 pp.

Grossman, G. M. and Helpman, E., 1990, "Trade, Innovation and Growth," *American Economic Review,* May 1990, pp. 86–91.

Hyman, J. B. and Wernstedt, K., 1991, "The Role of Biological and Economic Analysis in the Listing of Endangered Species," *Resources,* 104: 5–9.

Kneese, A. V. and Schultze, W., 1985, "Ethics and Environmental Economics," Chapter 5, in A. V. Kneese and J. L. Sweeney (eds), *Handbook of National Resource and Energy Economics.* North Holland, Amsterdam, 755 pp.

Lucas, R. E., 1990, "Why Doesn't Capital Flow from Rich to Poor Countries?" *Pap. Proc. Am. Econ. Rev.,* May 1990, pp. 92–96.

Mitchell, R. C., 1979, "Silent Spring/Solid Majorities," *Public Opinion*, 2 (4): 16–20, 55.

Nash, R., 1989, *The Rights of Nature: A History of Environmental Ethics*, University of Wisconsin Press, Madison, Wi, 209 pp.

Norgaard, R., 1984, "Coevolutionary Development Potential," *Land Econ.*, 60: 160–173.

Norton, B. G., 1987, *Why Preserve Natural Variety?*, Princeton University Press, Princeton, 281 pp.

Nozick, R., 1974, *Anarchy, State and Utopia*, Basic Books, Totowa, NJ, 418 pp.

Olson, M., 1965, *The Logic of Collective Action*, Harvard University Press, Cambridge, 176 pp.

Page, T., 1988, "Intergenerational Equity and the Social Discount Rate," in V. Kerry Smith (ed.), *Environmental Resources and Applied Welfare Economics: Essays in Honor of John Krutilla*, Resources for the Future, Washington, DC, 293 pp.

Rawls, J., 1971, *A Theory of Justice*, The Belknap Press of the Harvard University Press, Cambridge, Ma, 607 pp.

Rolston III, H., 1988, *Environmental Ethics: Duties to and Values in the Natural World*, Temple University Press, Philadelphia, Pa, 391 pp.

Romer, P. M., 1990, "Are Non-Convexities Important for Understanding Growth?" *Pap. Proc. Am. Econ. Rev.*, May 1990, pp. 97–103.

Sagoff, M., 1988. *The Economy of the Earth: Philosophy, Law and the Environment*, Cambridge University Press, New York, 271 pp.

Stone, C.D., 1988, "Moral Pluralism and the Course of Environmental Ethics," *Environmental Ethics*, 10: 139–154.

World Commission on Environment and Development, 1987, *Our Common Future*, Oxford University Press, Oxford, 400 pp.

12

LAWS OF NATURE VS. LAWS OF RESPECT

Non-violence in practice in Norway

David Rothenberg

Environmental pragmatism is about finding those philosophies which work in practice at clarifying and solving environmental problems. I believe environmental philosophy, conceived through this perspective, is fundamentally an applied philosophy, taking abstract questioning and demonstrating how it may be used to solve real disputes.

Here, I wish to examine the efficacy of putting Gandhian non-violent protest and communication to use. I identify a history of this approach in Norwegian environmentalism, specifically in Arne Naess' deep ecology, and suggest that this tradition might be applied to resolve the present whaling dispute in Norway. Against the Gandhian approach are other deep ecology activists like Paul Watson and the Sea Shepherds, who prefer a more confrontational strategy. But since both sides of this divide claim to act consistently with deep ecology principles, how can we resolve this situation from within the perspective of deep ecology? Environmental pragmatism, as defined by Andrew Light in this volume, directs us to choose that political strategy which most expediently results in solutions to environmental problems, and one of the tests of such effectiveness is whether it fits the political context where it is being applied. It is the pragmatist perspective that shows why Watson's tactics are wrong in this case, and why a Gandhian approach would be better suited to the Norwegian political context. This essay then is an exercise in using environmental pragmatist principles as a way of adjudicating the debates among the adherents of one kind of environmental philosophy.

Norway has been called the philosophical home of radical environmentalism. Most recently this is because it is the homeland of Arne Naess, the philosopher who invented the term "deep ecology" to refer to the growth of serious and hard-edged revolutionary ideas clustered around environmental concern. Before Naess, of course, Norway was known as a country that traditionally reveres and mythologizes its mountains and rivers, such that its entire history is interwoven with the land. Norwegian national identity is nothing without nature, and their natural resources are the basis of the country's current wealth, guaranteeing a pristine sense of civilization.

So are all its citizens environmentalists? It depends who you ask. Everyone wants to call themselves environmentalists these days, wherever they are on the political map. The term has become innocuous enough in the US, to serve as political capital for almost any public official. Carter, Bush, Clinton, probably everyone except Reagan would like to have the public know that they are the "right" kind of environmentalist, collecting the title while distancing themselves from whoever they decide is an extremist. Norway's prime minister Gro Harlem Bruntland has made an international reputation as a "Green Goddess" on her reputation as the organizer of the World Commission on Environment and Development and her outspoken positions at the 1994 Cairo Population Conference and similar events. But at home her reputation has never been consistent. As Minister of the Environment she was not particularly effective, and her support for the country to join the European Community (voted against 53–47 per cent in November 1994) has eroded her national stature.

Norway is *not* a country where everyone would want to accept the e-word. In rural areas environmentalists (it's even hard to say in Norwegian – *miljømennesker?*) are seen as stuck-up city types from the south who want to tell the hinterlands how to live. And the latest generation of Norse twentysomethings are as interested in the global grit of world culture as they are in the traditional mountain world of outdoor life and local control.

In our book *Wisdom in the Open Air*,[1] Peter Reed and I endeavored to present the history and prognosis of Norwegian ecological thought as a blend of the radical and the conservative, the traditional and the innovative. Following this outlook, I will first review the most salient and practical advice that the ideas of Arne Naess and his followers have brought to worldwide environmental activism. Specifically, I will examine how the country that brought the world deep ecology did so upon a firm Gandhian foundation (which is often forgotten in many descriptions of the development of deep ecology). Finally, I will discuss how this approach might be better applied to convince Norwegians to give up whaling than the confrontational approach of Sea Shepherd Captain Paul Watson.

NON-VIOLENCE IN ECOLOGY

Non-violent resistance is often an important part of environmental action: lying across the road to block the onslaught of bulldozers, chaining oneself to the floor of a valley as the dammed waters start to rise around you. These can be powerful forms of protest. The press will take notice, and the public will follow the news, so the world will learn of your case. If you are willing to lay your life on the line, they think, you must be quite convinced of the correctness of your position.

But does it work? Such demonstrations often seem to escalate quickly towards violence, or at least animosity. The fierceness of commitment to any cause can lead to overzealous fervor and a refusal to respect the views of the other side. This fervor is what motivated the militant wing of America's environmental movement, known for a while as Earth First!, to support the practice of violence towards property, if not people, in the name of a free and wild nature. There is a long tradition of violence against property belonging to the ruling class in the name of a greater common good. In Norway ecoactivist Sigmund Kvaløy had this to say about the application of controlled violence in environmental protest:

> People's will to defend themselves doesn't concern only their country, but also their living environment. . . . One has to defend what one feels close to. . . . We must get away from the idea that the dividing line between violence and nonviolence lies at the use of dynamite. *Yet no living thing should be harmed – not even a blade of grass.* If dynamite helps life to flow again where it has been stopped, this would be a truly nonviolent use of explosives, really a peace-promoting use of the name of Alfred Nobel![2]

This is a tactic for environmentalism that does tend to incite the enemy, maybe to something less than willingness to consider alien views.

A common opinion is that mainstream environmentalists are glad that extremists exist, to make themselves appear more reasonable, but there is a flip side: opponents of the ecological cause may tend to brand all environmentalists as unreasonable provocateurs. Paul

Bremer, US Ambassador-at-Large for Counterterrorism (during the Bush administration), sees little difference between ecoterrorists and other kinds:

> Like political terrorists, ecoterrorists start with a strong belief system centered on an uncompromising, utopian – even Messianic – vision. Such terrorists see their role as righting perceived wrongs, whether the "oppression of the Palestinian people" or "the rape of Mother Earth." But these worthy goals are then perverted by ideological extremism. . . . They seek to commit acts of violence which draw maximum attention to their cause with little concern for who gets hurt.[3]

To distinguish ecoterrorism from other acts of political terrorism, one needs to emphasize that it does uphold a separable moral code in kind from these other acts: Only harm things, not people. And to paraphrase the words of its literary inspiration, Edward Abbey: Never get caught. Never seek publicity. Just get the job done. Make sure no one knows who you are. The action speaks for itself.

Does this platform of protest offer much method for solution? This is the biggest problem with violent civil protest. It occupies an uneasy place between reform and revolution. It does not encourage a cooperative spirit towards the resolution of real and difficult problems. It seems better as symbol than reality. By giving in to violence the movement may gain a visible hard edge, but it loses the opportunity to foster constructive conversation between the sides of the conflict who will still disagree.

Those who support non-violence as political force find the strongest solace in the actions and words of Mahatma Gandhi. Although nature was not his primary cause, he offers specific insights appropriate to ecoactivism. Gandhi offers a total view of how an individual may act in an apparently selfless way, all the while pursuing a goal of self-realization in its widest sense.

It was this wide sense that most impressed Arne Naess. Before turning to ecological matters, Naess worked for many years to demonstrate that Gandhi's collection of aphorisms, life experiments, protests and meditations constituted a coherent philosophy, not a bag of contradictory assertions and actions. A Gandhian sense of Self-realization is the root, I would argue, of Naess' entire philosophy of deep ecology, conceived as personal philosophy in concert with respect for the human place in nature. Deep ecology, then, on Naess' account should in theory and practice be consistent with a Gandhian outlook.

The power of non-violence is built upon belief in the essential oneness of all life. "A drop torn from the ocean perishes without doing any good. If it stays part of the ocean, it shares the glory of carrying on its bosom a fleet of mighty ships."[4] Realizing the full potential of the self means recognizing that one's identity as an individual is expanded by embracing the concerns of those aspects of life first near and then far from your own particular place. Act selflessly, with detachment, and above all *without regard for the fruits of one's action*, an idea taken straight from the *Bhagavad Gita* itself, where Arjuna calls to Krishna from the battlefield, wondering whether to give in or to fight. Choice is essential, says the god. But the action should be performed because it is right, not for any gain, public or private, personal or political.

These spiritual principles seem quite far from a strategy for any issue-based campaign. It is precisely because of this distance that they have more to do with truth than with winning a game. Non-violence so defined becomes much more than a tactic for calling attention to environmental causes, but a strong platform from which to convince people of many different persuasions that the situation is severe. Gandhi conceived of non-violence as a philosophical quest, a mission towards the truth, rather than a method to fight for what you already are sure of.[5] The non-violent action should imply searching for the answer, not representing it in advance.

And the truth which it seeks is the answer most pertinent to the situation at hand. Here is Gandhi on trial by the Hunter Committee in 1919:

> *Council* However honestly a man may strive in his search for the truth, his notions of truth may be different from the notions of others. Who then is to determine the truth?
> *Accused* The individual himself would determine that.
> *Council* Different individuals would have different views as to truth? Would that not lead to confusion?
> *Accused* I do not think so.
> *Council* Honestly striving after truth differs in every case?
> *Accused* That is why nonviolence is a necessary corollary. Without it there would be confusion and worse.[6]

A diversity of truths, a plurality of right ways to live and fulfill the self, but each dictated by the situation. Gandhi believes non-violence will work only in situations where parties hold *honestly* their disparate views – where opponents truly believe in their differing positions, and no one is lying for political or other reasons. Opponents should not attack each other with straw-man caricatures or inflamed rhetoric designed cheaply to discredit the opposite side. Criticism should be truthful, direct and aimed at communicating with those who disagree. No exaggeration, no inflammation of the facts is appropriate. With action done just for its own sake, non-violence will inspire mutual respect among all involved parties.

People may honestly strive towards the truth but still have different views as to the content of that truth. Does the idea of truth specific to each situation sidestep the conflicts of relativism? The answer to such disparities lies in the sincerity with which different perspectives are presented. It becomes a contextualizing pragmatism guided by respect for the principle of paying attention to the full extent of each situation. Follow the action itself, here and now. Learn as much about the different sides of the case as you can. Be prepared to give up your principles if you discover information that challenges them. Yet if you are still sure you are right, steadfastness will also hold with greater knowledge. Above all, strive to meet the other side directly, on their own terms. If you do all this, you will be meeting the demands of non-violent environmental activism.

This approach works. It demonstrates the conviction of environmentalists and shows our willingness to convince people who disagree with us. First, though, we need to respect this disagreement, and not vilify the alternate view into an evil 'enemy' beyond possibility of reform.

These principles seem so calmly considered, so polite, so reflective, and far from the overblown pontificating we have come to expect in political statement. Non-violence may have worked wonders in India, but is it too idealistic for the arena of deception in which ideas are slyly marketed and packaged over here? I hope it can work, because unlike so many other forms of social critique, it does not begin with cynicism towards the general malaise of our time and place. It begins with the inhabitation of the other point of view. Before the divergence of your own interest comes a deep identification with the goals and aspirations appropriate to all fellow humans. There need be no hatred.

This is the aspect most often missing *even* from non-violent environmental protests: a real willingness and effort to communicate with the other side of the protest. Demonstrations are calculated to

256

bring media attention to an unjust situation, but they forget the importance of face-to-face, personal contact with those whom the grievance addresses. The US Forest Service may be initially unwilling to listen to those who attack their clear cutting practices. They will cite the need for jobs and criticize the overemphasis on the poor little spotted owl. The resolution of any genuine struggle is never as simple as finding better images to stand for your case. Gandhi told his biographer, "I am essentially a man of compromise, because I am never sure that I am right."[7] Here was someone who could labor tirelessly and at the same time question his whole campaign – such openness is essential to philosophy in action.

Non-violence can be a powerful but tough moral standard to uphold. Is there any reason it is uniquely appropriate to ecological struggles? There are many who believe nature itself to be the prime arena of non-violence, where all is a give and take balanced by moderation in all things. But one must be moderate in the recommendation of moderation as well. Nature may be calm, but there is also a backlash: a top-selling new video is entitled "The Violence of Nature," presumably depicting all those things that Mutual of Omaha would never let us see on "Wild Kingdom." This is the age when modern conventions are to be overturned, and the romance of placid nature may well be an idea whose time has come and gone. Nature changes through history from the human vantage point. It has been alternatively inclusive and unreachable, fearsome and embracing, with all the ambiguities in between. It remains a foil to all our attempts to limit it to one value or another.

Nature will always be more than something which has inalienable rights or inherent value from our expanding perspective of care. Nature is larger than what we can ever know. Limiting the purpose of this unknown is a dangerous fault of philosophy. This then is why protecting our natural home with Gandhian means of non-violence makes sense. Not because it pretends to mirror a process of the non-human natural world, but because the method can be solid and unwavering while still admitting skeptical searching.

Non-violent ecoactivism stands firm for respect for oppressed humanity *or* nature, not out of altruism, but as part of progress in self-realization, widening the grounding reaches of our identity. We learn to define truth not relatively, but absolute and precise to each specific situation. We become pluralists, but not relativists.[8] But as deep ecologists, we are holists: we think of responsibility to the greater whole before demanding individual rights. We try to face opposing positions without intimidation. And we will not be afraid to change our tack if new knowledge of the case appears which defies our ideology.

Non-violence is appropriate in the defense of the environment only if it truly aims to hurt no one. For the only kind of nature that emerges as being more important than people from the human point of view is the kind which everyone will eventually admit is deserving first of respect, and then of care. That will be the largest kind of truth that is specific to the world.

BY ANY MEANS NECESSARY?

Deep ecology encourages environmentalists to stand up for their feelings and convictions as a focal point for public debate.[9] Arne Naess has insisted that we should not be afraid to speak up to our opponents with the words, "this river is part of myself." When you dam its waters, you also close me off and shut me in. Norway confronted this issue first in 1970, when Mardøla Falls, a spectacular cascade over a thousand meters high, was threatened by a hydro-power project that would impede its flow above the cascade. Demonstrators from all over the country gathered to protest this injustice to one of Europe's highest waterfalls. Demonstrators including Arne Naess and Sigmund Kvaløy chained themselves to the earth to prevent bulldozers from marauding into the mountains. Civil disobedience brought environmentalism into national debate. A few years later a Ministry of the Environment was formed, today staffed by many of the protesters from the 1960s.[10]

Today, Norway's environmental climate is dominated by a different story. Norwegians have resumed whaling, despite a worldwide ban from the International Whaling Commission. They claim that one particular species, the Minke whale, is not endangered and may be hunted sustainably. The world's environmentalists do not concur, citing that Norway does not need whales for food or industry.

But for Norwegians, curtailing whaling also shuts off the self. In the north of the country some villages have survived for centuries on whaling and the practice is imbued in the culture as a whole as a symbol of its distinctive national character. Similar, to some degree, with the practices of whaling among indigenous North Americans, Norwegians whale for more than just food or industry. It is part of a way of life. If we believe in sustainable harvesting, and if there turn out to be enough of one or more whale species such that hunting can continue, why not allow a limited catch just like a hunting season on deer, moose and bear?

The strongest argument against whaling, I think, is in the realm of the sacred, upholding the belief that there is something to revere in whales that should require all traditions to hold back. But I won't elaborate on this argument here. The Norwegian way of life must also be taken seriously, and traditional ways of life may die easy, but change hard. As environmental concerns become increasingly multicultural, the challenge is to find ways to continue cultural diversity within a framework of world consensus, especially when it comes to the governance of nature that crosses national boundaries. Norwegian tradition will be a stronger thing once it can develop its own considered response to a now global problem. At present there is a tendency to defend troubled practices with a line like this: "You foreigners who don't know the first thing about whales have hunted too many, while we Norwegians represent a long-standing tradition that never killed too many or got greedy."

But can't a people's identity remain strong even through its traditions must change? Slavery was an important American tradition, and shaped part of some of its regional identities, and sure it died hard and reluctantly. Extermination of Native Americans was also part of the "Manifest Destiny" idea, crucial to the nineteenth-century American self-conception. These "institutions" characterized the United States for decades but they had to give way eventually to a more enlightened state of affairs. The ultimate interests at stake were greater than mere national identity. Similarly, Norway ought not become nationalistic about something so transnational in scope. This is not a philosophical argument against whaling, but a plea for a rethinking of the assumption that local tradition is always inviolate. But importantly (following a Gandhian path), it is important to do this by taking the argument of the traditionalists seriously.

Environmentalism has always been both radical and conservative. It involves a critique of current society along with a longing for the bucolic idylls of the past. The traditional Norwegian interest in whaling is somewhat at odds with its equally traditional love of nature. It has been said that Norwegians love their mountains as home but use the sea as resource, and this dual sense of nature, as something to be revered *and* harvested, has kept the country prosperous. But philosophically it may be a dual standard. Progressive ecological thought ought not simply come from the past. It makes use of traditional ideas and blends them with new findings to outline a path to the future.

So, how does an environmentalist approach this problem? More specifically, how does a deep ecologist take on the issue of whaling in the context of Norwegians' love of tradition, and their dual identity regarding nature. How in short, do we get Norway to listen?

The Sea Shepherds are a self-identified environmentalist group that claim to be deep ecologists, and who are directly involved with the controversy over Norwegian whaling. Like their brothers and sisters in Earth First!, they have taken the philosophical views of Naess (as translated through North American and Australian thinkers) and extended them into practice. But the confrontational tactics of extremists like the Sea Shepherds are not, I maintain, what Gandhi would recommend. Their view does not represent a version of Naess' original conception of deep ecological activism which is rooted in Gandhi's thought.

This activist organization advocates the "search and destroy" method in the struggle against ships that are violating the international ban. Leader Paul Watson has defended their attacks on Norwegian whaling vessels *Senet* and *Brenna* in the Lofoten Islands on the grounds that the evil of whaling is great enough to allow the destruction of private property, and even human lives.[11] Reviews of Watson are almost uniformly negative. He has been charged on criminal counts worldwide for his no-compromise efforts. Anti-environmentalists find Watson to be the archetypal extremist eco-villain that fulfills their worst fears. Some mainstream environmentalists are glad that a person like Watson exists, to make them look more reasonable. Others are worried that he gives environmentalism too much of an outlaw cast. Here's what Watson himself says:

> I do not feel like a freak. I feel normal. And sometimes I wonder if the rest of the world is normal, especially that part of it that goes around plundering nature. It is at such times that the opposing philosophies of violence and nonviolence tear at me. I know violence is morally wrong and nonviolence is morally right. But what about results? Nonviolent action alone has seldom produced beneficial change on our planet. I continue to fret over this point. I compromise by allowing myself violence against property but never against life, human or otherwise. . . .

Let's get something else straight. The killing of whales in the present day is a crime. It is a violation of international law, but more importantly it is a crime against nature and a crime against future generations of humanity. Moreover, whaling is a nasty form of anti-social behavior and an atrocity which should be stamped out. So I don't want any crappy letters about tradition, livelihood, or [Norwegian] rights.[12]

So who knows best the definition of crime here, and who is qualified to mete out the punishment? Watson's defense of the legality of his organization is the most troubling part of his argument, blurring the rules agreed upon by human society and the principles we humans have observed at work in the natural world:

The Sea Shepherd Conservation Society is a law-abiding organization. We rigidly adhere to and respect the laws of nature, or *lex natura*. We hold the position that the laws of ecology take precedence over the laws designed by nation states to protect corporate interests. If Norwegian laws were more respectful of international agreements, then there would not have been cause for Sea Shepherd intervention.[13]

This is not the kind of argument that will lead Norway to respect the intent of foreign environmentalists, deep ecologists or not. More importantly, it is not the kind of action that will win over a national population with a history of experiencing deep ecology in the context of non-violent resistance. The attack by Watson and his crew is only being acknowledged as foreign intervention, and the principles they may share with home-grown deep ecologists gets lost. No wonder the whole country is defending whaling as patriotically as ever! By and large, they have come to see the global environmental movement as a violent threat. And this is a sad consequence of Watson's uncompromising militancy.

The plot thickens, the Sea Shepherds only get less popular. In July 1994, there was a confrontation between Watson's M/Y *Whales Forever* (registered in Belize) and a Norwegian coast guard vessel. The coast guard says it took place within territorial waters, Watson says it was outside. The Norwegians threatened to fire upon the activists, and they sought assistance via radio from the US State Department. None was forthcoming. The boats collided (each side claims it was the other's fault), and Watson beat a hasty retreat to the Shetland Islands.

Both sides claimed victory, no one admits to be the aggressor and there were clamors of outrage all around. Here is why Watson thinks he won:

> The campaign was more successful than anticipated. Our objective was to engage in a political confrontation with the government of Norway. We anticipated an encounter with a whaling vessel outside of the Norwegian territorial waters which would provoke a Norwegian military response. Norway's premature provocation led to a confrontation some 17 miles from the closest point of land. Our hoisting the yellow Q flag gave legitimacy to our intentions to enter Norwegian territorial waters, although we had not yet done so at the time of the unprovoked attack on the M/Y *Whales Forever* by the Norwegian warship K/V *Andenes*.
>
> The confrontation between the K/V *Andenes* and the M/Y *Whales Forever* was a classic example of the successful application of non-violent resistance. The crew of M/Y *Whales Forever* faced the guns and violence of the heavily armed warship K/V *Andenes* and won the day! The crew of the M/Y *Whales Forever* displayed great courage and demonstrated their dedication and their commitment to the protection of the whales.[14]

The situation looked quite different to the Norwegian government. In Watson they perceive an "eco-terrorist," a man and an organization that has avowedly destroyed private Norwegian property in the name of defending a part of nature, quoting the "laws of nature" in his defense. They did not consider him a probable innocent and, in fact, many Norwegians consider him a fugitive of justice for the sinking of the *Branna* in 1993. They are surprised at how conservative the coast guard was – they just let this criminal go![15]

On the other hand, Greenpeace, so upstaged in recent events by the more aggressive Sea Shepherds, toughened up their act and cut a dying whale loose from a ship! The whalers will sue, and the Commissioner of Whaling is condemning Greenpeace for interrupting the "humane killing" of an animal to let it suffer unnecessarily in the cold ocean depths.

Each side takes the event and turns it into propaganda for their point of view. How do we evaluate this dilemma from a pragmatist viewpoint? How do we make a claim for the correct deep ecological approach to this now embroiled situation?

The first pragmatist question is: What are the other options consistent with a deep ecological perspective? Gandhi would most likely choose a different tact from Watson. He would say: sit down with the hunters, go with them on a hunt, learn about their way of life. Show them you respect their ways, and gradually teach them why you think they should change. To come from outside with a message of transformation is the hardest stand to take, but it is a necessary one. The change must be phrased as a positive thing, based upon genuine care for the other peoples' way of life as much as a desire to save the whales, or any other part of nature. This means taking the argument of Norwegian tradition seriously.

Perhaps this is what Arne Naess has in mind when he writes that he is against whaling "not because there is a possibility for extinction or because whales hold special status among mammals. . . . The most important argument for me is based on complete richness and diversity of life forms on Earth."[16] Our culture does not need to hunt whales to further its development. The nations of the world have agreed to ban whaling. This is a form of eco-cultural progress. Given such a backdrop, there is reason, it seems according to Naess, to abandon whaling not because of any philosophical argument, but because the argument for tradition and national identity can be mollified in certain situations. And to take on the Norwegians on this ground, the ground they choose to defend their policies, is much more pragmatically consistent with the context of nature in their culture, than to take the high hand of moralism as Watson does – to claim to act according to the laws of nature. Why criminalize the Norwegians for breaking these laws, when consistent with the same underlying principles of deep ecology that Watson embraces, we can mount an argument that there is a much better chance the Norwegians will hear. Isn't the point, from an environmentalists' perspective, to get the practice to stop? And finally, if action is needed, why mask violence in the rhetoric of non-violence? The true principles of Gandhian non-violence seem to fit more organically with the deep ecology of Naess and his Norwegian followers. And importantly, this legacy also represents a tradition to Norwegians, a legacy of environmentalism that could be as rich as its legacy of harvesting the sea.

By the time you read this, the political situation in Norway, and in the Arctic Ocean in general, may very well have changed. Perhaps a backlash against environmentalism will lead the world to consider the commercial viability of whaling once more. Or maybe respect for cetaceans may be well on the way to being a universal human belief, on a par with racial and sexual equality, both ideas which have gained widespread acceptance only in the twentieth century. Or a compromise solution may evolve, whereby whaling is permitted in a very restricted manner, with the whale hunters learning to act in a responsible manner, deferential to international law and global environmental realities. Andrew Light's metaphilosophical pragmatism directs us to acknowledge certain "rules for the game" for deliberating between different approaches to environmental problems. These rules include pluralism, respect for the context of a situation, and an attempt to come to an underlying compatibilist solution with other environmentalists *and* with the other stakeholders in a controversy.[17] Here I have shown how one kind of environmentalism – deep ecology – must be sensitive to the specific contexts in which it finds itself, and engage in those solutions that will best make a difference in the here and now. Whatever the outcome of the whaling controversy, I believe that environmentalists will only play a *constructive* role if they pay more attention to the context in which they find themselves and then adapt their strategy to that context. A pragmatist deep ecologist would choose the Gandhian/Naess approach and not the Watson approach to Norwegian whaling because of its underlying compatibilism with the Norwegian position. This is not to say this perspective is more likely to give up its resolve against whaling, but instead is more likely to make progress *with* the Norwegians, rather than against them.

In the end, or in the beginning of the struggle for a new way of life, we may all sit down, discuss our differences, and realize that we will all have to change as the world gets smaller and smaller. Environmentalism is only radical when it gets to the root of a problem, not when it is extreme for the sake of difficulty. Intelligent people are not afraid of change, in their food, or their ideas. The non-violent tradition, which has been articulated in Norway as relevant to the environmental crisis, can teach us a thing or two about how to talk to one another. Let us hope the twenty-first century will bring a revolution in conflict resolution as well as respect for the value of the Earth to live and flourish with humanity within it, not above.

NOTES

1   Peter Reed and David Rothenberg, eds, *Wisdom in the Open Air: The Norwegian Roots of Deep Ecology* (Minneapolis: University of Minnesota Press 1993).

2 Ibid., p. 115.
3 Paul Bremer, "Eco terrorism. The Rise of the Green Revolutionaries," *CEO/International Strategies*, 5 (3), 1992, p. 79.
4 Gandhi quoted in Arne Naess, *Gandhi and Group Conflict* (Oslo: Scandinavian University Press 1974).
5 See Joan Bondurant, *Conquest of Violence* (Berkeley, CA: University of California Press, 1971) [1958].
6 Naess, *Gandhi and Group Conflict*, op. cit., p. 43.
7 Ibid., p. 50–51.
8 See Gary Varner's "No Holism without Pluralism," in *Environmental Ethics* 13:2, Summer 1991, for the intricacies of this distinction. Varner does not carry pluralism as far as I do.
9 See Arne Naess, *Ecology, Community and Lifestyle*, tr. and ed. David Rothenberg (Cambridge: Cambridge University Press, 1989).
10 See David Rothenberg, "Individual or Community: Two Approaches to Ecophilosophy in Practice," in *Ecological Prospects*, ed. Christopher Chapple. (Albany: SUNY Press, 1994) pp. 83–92.
11 *Norway Times*, "Norway calls on the US to take action against whaling activist group," 104 (6), 1994: 1–2.
12 Paul Watson, "Goodbye to Greenpeace" and "Raid on Reykjavik," *Radical Environmentalism: Philosophy and Tactics*, ed. Peter List (Belmont, CA: Wadsworth Publishing, 1993), p. 170.
13 Paul Watson, "An Open Letter to Norwegians," *Sea Shepherd Log*, First Quarter, 1993, p. 5.
14 Paul Watson et al 1994, "NEWS: Report on Norwegian Attack," downloaded from the internet. [Contact <dcasmedic@aol.com> to be on the Sea Shepherd electronic mailing list.]
15 *Norway Times*, "Norway lets Watson go!" 104 (28), 1993: p. 1.
16 Arne Naess, "Hvem er hvalsakens tapere?" [Who loses in the whaling debate?], unpublished manuscript, University of Oslo: Center for Environment and Development, 1993. See also David Rothenberg, *Is It Painful to Think? Conversations with Arne Naess* (Minneapolis: University of Minnesota Press, 1993), pp. 137–139.
17 See A. Light, "Compatibilism in Political Ecology," in this volume.

13

TEACHING ENVIRONMENTAL ETHICS AS A METHOD OF CONFLICT MANAGEMENT

Gary E. Varner, Susan J. Gilbertz and Tarla Rai Peterson

> "Do we know what we're doing and why?"
> "No."
> "Do we care?"
> "We'll work it out as we go along. Let our practice form our doctrine, thus assuring precise theoretical coherence."
>
> Edward Abbey, *The Monkey Wrench Gang*

Environmental activists commonly express impatience with professional environmental ethicists. Describing his conversion from "a leading moderate among New Mexico conservationists" to a co-founder of Earth First! Dave Foreman reckoned that "Action is more important than philosophical hair-splitting or endless refining of dogma (for which radicals are so well known). Let our actions set the finer points of our philosophy."[1] And later added:

> Too often, philosophers are rendered impotent by their [in]ability to act without analyzing everything to an absurd detail. To act, to trust your instincts, to go with the flow of natural forces, *is* an underlying philosophy. Talk is cheap. Action is dear.[2]

Christopher Manes went further and opined that "most radical environmentalists look at systematic philosophy as the problem" rather than a part of the solution.[3]

It is not just activists who doubt the relevance of academic environmental ethics to real world disputes. John Lemons, a scientist with a keen interest in environmental ethics who has himself

266

contributed several papers to *Environmental Ethics*, complained in a 1985 comment that "too many papers [in that journal] may lack relevancy to environmental affairs."

> As an environmental scientist, I believe that questions about empirical data are just as important as metaethical and normative questions, and that the field of environmental ethics will not make much practical progress until it recognizes the validity of this point.[4]

A number of environmental philosophers have expressed similar concerns about the emphasis on theory in environmental ethics, as some of the essays in this volume will attest.

In contrast to the Monkey Wrench Gang's wholesale rejection of theory, however, both Lemons and the more pragmatically oriented philosophers see a role for theory. Bryan Norton, for instance, urges us to let theory arise out of, rather than precede and then be applied to, real world controversies. He calls this "practical philosophy" to distinguish it from business-as-usual "applied philosophy."[5] And Paul Thompson sees a role for theory in the "diagnosis" of environmental disputes like water rights controversies.[6] What such pragmatically oriented philosophers are impatient with is the obsession, typical of philosophical environmental ethics, with very abstract controversies over anthropocentrism vs. holism, intrinsic value vs. instrumental value, and so on.

While recognizing that obsessing on theoretical debates will not solve any actual environmental problems, we wondered if these debates might not be useful in an unrecognized way. A discussion of philosophical abstractions like the anthropocentrism vs. holism and intrinsic vs. instrumental value debates could, we reasoned, be used to cultivate a calm, reflective atmosphere which is the antithesis of that created by media treatments of controversial issues and, perhaps, by public hearings. The latter encourage truncation and radicalization of position statements. By contrast, traditionally philosophical discussions encourage lengthy, careful reflection using a common vocabulary. We therefore expected that a workshop beginning from traditionally recognized theoretical distinctions could become an occasion for participants to achieve a better understanding of their own and others' views, and possibly, as a result, to defuse conflict and facilitate constructive management of environmental problems.

To get a better idea how people involved in real world environmental disputes would be affected by exposure to arguments and analyses typical of the environmental ethics literature, we used funding from the US Environmental Protection Agency[7] to conduct two-day workshops on environmental ethics modeled after typical college philosophy courses on the subject. We recruited representatives of divergent interest groups in the Rockport and Brownsville, Texas areas. Rockport is in a region traditionally dependent on agriculture, with growing petro-chemical and tourism industries. The nearby Arkansas National Wildlife Refuge is the winter home of the endangered whooping crane. Brownsville was thrust into the national media spotlight when a number of cases of neural tube birth defects occurred during 1991. Controversy has centered around the cause of these birth defects, with various groups blaming agriculture, transnational manufacturing industries, and the health care system. We attempted to include representatives of agriculture, business and industry, tourism, environmental groups, the health care professions, state and federal regulatory agencies and educators (broadly construed to include extension agents in addition to teachers).[8]

Our workshops began with introductions to some basic concepts in ethics generally: utility, rights, and intrinsic vs. instrumental value; and some basic theoretical frameworks in environmental ethics specifically: anthropocentrism, animal welfare/rights, biocentric individualism and holism. We then moved to two topic areas in environmental ethics: the private property "takings" issue and the use of cost-benefit analysis in setting environmental policy. A workbook was designed to provide participants with skeletal notes, accompanied by copies of the visual aids used. The workbook also included diverse interactive exercises designed to serve three purposes:

1 They facilitated participants' understanding of various concepts;
2 They exposed differences and similarities among members of similar interest groups; and
3 They facilitated understanding of positions taken by participants from other interest groups.

Sometimes, participants were asked to describe themselves individually in terms of the concepts being presented. Other exercises asked members of the same interest group to characterize their group's views using those concepts, and still others brought together members of divergent interest groups for comparisons of views. (For a copy of the workshop materials, contact the authors.)

Interviews with participants were conducted three months before and after to determine if the workshops had any meaningful impact on the participants' expressions of concerns and/or characterizations of opposing groups' concerns. In evaluating the utility of our workshops, it is important to keep in mind that informants were never prompted to discuss any particular topic or even ethics in general. Nor were they prompted, after the workshop, to discuss it or to use any of the concepts presented there. The workshops and both interviews were represented to participants as information-gathering exercises on our part, rather than as an attempt to evaluate the workshops' impacts on participants' thinking about environmental issues. Only if an informant used an ethical concept or raised an issue from the workshop would the interviewers ask for clarification.[9] Nevertheless, in follow-up interviews, a number of participants incorporated concepts presented in the workshops into their explanations of their own perspectives on environmental issues. Less obvious, yet still clearly indicated, was the impact of the workshops on participants' understanding of others' perspectives and increased recognition of opportunities for partnership building.

A particularly striking example of an effective exercise was the one we designed around the so-called "takings" issue. The issue is quite complex, conceptually. "Taking," "regulatory taking," "eminent domain" and "police power" are all key concepts with checkered histories in American jurisprudence. Nevertheless, we were able to design a simple exercise which clarified participants' views and framed the resulting discussion in a constructive way.

Participants were first asked to fill in, on an individual basis, a brief questionnaire. It asked, "Without compensation, would the government be justified in taking property or restricting use rights in any or all of the following cases?" The first two cases, (1) prohibiting farming practices which result in polluted runoff, and (2) taking property to build a highway, were chosen to be paradigm examples of cases in which our legal tradition (respectively) does and does not recognize a government's right to take without compensation. As the US Supreme Court put it in recent cases, the common law recognizes no right to use one's property in a way that results in harm to others, so government is not taking any use rights when it prohibits a practice which, like that described in case (1), causes harm to others. However, when the taking cannot be so justified, as in case (2), compensation is required. These are sometimes distinguished as exercises of *police power* and of *eminent domain* respectively.[10]

269

Additionally, two other cases, (3) prohibiting the filling of wetlands lying entirely on private land, and (4) prohibiting modification of the habitat of an endangered species, were chosen because we expected that some participants (those with sympathies in the "wise use" or private property protection camp) would analogize them to the second case, arguing that if a taking occurred, it would have to be compensated, while others (those with more environmentalist leanings) would analogize them to the first case, arguing that here takings could occur without compensation.

After filling in questionnaires individually, participants were randomly organized into groups of four to six, a group scribe was appointed, and they were given 15 to 20 minutes to discuss the four cases, with the scribe recording every reason the participants had for answering "yes" or "no" in each case. During the small group discussions, universal agreement emerged about the first two cases. Everyone agreed that the government could prevent practices that result in polluted runoff without compensating the landowner, and no one thought the government could condemn one's property in order to build a highway without at least compensating the landowner. Participants differed, however, in their treatments of cases (3) and (4). As expected, some analogized them to case (1), arguing that harm to other people results when landowners adversely affect endangered species or ecological processes, whereas others analogized them to case (2), arguing that no harm to others was caused in such cases.

After about 20 minutes, each of the small groups was asked to report to the whole workshop their reasons "pro" and "con" in each case, and the resulting discussion was facilitated in a way that placed the takings controversy in the context of US common and constitutional law. At that point (but not before) participants were given prepared note pages with copies of the slides used in the facilitator's summary presentation.

This exercise generated lively, but focused and constructive discussion during the workshops. By making the distinction between taking to prevent harm to others (the exercise of *police power*) and taking to secure a public good (the exercise of *eminent domain*), and then comparing environmental regulations to clear examples of each, we helped participants to frame their disagreements more precisely than they would in a less structured discussion of property rights in general. That the exercise had a lasting influence is borne out dramatically in the transcripts of two participants who were particularly hostile to government takings.

Bill and Elizabeth Blake[11] were participants in the Rockport-area workshop. The Blake family owned and ranched 19,000 acres on Matagorda Island – a barrier island off the Texas coast – prior to the Second World War. All their land was taken to build an Air Force pilot training base in 1940. The land was never given back after the war as originally promised and was in 1975 transferred to a very different use as a wildlife refuge. The Blakes were permitted to lease back grazing land from the government after the war, but with increasing restrictions and finally, in the early 1990s, cattle grazing was banned entirely from the island on the grounds that it was incompatible with its status as a wildlife refuge. Although the Blakes were belatedly compensated for their land, they remain indignant over the fact that other landowners in the region were compensated at a higher rate when their lands were taken for similar purposes, as well as over the fact that it was permanently dedicated to use specifically as a wildlife refuge.

The first questions in our interview protocol were: "How long have you lived in this region?" and "What changes have you noticed since?" Mr. Blake's response was (and Mr. Blake did almost all of the talking in their interviews):

> Our biggest change is the government interference in private rights. And that's our biggest concern over this environmental binge that's expounded by people that don't have any idea what they're talking about – extremists in other words (Response #21).[12]

And he gave their situation as "the biggest example" of this "interference" that "extremists" are guilty of.

> Well I'll give Matagorda Island as the biggest example. There's 19,000 acres that the Air Force condemned and took for building an Air Force base for training World War II pilots. And they turned around and ended up making a wildlife refuge out of it. In other words, that wasn't the purpose of taking whatsoever. It was not taken for that purpose (Response #22).

Three themes, implicitly but not explicitly interrelated, emerge in these two early passages. One is that too much taking is occurring in the name of the environment. A second is that the advocates of these takings are scientifically naive. In the first passage quoted above, Mr. Blake states that the advocates of "this environmental binge . . . don't

271

have any idea what they're talking about." This view – that environmental policies are too often uninformed by relevant science – surfaces regularly in the initial interview. He repeatedly mentions, for example, agency-commissioned studies which indicated that cattle grazing on Matagorda Island was not incompatible with the presence of endangered whooping cranes during the winters. The third implicitly related theme is that if the Blakes' land had been taken for the original purpose of training pilots during the war and then returned, it would have been justified. As Mr. Blake puts it elsewhere, "people in them days was patriotic" (response #54) and would let government agencies use their land for certain purposes.

However, nowhere in the first interview does Mr. Blake articulate a principle which explains how the scientific studies he cites relate to the takings issue. And although he seems to believe that a temporary taking of their land for use as a war-time training facility would have been justified, nowhere does he state any general principle implying that when it was retained after the war and put to a very different use (as a wildlife refuge), the taking became unjustifiable.

By contrast, in the follow-up interview, Mr. Blake explicitly and repeatedly articulates a general principle, which he believes should govern takings decisions, and which clearly interrelates these three themes. Early in the interview he says:

> Unless they can show that you're harming the environment or something else, on your property you oughta be able to do what you want. If you were harming a neighbor or anybody else, they've already got laws against it (Response #36).

As he puts it later: "[T]o me unless they can show I'm harming something – actually harming it – I don't think that's right [to prevent grazing]" (response #44). And again: "[I]f some agency can show that you're gonna harm something, there's a good reason for that [garbled word] to study and so on and so forth. But they're [garbled word] holding back . . . progress" (Response #62).

Mr. Blake appears to be saying that private uses of land may justifiably be restricted only when they can be shown to cause harm to others. This is a more restrictive position than the US common law's. Mr. Blake does not explicitly address the question of whether and when the government may take land for other purposes, but if taken literally his principle would rule out all exercise of eminent domain. Nevertheless, he is acknowledging the police power justification for taking without compensation. The philosophical significance of the principle articulated in Mr. Blake's post-workshop interview is that it clearly interrelates the three themes identified above. He takes pains to indicate that his family's use of Matagorda Island for ranching caused no harm to anyone, and this, in his view, implies that the taking went too far:

> No one . . . can ever say . . . that we've ever contributed to any pollution or anything environmentally unsound. They never have said it – even on the island with cattle raising. There's not one agency that's ever said we did anything to hurt the environment [garbled word] Matagorda Island. Nothing (Response #92).[13]

In the post-workshop interview it is also clear how environmentalists' alleged scientific naïveté relates to the takings issue: Mr. Blake cites scientific studies, commissioned by the very agencies which managed the island, showing that their cattle business was not harming the endangered whooping cranes; environmentalists advocate going too far in regulatory takings because (in Mr. Blake's opinion) they are ignorant of relevant scientific findings.

In summary, it is not that Mr. Blake's position has changed. As before the workshop, he believes that taking the Blakes' land on Matagorda Island for a wildlife refuge was unjust. But after the workshop he repeatedly articulated the principle that uses of private land may justifiably be restricted only when they can be shown to cause harm to others, a principle which makes clear the relevance of the scientific studies he cites to the takings issue and makes clear why, in his opinion, an "environmental binge" of takings is occurring. So he is articulating his own position better. But in developing his position in this way, he is also laying down a clear challenge to the environmentalists and focusing the disagreement between them in a constructive way. "Show me," he is in effect asking, "how cattle ranching on Matagorda Island was harming our communal environment. If you can't demonstrate harm to others, then seizing my family's property was unconstitutional and unjust."

Our property rights exercise illustrates the way conceptual analysis characteristic of contemporary academic philosophy can help participants frame issues more clearly. Although our exercise was highly conceptual in nature, the takings issue is, admittedly, very immediately related to many participants' practical affairs. However, our workshops began with sections introducing participants to several more abstract concepts in ethics (including the distinction between intrinsic and instrumental value) and to a very abstract taxonomy of views in environmental ethics (including anthropocentrism, animal rights/welfare views, and holism or ecocentrism). Without prompting during their follow-up interviews, a number of participants invoked these more abstract concepts and distinctions in expressing their own views and those of others or expressed those views more clearly, apparently as a result of the workshop experience.

Hal Ketchum is a striking example. Ketchum is a wildlife biologist with a federal agency stationed in the Brownsville area. Ketchum devoted much of his pre-interview to the unique plants and animals of the region. In her analysis of Ketchum's November interview (which she had conducted), one of the co-authors wrote: "I think he values these plants and animals for their own sake [i.e. intrinsically], although he never says so quite explicitly in this [pre-]interview." In his follow-up interview, Ketchum more clearly and explicitly sided with a non-anthropocentric viewpoint.[14] He says:

> I guess I have more of holistic type of thinking in that y'know all this is interrelated . . . so y'know I don't have a problem with coming up with ways to help people understand why should they preserve or protect or whatever (Response #79).

The terms "holist" and "holistic" do not occur in Ketchum's pre-interview. Taken by itself, however, this quoted statement is ambiguous. "Holistic type thinking" could either refer to *ethical* holism (to the view that ecosystems themselves have moral standing), or it could just refer to a holistic *management strategy*, which self-professed anthropocentrists can embrace. In fact, two of our most explicitly anthropocentric participants, Ned Holden and Mike Jackson (both land managers simultaneously involved in education), both stressed the need for taking a systems approach to land management.

An analogy from business helps here. Suppose one believes that a business has only instrumental value to its stockholders. That is, suppose one believes that one's only moral obligations regarding the business are to the stockholders for whom it makes a profit. Taking a systems approach to managing that business would not commit one to thinking that one has any obligations to the business itself, above and beyond its stockholders. Similarly, Holden and Jackson are both explicitly anthropocentrists in their thinking about ethics. They do not think in terms of obligations to the ecosystems, but only obligations to human beings regarding their management of ecosystems. Nevertheless, both say that they use "holistic type thinking" when making management decisions. Ketchum might be thinking similarly.

However, Ketchum repeatedly characterizes anthropocentric arguments as useful for convincing people, and he says this in ways that suggest he thinks other, non-anthropocentric arguments are at least relevant, if not more important. The role of endangered species as indicators of the health of biotic systems and as warnings about problems for humans was mentioned in his pre-interview (at response #142), but it was not tied into any ethical framework the way it is in his follow-up interview. Here he says that "maybe that's something that would help people understand" (response #73), "understand why they should want to do a certain thing . . . even though it may be different than what your particular goal or um responsibility is" (response #79). Here Ketchum recognizes, as he puts it elsewhere, that this is "an argument we use" (response #72) whether or not we agree that it is the most important one. Similarly, he says that endangered species serve "human needs as well . . . maybe that's something uh that would help people understand y'know what's involved. Maybe it will hit home a little bit better" (response #73).

Asked specifically what "[his] own perspective" is, Ketchum says, "Well y'know it depends on what day you're asking me" (response #80), and he goes on to say that in his job capacity "aesthetics or whatever" are important. This suggests that he indeed believes that non-human nature has something more than instrumental value. He then adds: "[P]ersonally I believe we should do it because I think we have that obligation as a human race. But also I believe that we need to do it for [our] fellow man y'know (response #83)." Saying "we have that obligation as a human race" could be an expression of anthropocentrism, but when it's followed immediately by "But also I believe that we need to do it for [our] fellow man," which is clearly an expression of anthropocentric reasoning, the obligation of "our race" sounds like an obligation to other races [species] on the earth.

Ketchum is more specific in his descriptions of both his own and others' views in his follow-up interview. More complicated is the case of Tony Jordan, a junior college biology teacher from the Rockport area. Jordan appears to have changed his views as a result of his exposure to conceptual analysis at the workshop. In his pre-interview, Jordan says:

the other pet peeve that I have is the "save the planet" deal y'know – because the planet is gonna be fine y'know even if we take and nuke ourselves out of existence. The planet will recover in a few thousand years y'know and it will be better than before. Y'know it should be "save the people" (Response #64).

He goes on to say that he is a vegetarian because "it's better for people and it's better for the environment" in terms of resource utilization, rather than "animal rights or anything like that" (response #72). Here Jordan's language seems to imply that he embraces an anthropocentric stance. Neither ecosystems nor animals are of direct concern to him. For him, our treatment of both should be a function of their usefulness to humans.

In his follow-up interview, Jordan appears to have reconceptualized his own views. When asked if he considers himself an environmentalist, he says that he dislikes the "connotations that it has" (response #37) but he refers to the workshop in positive terms, saying the "discussion" there "really y'know got me thinking" (response #38) and that this led to him now characterizing himself as a holist (response #39). He says that he now sees himself as

an environmentalist – from the point of view that I would like to see the earth itself preserved to a certain extent, okay. Just for the sake of having that system preserved, not so much . . . for people so that it will be good for people and it will be good for future generations (Response #38).

This contrasts dramatically with his pre-interview, where he said that the rallying cry of the environmental movement should not be "Save the planet!" but "Save the people!"

The evidence is inconclusive, but it *may* be that the comparison made at the workshop between holistic or ecocentric views on the one hand, and all individualistic views on the other, is what inspired Jordan to redescribe his views. He says, "I'm not as concerned as some people are about – about the people. And I'm not that concerned about the individual animals either" (response #38). In his pre-interview, Jordan said that he didn't care much about individual animals, even though he thought they should not be abused or suffer needlessly. In this follow-up interview, he seems to have generalized this de-emphasis on individuals to humans. If so, the probable inspiration was the comparison, made in the first part of the workshop, between emphasizing eco*systems*, as ethical holism does, and emphasizing *individuals*, as both anthropocentric and animal rights views do. After the workshop, but not before, Jordan makes a distinction between holistic views on the one hand, and individualistic views on the other, including both anthropocentrism and animal rights views. Greater clarity about this distinction may be what allowed him, after the workshop, to redescribe himself as a holist rather than an anthropocentrist.

In other cases, there were less dramatic differences in participants' articulation of their own and others' views in post-workshop interviews, but differences which can plausibly be attributed to the workshop experience. Bill Coining, a Rockport-area rancher, appeared to go from describing his perceived opponents as purely evil people to seeing them as thoughtful individuals with positions that he could understand while continuing to disagree with them. As one of the co-authors put it in an analysis of his transcripts, in his pre-interview Coining "really has only one theme – good versus evil, or right versus wrong." There Coining cast his perceived opponents in an evil light, describing one agency representative in these terms: "he didn't give a damn" (response #42), "he's a cold-blooded, ruthless son-of-a-gun" (response #51), and "he'd sell out his soul . . . to a pressure group . . . he would sell you out" (response #52). Analogously, he described one business' managers as "absolutely bad people . . . cold-blooded" (response #72). While Coining did not retract his descriptions of any of these specific individuals during his follow-up interview, he did switch to describing his perceived opponents as reasonable individuals. In his follow-up interview Coining described agencies involved in management of the whooping crane as simply spending more money on the project than he thinks is necessary (responses #24 and #48). And he described one environmentalist whom he had known for some time as if interacting with her at the workshop had, coupled with the introductory materials on non-anthropocentric views in ethics, changed his understanding of her.

> You know what that Ms. Applethorp [Fran Applethorp] told me? I was trying to pin her down up yonder [at the workshop] and make a choice between the owl and the human y'know. And when it comes right down to it, I believe she's gonna go with that owl. And like I said, nobody takes care of wildlife better than I do. I'll guarantee you that. But if it come right

down to it – of saying "Bill, you gotta either let them deer go or let y'all go" – when it come down to that, don't you see I'd have to stick with the human race? It's just that simple (Response #30).

Here Coining describes the difference between himself and the environmentalist, not as a difference between someone who has ethical principles and someone who is immoral or amoral, but as a difference between people with very different ethical principles. This is, we believe, an artifact of the workshop.

Similarly, after the Rockport workshop, ranchers Bill and Elizabeth Blake switched from describing government agencies and officials as simply evil or bad people, to describing them as relatively non-anthropocentric in their thinking. In their pre-interview, the Blakes repeatedly described the government as deceitful (e.g., responses ##22, 53, 62, 63 and 202), unfair (e.g., responses ##44 and 188), and power-hungry (e.g., responses ##202 and 205). In their follow-up interview, the Blakes continue to describe government agencies as deceitful, but these condemnations are tempered with descriptions of regulations as "unclear" or not well explained (e.g., responses ##11–12, 63 and 69), and with calls for a less anti-anthropocentric stance in agencies' thinking: "They're really bringing it down to the environment against the people. That's not the way it oughta be. . . . They're gonna have to put people back in the picture" (responses ##62 and 95).[15]

In these ways, the transcripts bear out our hypothesis that workshops modeled on college philosophy courses in environmental ethics can improve participants' articulations of their own views and those of their perceived opponents. In his contribution to this volume, Paul Thompson worries that

the practical effect of philosophical analysis is to provide disputants with robust accounts of their earlier opinions, viz. to freeze the actual political debate into stasis. . . . If this is what applied philosophy yields, we might be better off with lawyers![16]

But we hoped that our workshops would do more than sharpen participants' articulation of their own and others' views. We hoped that bringing members of opposing interest groups into a philosophical discussion would defuse conflict and facilitate cooperation by helping them to see each other as reasoning individuals.

278

Although less obvious than the workshops' impact on participants' articulation of views, this does appear to have been a result of the workshop experience. As detailed above, some participants who described government agencies and their perceived opponents as thoroughly evil in pre-interviews instead described them as strongly non-anthropocentric in their follow-up interviews. Although these participants still characterized government agencies as difficult to work with, the switch in rhetoric is not insignificant. To see agencies or individuals as thinking differently about ethics is to see more possibilities for cooperation than where the agencies or individuals are perceived as immoral or amoral and simply not to be trusted.

There were also clear cases where the Brownsville workshop helped participants to see previously unrecognized opportunities for partnership and cooperation. Wildlife professional Hal Ketchum recognized in his follow-up interview, but not his pre-interview, that the land management professions could build bridges with the health care community (response #74). And Enriqué Gonzales, a doctor, and Suzanne Ramirez, a school nurse, noticed the parallel point that the health care professions need to build bridges with the environmental community (Gonzalez: responses ##21–23; Ramirez: responses ##50, 52 and 71).

Relatedly, a recurrent theme in interview transcripts was the need for members of opposing interest groups to communicate accurately and interact in a calm, reflective setting. In follow-up interviews, several participants specifically mentioned the workshops and associated materials in this vein (again, without prompting from the interviewers). For instance, Hal Ketchum cited a need for "better understanding – overall understanding of why they [various interest groups] y'know why they think the way that they think" (response #122) and for "An educational program that anybody could give, y'know whether it's us or them or a third party – that combines" education and outreach in a way that fosters mutual understanding among interest groups (response #133). He then concluded his follow-up interview with remarks on how valuable he and a colleague have found "every piece of material that [they have] received from [the project team]" and how they are sharing all of it with other people (response #158): "I've been able to use some of the things that I've learned and some of the information that we've received and had discussions about with other people in my agency as well as others" (response #156). Our participants appear to have valued the workshops for precisely the characteristics we attempted to build into them.

We undertook this project because we believed that when partisans are stuck in intractably opposed positions, retreating temporarily to a higher level of abstraction can facilitate communication among interest groups in a way that contributes to constructive management of the conflict. Our workshops used the study of philosophical debates in environmental ethics as a vehicle for encouraging calm, interactive reflection, rather than posturing and confrontation. Transcript analysis suggests that study of these debates *can* help participants frame their own and others' views more clearly and precisely, and that it *can* help them perceive opportunities for cooperation and partnership rather than confrontation. While our study was based on a small number of participants,[17] we think this approach deserves further attention and development.

Environmental activists have good reason to be impatient with endless philosophical debates over anthropocentrism vs. holism, intrinsic vs. instrumental value, and so on. Foreman is right that "philosophical hairsplitting" and "endless refining of dogma" will never by themselves solve an environmental problem. On the other hand, the politically supercharged environments of media presentations and public hearings on environmental problems are suboptimal in part because they tend to stifle philosophical reflection on basic issues in environmental ethics. For as our study suggests, a workshop that fosters such theoretical reflection can improve interest groups' understanding of each other and help foster cooperative problem solving rather than conflict. One of us (Varner) defends a specific theory of environmental ethics in published and forthcoming work.[18] However, when we introduced workshop participants to academic debates in environmental ethics, it was not in order to defend a particular theory against all comers, or even to encourage our participants to do this. We used theory as a vehicle for less confrontational discussion of regional environmental disputes, but we left the totalizing vision of academic philosophy debates in our offices. In this way our approach was roughly what Andrew Light labels "metaphilosophical pragatism."[19] Our study suggests that the optimal climate for constructive resolution of environmental controversies will be one which tempers the inspiration of activism with philosophical reflection on basic theoretical issues.

NOTES

1 Dave Foreman, "Earth First!" *The Progressive* (volume 45, October 1981), pp. 39–42, at pp. 40 and 42.

2 Dave Foreman, "More on Earth First! and The Monkey Wrench Gang," *Environmental Ethics* 5 (1983), pp. 95–96, at p. 95.

3 Christopher Manes, *Green Rage* (Little, Brown and Company, 1990), p. 21.

4 John Lemons, "Comment: A Reply to 'On Reading Environmental Ethics'," *Environmental Ethics* 7 (1985), pp. 185–188, at pp. 186, 187.

5 Norton makes the distinction between "applied philosophy" and "practical philosophy" in "Urging Philosophy to Become Practical," in Frederick Ferré and Peter Hartel, eds, *Ethics and Environmental Policy: Theory Meets Practice* (Athens, GA: University of Georgia Press, 1994), pp. 235–241. See also his *Toward Unity Among Environmentalists* (Oxford: Oxford University Press, 1991).

6 Paul Thompson, "Pragmatism and Policy: The Case of Water" in this volume.

7 EPA assistance ID #MX822144–01–0, "Addressing Coastal Challenges Through Environmental Ethics Education."

8 Difficulty was encountered securing participation from some regulatory agencies in both locations and, in Brownsville, from transnational manufacturing industries. Also, participants found it difficult to devote most of a weekend to a workshop. Indeed, although over 60 agreed to participate in the project and about 40 were interviewed prior to the workshops, only 16 completed the whole two-day workshop and participated in follow-up interviews. For this reason, future projects of this sort should explore scheduling alternatives and should use a significant economic incentive for full participation.

9 We used McCracken's "long interview" technique (G. McCracken, *The Long Interview* [Thousand Oaks, CA: Sage Publications], 1988), which relies on a few questions, phrased in a general and non-directive manner, with the sequence of questions determined by the informant's responses. The relatively unstructured format encourages informants to tell their stories in their own idiolect. This approach has been used by a variety of researchers to uncover meaning systems within interest groups similar to those convened for this project. See, for instance, Tarla Rai Peterson, "Telling the Farmers' Story: Competing Responses to Soil Conservation Rhetoric," *The Quarterly Journal of Speech* 77 (1991), pp. 289–268; Peterson, Kim Witte, Ernesto Enkerlin-Hoeflich, Lorina Espericueta, Jason T. Flora, Nanci Florey, Tamara Loughran and Rebecca Stuart, "Using Informant Directed Interviews to Discover Risk Orientation: How Formative Evaluations Based in Interpretive Analysis can Improve Persuasive Safety Campaigns," *Journal of Applied Communication Research* 22 (1994), pp. 199–215; and Cristi Horton, *An Interpretive Analysis of Texas Ranchers' Perceptions of Endangered Species Management Policy as Related to Management of Golden-cheeked Warbler Habitat,* master's thesis in Speech Communication, Texas A&M University.

10 See Gary Varner, "Environmental Law and the Eclipse of Land as Private Property," in Frederick Ferré and Peter Hartel, eds, *Ethics and Environmental Policy: Theory Meets Practice* (Athens, GA: University of Georgia Press, 1994), pp. 142–160.

11 To preserve anonymity, all informants' names have been changed. Except as indicated with bracketed words or phrases, responses are quoted verbatim without grammatical or other corrections.

12 Response numbers refer to interview transcripts on file with the authors.

13 Mr Blake recognizes that an implication of his view is that if there *were* appreciable harm to others, then the practice in question could be banned, and without compensation. For he contrasts his own benign ranching practices to certain farming practices, saying "I don't use any herbicides, insecticides or [garbled word]. Nothing. . . . But the farmers do, and they use too much. . . . But that's what they were raised to do. And I think there should – they've already got restrictions on it of course. And it's – might put the rice farmers out of business because they have to turn that water loose after they've used it on their crop and it goes directly into your watersheds, bayous and bays" (Responses ##37–40).

14 In our workshops, "anthropocentrism" was defined as the view that, morally speaking, only humans count; "intrinsic value" was defined as the value something has independently of its utility to other things. An anthropocentrist was further characterized as one who believes that non-human nature has only instrumental value.

15 Transcripts of interviews with Mike Jackson (a land manager) and Enriqué Gonzales (a doctor) are also noteworthy for changes in self-descriptions of their value commitments, but space precludes discussion here.

16 Thompson, "Pragmatism and Policy," pp. 199–200, this volume.

17 See note 8 above.

18 In a book manuscript (*In Nature's Interests? Interests, Animal Rights, and Environmental Ethics*, Lanham, MD) Varner labels his position "biocentric individualism" and questions the widely espoused views that non-holistic and/or relatively anthropocentric theories provide inadequate philosophical grounding for the environmentalist agenda. Related work already published includes "Biological Functions and Biological Interests," *Southern Journal of Philosophy* 27 (1990), pp. 251–270; and "Can Animal Rights Activists be Environmentalists?" invited paper in Donald Marietta and Lester Embree, eds, *Environmental Ethics and Environmental Activism* (forthcoming from Rowman & Littlefield). Both papers are reprinted in Donald VanDeVeer and Christine Pierce, eds, *People, Penguins, and Plastic Trees*, second edition (Belmont, CA: Wadsworth, 1994).

19 Andrew Light, "Compatibilism in Political Ecology," this volume.

Part 4

ENVIRONMENTAL PRAGMATISM: AN EXCHANGE

14

BEYOND INTRINSIC VALUE
Pragmatism in environmental ethics
Anthony Weston

I INTRODUCTION

"Pragmatism" sounds like just what environmental ethics is against: shortsighted, human-centered instrumentalism. In popular usage that connotation is certainly common. *Philosophical* pragmatism, however, offers a theory of values which is by no mean committed to that crude anthropocentrism, or indeed to any anthropocentrism at all. True, pragmatism rejects the mean–ends distinction, and consequently rejects the notion of fixed, final ends objectively grounding the entire field of human striving. True, pragmatism takes valuing to be a certain kind of desiring, and possibly only human beings desire in this way. But neither of these starting points rules out a genuine environmental ethic. I argue that the truth is closer to the reverse: only these starting points may make a workable environmental ethic possible.

One charge of anthropocentrism should not detain us.[1] Pragmatism is a form of subjectivism – it makes valuing an activity of subjects, possibly only of human subjects – but subjectivism is not necessarily anthropocentric. Even if only human beings value in this sense, it does not follow that only human beings *have* value; it does not follow that human beings must be the sole or final objects of valuation. Subjectivism does not imply, so to say, subject-*centrism*; our actual values can be much more complex and world-directed.

Pragmatism insists most centrally on the *interrelatedness* of our values. The notion of fixed ends is replaced by a picture of values dynamically interdepending with other values and with beliefs, choices, and exemplars: pragmatism offers, metaphorically at least, a kind of "ecology" of values. Values so conceived are resilient under stress, because, when put to question, a value can draw upon those

285

other values, beliefs, etc. which hold it in place in the larger system. At the same time, though, every value is open to critical challenge and change, because each value is also *at stake* precisely with those related values, beliefs, etc. which on other occasions reinforce it. We are thus left with a plurality of concrete values, in which many different kinds of value, and many different sources of value, can be recognized as serious and deep without requiring further reduction to some single all end in itself. And there is every reason to think that respect for other life forms and concern for natural environments are among those values. The problem is not to devise still more imaginative or exotic justifications for environmental values. We do not need to *ground* these values, pragmatists would say, but rather to situate them in their supporting contexts and to adjudicate their conflicts with others – a subtle enough difference at first glance, perhaps, but in fact a radical shift in philosophical perspective.

II INTRINSIC VALUE AND CONTEMPORARY ENVIRONMENTAL ETHICS

We seem to be compelled to distinguish means and ends almost as soon as we begin thinking about environmental values. Nature has certain obvious appeals: recreational and aesthetic satisfactions, "ecosystem stabilization" values (seemingly useless species may play a role in controlling pests, or fixing nitrogen), research and teaching uses, the attraction of natural objects and lifeforms simply as exemplars of survival, and so on.[2] In making these appeals, however, we value nature not "for its own sake," but for a further end: because it is necessary, useful, or satisfying to *us*. Even aesthetic appreciation does not necessarily require valuing nature for itself, since we might be tempted to say that only aesthetic *experience* is valued intrinsically. Beauty is in the mind of the beholder: aesthetic objects are only means to it.

The familiar next step is to ask whether nature could also be valuable in its own right. Could nature have *intrinsic* value, could it have worth as an end in itself, and not just because it serves human ends?[3] This question, of course, frames much of the debate in contemporary environmental ethics. If human beings, or some particular and unique human characteristics (e.g., a certain kind of conscious experience), are the only ends in themselves, then we have, for better or worse, "anthropocentrism." If some broader, but not universal class of beings has intrinsic value, and if, as usual, this class is taken to be the class of sentient or (even more broadly) living beings, then we have what might be called "sentientism" or (more broadly) "biocentrism." If *all* ("natural"?) beings, living or not, have intrinsic value and must not be treated merely as means, then we have what might be called "universalism." There is a continuum of possible ethical relations to nature, then, ranging from views which limit the bearers of intrinsic value strictly to human beings through views which progressively extend the franchise until finally it is (nearly?) universal.[4]

This much seems perfectly innocent. No views are actually endorsed, after all: only a range of possibilities is set out. In fact, however, I think that this "frame" is far from innocent. This seemingly uncommitted range of possibilities is in fact narrowly restricted by the underlying notion of intrinsic value itself.

Consider, after all, how that range of possibilities is determined in the first place: each option is defined precisely by the set of beings to which it attributes intrinsic value. Richard and Val Routley, for instance, argue that anthropocentrism represents a kind of moral "chauvinism," as egregious as the egoist's blindness to values beyond his or her self or the racist's failure to look beyond his or her race;[5] they insist upon the existence of *other* intrinsic values besides conscious human experience, values which deserve similar respect. Tom Regan *defines* an environmental ethic as a view which attributes "inherent goodness" to at least some non-human natural objects, where "inherent goodness" is an "objective property" of objects which compels us to respect its bearers.[6]

That notion of intrinsic or "inherent" value, however, is itself extremely specific and demanding. A great deal of philosophical baggage comes with it. Regan already weighs in with some of it, as Evelyn Pluhar points out, by construing inherent value as a "super-venient," "nonnatural" property, notions whose Moorean ancestry and problematic metaphysical commitments are plain to see.[7] But there is more to come. Let me try to set out the traditional requirements for intrinsic values more systematically.

(1) To qualify as intrinsic a value must be *self-sufficient*. G. E. Moore – the patron saint of intrinsic values – wrote that "to say that a kind of value is 'intrinsic' means . . . that the question whether a thing possesses it . . . depends solely on the intrinsic nature of the thing in question."[8] In his famous thought experiment in *Principia Ethica*, Moore says that to decide what things have intrinsic value "it is necessary to consider what things are such that, if they existed *by themselves*, in absolute isolation, we should yet judge their existence

to be good."[9] While everything else is dependent and, by itself, value-less, intrinsic values hold the sufficient grounds of their worth within themselves.

Moore appears to find it conceivable that anything at all could be valued intrinsically. In practice, however, self-sufficiency may not be such a neutral requirement. Even Moore came in the end to the conclusion that nothing but an experience can be intrinsically good; his argument turns on the claim that only experiences can be "worth having even if [they] exist quite alone."[10] Here Moore invokes a fundamentally Cartesian outlook. Consciousness is aloof from, not implicated in, the failures and ambiguities of actual objects and states of affairs in the world. Descartes argued that while my beliefs may or may not correspond to something in the world, I am sure at least that I have them. Perhaps Moore is arguing that while my acts too, in the world, may be incomplete, damaging, or uncertain, at least my conscious enjoyment of them, taken by itself, is solid and unquestionable. Just as Descartes' way of setting up the problem of knowledge made consciousness the natural and necessary standard-bearer against skepticism, so the demand that intrinsic values be self-sufficient may make consciousness the natural and necessary standard-bearer of the intrinsic. Only a commitment to a philosophical "paradigm" of this sort, I think, can explain the strikingly *unargued* insistence, even by such careful writers as W. K. Frankena, that "[no]thing can have intrinsic value except the activities, experiences, and lives of conscious, sentient beings."[11] Frankena just "*cannot see*" that "we ought morally to consider unconscious animals, plants, rocks, etc."[12]

(2) Philosophical tradition also demands, at least by implication, that intrinsic values be *abstract*. Intrinsic values are, after all, special: not everything can be intrinsically valuable. But the distinction between special ends and ordinary means, perhaps innocent enough at first, sets in motion increasingly radical demands. Everyday values are integrated as means under fewer and somewhat more general ends. On the next tier these still proximate ends become means themselves, to be unified in turn under still fewer and more general ends. Already this is a kind of "slippery slope" – upward, as it were. The supercession of each proximate end seems to deprive it of any independent value at all: now they are only means to the ends on a still higher tier. But these ends too may be superseded. Nothing will stop this regress, we say, except the most general, not-to-be-superseded ends in themselves: traditionally, values like "happiness"

or respect for persons. Having reached this point, moreover, there is a familiar and strong impulse towards erecting a *single* end on the first and highest level. Traditional value theory tends towards a kind of monism. We are not inclined to leave two or five values at the top of this pyramid when we might abstract down to one: on the most general level we want unity. Respect for persons might be reinterpreted as another source of happiness; happiness might be reinterpreted, as in Aristotle or Rawls, as valuable insofar as it represents the self-actualization of autonomous persons; but in any case, as Kenneth Goodpaster puts it, "one has the impression that it just *goes without saying* . . . that there must be some unified account of our considered moral judgments and principles," some sort of "common denominator."[13]

This monism too, moreover, may not be so neutral in practice. Conscious experience is supposed to be a single, unified sort of thing, abstract and self-sufficient enough, given Cartesian presuppositions, to be a bearer of intrinsic value. Adding a second sort of thing as another bearer of intrinsic value would destroy this tight unity. Thus, the implicit demand to reduce intrinsic values to a single common denominator may incline us once again towards the anthropocentric-sentientist end of the range of possible environmental ethics. Goodpaster reminds us, for instance, that many philosophers have been tempted to underwrite environmental values by extension from familiar "interest" or "dignity" ethics, respectively Humean or Kantian. Both are monistic models, tied at least historically to human beings as exemplars, and therefore run the risk of "constraining our moral sensitivity to the size of our self-wrought paradigms," just as they gain plausibility from the very same appeal.[14]

On the speculative side, some metaphysical consciousness monisms have become attractive. Some environmental ethicists want to attribute conscious experience even to the seemingly inanimate world: Po-Keung Ip, for example, uses a panpsychic Taoism to vindicate the intrinsic value of nature; Jay McDaniel uses a Whiteheadian reading of quantum mechanics.[15] Christopher Stone suggests that we regard the whole planet as a conscious entity.[16] Nature itself is thus animated, and all of us can enter the Kingdom of Ends together. At this extreme, then, a monism of intrinsic values is perhaps compatible with a powerful environmental ethic after all. The cost, however, is a radical revision of our metaphysics – in itself not unattractive, perhaps, but in the process we must also reaffirm, rather than escape, the absolute ethical centrality of sentience.

ANTHONY WESTON

(3) Intrinsic values demand *special justification*. Given their supposed self-sufficiency, they cannot be justified by reference to other values. Given their abstractness, they are too special, too philosophically fragile, to exist unproblematically in the world. But merely to assert them is insufficient: that would make them arbitrary, or condemn us to speechlessness about them, and so would cast our whole system of values adrift. Justification, we say instead, must take a special form: a "grounding" of intrinsic values is called for. Value as such must be derived, ontologically, from something else. Thus, intrinsic values have been construed as God's commands, as *a priori* truths about a special moral world revealed by intuition, as deliverances of Pure Reason, as aspirations fundamental to "human nature," and so forth. It is not surprising, then, that when Regan tries to ground his "inherent values," he feels driven to an ontology of "nonnatural properties" – despite the irony of appealing to "nonnatural" properties precisely in order to vindicate the value of *nature*! Some such ontology seems necessary. David Ehrenfeld holds that only the religious tradition will do: only a transcendental perspective can transfigure nature into "the present expression of a continuing historical process of immense antiquity and majesty."[17]

Many philosophers, however, no longer accept any of the traditional ontologies of values. Once again the result is to make some form of anthropocentrism or sentientism seem the only live option. Human concerns can always be counted upon to motivate, and the intrinsic value of conscious experience is often accepted without a fight. Thus, the temptation is to eschew the traditional ontology and to try to "build out" from these readily available anthropocentric starting points. Bryan Norton, for instance, proposes what he calls "weak anthropocentrism," a view which countenances not only occurrent human desires but also "ideals," like living in harmony with nature, which represent patterns of *considered* desire. Norton explicitly "avoids attributing intrinsic value to nature" because of the "questionable ontological commitment" that attribution would involve.[18] "Strong" anthropocentrists are often similarly motivated. Some utilitarians argue that cost-benefit analysis can accommodate environmental values more effectively than they have so far.[19] Here dubious ontological claims are avoided because only human interests are considered: utilitarianism is the epitome of an ontologically unadventurous theory of values. Mark Sagoff holds that we may value in nature expressions of things we value intrinsically in our own lives: freedom, nobility, etc.,[20] and, in a similar way, Thomas Hill, Jr. argues that the

290

best moral attitudes towards persons – humility, self-acceptance, gratitude – are mirrored and promoted by more respectful environmental values.[21] Both Sagoff and Hill, however, are still "building out" from human-centred value systems, from expressions or personal qualities which we value in our own and other *human* lives.[22]

Regan has argued effectively that no strong anthropocentrism can vindicate environmental values to the extent that our convictions demand.[23] Sagoff, Hill and others may well disagree, but all the same they often convey a sense that they consider even their own approaches somewhat "second best." Hill writes at one point that "even if there is no convincing way to show that [environmentally] destructive acts are wrong . . . we may find that the willingness to indulge in them reflects an absence of human traits that we admire and regard as morally important."[24] *Even if* . . . we *may* find: the suggestion seems to be that modified anthropocentrism is the best we can do, though definitely not the best we might wish. Regan, meanwhile, according to Pluhar, draws the opposite conclusion from the same premise: Regan, she says, "seems to find it preferable to make the commitment to dubious property instances and thus salvage the possibility of the kind of ethical justification he wants. The possibility is remote, but he may reason that it is better than nothing."[25] So "better than nothing" is the bottom line on both sides. We are in a sorry state indeed.

Only occasionally are there hints of anything truly different. Some of these are attempts to formulate a new language for values in nature. Holmes Rolston's essay "Values Gone Wild," for instance, is striking in this regard for its plays on "source" and "resource," "neighbor," etc.[26] Later I will suggest that Rolston's promising start too is partially undercut by his attempts to meet the demands of intrinsic value: what *is* promising, I hold, is precisely the part that has worked free of those shackles. So far I am only trying to show how confining those shackles are. In short, not only has environmental ethics taken over from philosophical ethics an extremely specific and demanding notion of intrinsic value, rooted in various ways in Cartesian metaphysics and in time-honored philosophical temptations to abstraction and special justification; those very roots in turn put extraordinary constraints on any attempt to demonstrate intrinsic values in nature. At the deepest level, non-anthropocentric environmental ethics may simply be impossible within the inherited framework of intrinsic values. In itself, of course, this is not necessarily an objection to the tradition: maybe environmental ethics finally *is* impossible. But it is time to ask whether that tradition has any compelling defence.

III AGAINST INTRINSIC VALUE

Moore argues that some notion of "valuable for its own sake" or "valuable in itself' is required simply to *understand* the notion of "valuable for the sake of something else," the everyday notion of instrumental value which we usually take for granted. If we speak of means, then logically we must also be able to conceive of ends, since an end seems to be implicated in the very concept of a means. Thus Moore reads the phrase "good as a means" as equivalent to "a means to good," where the "good" in the second case seems to be intrinsic.[27]

This rationale fails, however, for a simple reason. We can also understand the notion of instrumental value by reference to further, but non-intrinsic values. Values may refer beyond themselves without ever necessitating a value which must be self-explanatory. The value of a day's hike in the woods need not be explained either by the intrinsic value of my appreciation of the woods or by the intrinsic value of the woods themselves; instead, both the appreciation and the woods may be valuable for further reasons, the same may be true of *those* reasons, and so forth. Appreciation may be valued, as Hill points out, partly because it can lead to greater sensitivity to others; but greater sensitivity to others may in turn make us better watchers of animals and storms, and so on. The woods may be valued not only as an expression of freedom and nobility, but also as a refuge for wildlife, and both of these values may in turn be explained by still other, not necessarily human-centred values.

Someone may respond that explanations such as these must still have stopping points somewhere. If X is valuable because it leads to or enhances Y, we might seem to be required to say that X's value is "passed on" from Y. Y's value in turn may be passed on from Z. But – the argument goes – there must be some origin to the value which is thus "passed on." Like a bucket of water in a fire chain, it must have started in some reservoir which is not merely another bucket. Monroe Beardsley likens this argument to the first cause argument for the existence of God: ". . . the existence of any instrumental value [is supposed to] prove the existence of some intrinsic value just as the occurrence of any event is said to prove the existence of a First Cause."[28]

Beardsley's analogy, however, suggests an initial objection. The "first value" argument may beg the very question it is trying to answer. Just as the first cause argument must assume that the chain of causes it invokes cannot be infinite, so the "first value" argument assumes that the long process of tracing means back to ends must have a final stopping point. But actually this is just what it was supposed to *show*.

Most importantly, however, there are many ways of not having a stopping point. We need not think of an endless series of means each necessitating the next like a long line of falling dominoes. It is more appropriate to think in quite different terms. Consider a more holistic picture conception according to which values are connected in a weblike way, so that any value can be justified by referring to those "adjacent" to it. On this model there is no ultimate reference or stopping point simply because the series of justifications is ultimately, in a sense, circular: to justify or to explain a value is to reveal its organic place among our others. These justifications need not wind their way only in a single direction or even towards a single type of value. If sometimes I value the mountain air because in it I feel (and *am*) healthy, other times I value health because it enables me to reach the mountains. If sometimes I value the melancholy glory of the autumn because it mirrors the closure of my own year, other times I value the rhythms of my yearly schedule because they mirror the glories of the seasons. The web image also emphasizes the *multiple* "adjacencies" of most values. To explain why I climb mountains may take hours; Henry Beston took a whole book to chart the riches of a year spent living alone on Cape Cod. By extension we may think of multiple circularities and feedback loops, multiple arcs returning to completion, so that the summation of those arcs is a rough map of one's whole system of values. To explain why I climb mountains may take hours, but it is not an endless task: although the story has no final stopping point or ultimate appeal, it is *complete* when I have articulated the manifold connections between mountain climbing and the other values, beliefs, etc. which make up my self.

Conceiving values in this holistic way undercuts the very center of the traditional notion of intrinsic value. Self-sufficiency, in the first place, is just what we should *not* want in our values. Beardsley argues that the notion of "intrinsic value" is almost a *contradiction* precisely because it insists on cutting values off from their relations with others in order to consider them "just in themselves." Following Richard Brandt's suggestion that the statement "X is desirable" means something like "desiring X is justified," Beardsley argues:

293

What "desirable" adds to "desired" is this claim to justifiability. But the only way this claim can be made good is by considering X in the wider context of other things, in relation to a segment of life or of many lives. Thus the term "intrinsic desirability" pulls in two directions: the noun tells us to look farther afield, the adjective tells us to pay no attention to anything but X itself.[29]

What would it actually be like, after all, to value a conscious experience for itself, "in absolute isolation"? Clearly it could qualify only in so far as it approximates the Cartesian self-sufficiency of dreams or visions: it could not matter whether the experience is connected to anything else in the world. But it is not obvious that this self- sufficiency makes an experience good at all, let alone good intrinsically – and the reasons are precisely the considerations that the self-sufficiency criterion requires us to rule out. What can exist and attract in isolation from everything else may be, for just that reason, *bad* : like the dream world of the drug user, it seduces us away from the complexity of our lives, substitutes solipsism for sociality, divides certain parts of our lives from the rest. We should prefer a conception of values which ties them to their contexts and insists not on their separability but on their relatedness and interdependence.

Beardsley himself has a somewhat different line of response to the "first value" argument. It is not so much a challenge to the alleged self-sufficiency of intrinsic values as a challenge to their abstractness. He begins by recalling Hume's response to the first cause argument. In ordinary life, Hume points out, we are not only familiar with specific causal relations, but are entirely capable of dealing with them concretely. The ultimate nature of causality, by contrast, is neither knowable nor important: it is "merely speculative," as Hume put it, both in the sense that it is endlessly debatable and in the sense that it is irrelevant to practical purposes. Beardsley makes just this argument with respect to intrinsic values. "We have a good deal of sound knowledge about instrumental values," he writes, "but we are in considerable doubt about intrinsic values."[30] In ordinary life we are not only familiar with specific values, but are eminently capable of dealing with them concretely. We know that it is better to be healthy than to be sick, better to live amidst beauty than monotony or ugliness, better to walk in a virgin forest than along the median strip of Interstate 84, and so on. But we do not know whether these things are good because they maximize our net hedonic quality, or good because they cultivate a good will, or what. So far from being the absolutely central project of any philosophy of values, the search for an ultimate end seems "merely speculative." It is better to think of values more concretely, in all their richness and plurality.

Besides, why *should* there be something which all values have in common? It is more plausible to deny that there *is* any final end from which all the others flow and which plays end to all the others' means. We have instead an irreducibly pluralistic system of desires. Some are straightforwardly biological, others culturally rooted, others more personal, and many are mixtures of all three. If anything we are doomed to hopelessly *conflicting* desires. Neither our biological predispositions nor our cultural heritage are even self-consistent, let alone fully compatible with the other.

These last points, however, may lead us to a third and final argument for intrinsic values. It may be urged that, in fact, intrinsic values *can* be concrete, plural, and possibly even inconsistent. This is Holmes Rolston's view, and a version of it has been held even by some pragmatists, such as C. I. Lewis. There are times, Rolston or Lewis would say, when we apprehend value concretely and directly, without having to look farther afield or into the future in order to recognize it. Lewis echoes Moore by comparing this recognition to the way we see redness or hear shrillness.[31] Rolston speaks of the intrinsic value of "point experiences," like the warmth of the spring sun, calling it "as fleeting and plural as any other kind of value."[32] Rolston's intrinsic values need not be abstract, then, and they need no justification at all, let alone "special" justification. A day's hike in the woods is worthwhile even if it does not contribute to peace of mind or animal-watching ability or job performance: the experience, as well as the woods itself considered even apart from my experience, is simply good "for what it is in itself."[33]

Undeniably, Lewis and Rolston are pointing to a real kind of experience; the question is what this kind of experience shows. It is, at least, an experience of what we might call *immediate* value. John Dewey argued, however, that "to pass from immediacy of enjoyment to something called 'intrinsic value' is a leap for which there is no ground."[34] When we do endorse something in an immediate and non-inferential way, according to Dewey, we do not usually make a judgment of value at all, and so *a fortiori* do not make a judgment of intrinsic value. Instead, that endorsement is a "statement to the effect that no judgment is required, because there is no conflict of values, no occasion for deliberation and choice."[35] Even obviously instrumental activities – doing the dishes, driving the highways – are sometimes appreciated in this immediate and non-referential way. Even something that destroys, a virus or a tornado, can sometimes be arrestingly beautiful. *Arresting* is the right word, too: our response to them precisely *disconnects* the frame of reference in which value questions even arise.

When values do become problematic, when choice is required, then they need articulation and defense. But to call them "intrinsic," in Rolston's sense, now offers no help. Since we have to disconnect objects and actions from their contexts in order to value them just "for what they are in themselves," what they are in relation to everything else is pushed out of focus. If I lose myself in the beauty of the tornado, I may not reach shelter in time. Rolston insists that immediate values must be put in context, like any others, and that they are sometimes ambiguous or even downright bad when contextualized. The upshot, however, is that the attribution of intrinsic value, in his sense, carries no special force in the real world. A thousand other "point experiences" of values press in upon us from every side, just as ordinary values have always pressed in upon us, and what we *do* will and should be determined, just as it has always been determined, by the balances and synergies and trade-offs between them. By all means let us remember that this is a world lavish with its moments of beauty and preciousness – but let us honor those moments without cutting them off from the practical living of our lives.

Earlier I called into question the traditional demands for self-sufficiency and abstractness in intrinsic values. Here, finally, the task of justification too is reconceived. It is not the task of "grounding" values: what Rolston's defense of the notion of intrinsic values may finally illustrate, in fact, is the way in which the project of "grounding" natural values (or, perhaps, any values) finally cuts itself off from the real-life task of assessment and choice. For assessment and choice we must learn, again, to *relate* values. Any adequate theory of valuation must recognize that valuation involves desires with a complex internal structure, desires interlinked, and mutually dependent with a large number of other desires, beliefs, exemplars and choices.[36] Love, for example, interlinks with a wide range of desires and beliefs, from the tenderness of "being with" to sexual desires, from one's complex understanding of the other person to the culture's images and exemplars of love, and so on. Justification draws on these interdependencies. We justify a value by articulating the supporting role it plays with respect to other values, which in turn play a supporting role with respect to it, and by referring to the beliefs which make it natural, which it in turn makes natural by re-affirming those choices and models which link it to the living of our lives. Precisely this is Beardsley's "wider context of things."

Interdependent values are not closed to criticism: it may actually be this sort of interdependence, indeed, which makes the most effective criticism *possible*. Criticism becomes an attempt to alter certain desires by altering something in the constellation of other desires, beliefs. choices, etc. to which they are linked.[37] Some of the beliefs in question may be false, desires artificial or shallow, and so forth. Norton is right to point out that "felt preferences" exploitative of nature can often be criticized on the basis of "considered prefer-ences." Too often we are simply thoughtless, or not thoughtful enough. But the power of this sort of criticism goes far beyond the dialectic of "ideals": only Norton's wish to set up shop on the edge of the concept of intrinsic value, I think, leads him to conceive considered preferences on the model of ideals, thus making them seem far more marginal than they are.[38] As Pluhar writes:

> It is amazing how much prejudice and ignorance fuel ethical disputes, not to mention bad reasoning. . . . How much lack of impartiality and empathy underlie common attitudes towards animals . . . ? How much greed (a prime source of partiality), ignorance, and muddled thinking fuel common attitudes about ecosystems and natural objects?

As she points out, visiting a meat factory makes many vegetarians! Although Pluhar, oddly, regards this pragmatic sort of criticism as an alternative way of defending Regan's "inherent values," she offers no argument that the values which might emerge from this procedure are in any sense "inherent" or intrinsic.[40] I suspect that no such arguments can be found. It is time to abandon the old preoccupation with intrinsic values entirely: let practical criticism be practical.

Not even radical criticism is excluded. The culture to which we owe so many of our explicit desires and their interlinkings also includes an attic full of latent ideals, inconsistent perhaps with its main tendencies, but still there waiting to be drawn out. God may have given us dominion over land and sea, but He also gave us St Francis; against the swashbuckling exploitation of the Industrial Revolution we have the romantic poets, landscape painting, Rousseau, Emerson, Thoreau; against factory farms we have the still

297

compelling image of the solitary farmer close to the soil. The wide-ranging recent debates about Christian and Judaic attitudes towards nature underscore this fundamental dissonance.[41] It is a mistake to try to find *the* Christian (or *the* American, etc.) attitude towards nature: there are many. Our traditions, I want to suggest (I have tried to argue this general point elsewhere[42]), contain their dialectical opposites within themselves. Even our biologically rooted desires are far from monolithic and static. Sometimes criticism simply needs the time and the patience to draw these latent elements out.

IV PRAGMATISM IN ENVIRONMENTAL ETHICS

The real power of the pragmatic approach lies in what it does *not* say, in what it has removed the need to say. Thus my concern here is emphatically not to devise new arguments for environmental values, but instead to show that the familiar ones are laboring under needless constraints. Still, this may be a modest, if unexotic, bit of progress, and I expect that it will be controversial all the same. I think that if values are conceived along the lines just sketched, then the case we can already make for environmental values – and in quite simple terms – is far stronger than most environmental ethicists themselves seem to believe.

We know that the experience of nature can awaken respect and concern for it. We know indeed that these feelings can become deep and synergistic desires in some lives, and we have before us exemplars of such lives in Muir, Thoreau, Leopold and others. Most of us are not so single-minded, but we too know how essential a return to nature can be, how Thoreau felt returning to Walden Pond from town, and why Yeats yearned for the bee-loud glade. While there are varied motives behind the recent boom in backpacking, cross-country skiing, canoeing, camping, and the like, at least part of the cause is surely a growing appreciation of nature, not just as another frame for our exercise and relaxation, but for its own unique voices, from the silence of the winter woods to the roar of waterfalls in spring.

These feelings are essential starting points for a pragmatic defense of environmental values. They are *not* "second best," "weak" anthropocentric substitutes for the intrinsic values philosophers want but cannot find. They do not need a philosophical "grounding." The questions that arise for us are of quite a different sort. Again, we need to know how to articulate, to ourselves and to others, the *relation* of these values to other parts of our system of desires, to other things that are important, and to the solution of concrete problems. For ourselves we want to understand and strengthen these values; in others we want to nourish and extend them. Nor, finally, need we start by trying to assimilate environmental values to our other values. Even our respect and concern for each other may be of quite a different type, and have entirely different sources, from our respect and concern for the environment.

The articulation of these values is not the province of philosophy alone. Poetry and biography are just as vital. Think of Wordsworth:

> And I have felt
> A presence that disturbs me with the joy
> Of elevated thoughts; a sense sublime
> Of something far more deeply interfused,
> Whose dwelling is the light of setting suns,
> And the round ocean and the living air . . .
> Therefore let the moon
> Shine on thee in thy solitary walk;
> And let the misty mountain winds be free
> To blow against thee. . . .[43]

We must not read this as an incomplete statement of pantheism, in need of philosophical clarification. Maybe Wordsworth was a closet metaphysician, but the possible linkage to Spinoza is not what makes us ache to feel those winds. Wordsworth offers a way to begin to describe a kind of experience which for our purposes may not need a stricter formulation. It is not a "grounding": it is a kind of *portrait*. Likewise, what is finally important in *Walden* is not Thoreau's misanthropic philosophizing, but the way in which he shows us, in his own person, how a human being can meet the evening, between the squirrels and the shadows, or how to look at a lake:

> A lake is the landscape's most . . . expressive feature. It is earth's eye, looking into which the beholder measures the depth of his own nature. The fluviate trees next to the shore are the slender eyelashes which fringe it, and the wooded hills and cliffs around are its overhanging brows.[44]

Nietzsche suggests more than once that philosophers are too clumsy to handle real values. He may exaggerate, but all the same we do know that philosophy has too long failed to take seriously what it cannot itself fully articulate. By rejecting the demand to "ground" these values, then, pragmatism also begins to undercut the demand that we articulate them in philosophy's peculiar, epistemically oriented way.

Still, on the whole, many philosophical arguments fare well in terms of the new set of questions I am advancing. Indeed many of them fare *better* when measured against this new set of questions than against the set of questions that they are actually trying to answer. Let us first return to Rolston's "Values Gone Wild." Rolston begins with a critique of the idea of nature as a "resource." The idea that "everything is a resource," he argues, like the idea that "everybody is selfish," becomes simply trivial at the extremes, "eating up everything, as if humans had no other operating mode *vis-à-vis* wilderness." In fact, we must enter wilderness "on its own terms" – not, or not primarily, as a means to "high quality experience." In this way, he argues, "one is not so much looking to *resources* as to *sources*, seeking relationships in an elemental stream of being with transcending integrities."[45] At this point, however, Rolston goes on to suggest that nature is intrinsically valuable because it is a source, in this sense, of whatever (else) we intrinsically value. This seems to me to add nothing; it only *weakens* the evocative force of the notion of "sourcehood." Although "elemental . . . transcending integrities" make a certain ecosystemic sense, trying to make their *value* transcendental either introduces an extremely problematic ontology, as I argued in part two, or represents only one way of talking, as I argued in part three, with no special force in actual moral thinking. "Sourcehood" is a perfectly understandable and powerful model of value in its own right: why force it into the mold of intrinsic values?

Consider one other example. Rolston writes of "sympathetically turning to value what does not stand directly in our lineage or underpinning" – our "kin" and "neighbors" in the animal world.[46] This too is genuinely perceptive: we do have a latent sense of community with animals which close acquaintance may bring out. But here too Rolston tries to wring intrinsic values out of facts which are better left alone. He argues, for instance, that the similarity between our reactions and those of animals suggests that we should take their reactions to express imperatives – values – as well, presumably including intrinsic values. Why these imperatives also bear on *us*, however, is not clear, and the claim that they do bear on us involves analogic arguments problematic in both philosophy of mind and moral theory. Once again Rolston's concrete notions, here of "kinship" and of being "neighbors," capture the values at stake much more freshly and directly than the philosophically problematic analogies necessary to make them over into intrinsic values. Moreover, as Rolston also points out, even within the animate world the notion of kinship eventually stretches beyond the breaking point: certainly we have little kinship with spiders. If another kind of value must be invoked for such "aliens," then it is not clear why this should not be so even for "neighbors." There is no need to fit all values into a single model.

Even more standard philosophical arguments – or at least their basic intentions – fit naturally into this framework. Recall Sagoff's argument that we may value in nature expressions of things that we value intrinsically in our lives: freedom, nobility, etc. Critics have pointed out that this cannot demonstrate the intrinsic value of nature itself.[47] Pragmatists, however, want to know simply how this value relates to others and can form an organic part of our lives. This is exactly what Sagoff helps to show us, locating it partly in the orbit of the desire for freedom. Or again, the persistent inclination to attribute "rights" directly to nature might now be reapproached and understood. In part, certainly, that attribution is a straight-forward political attempt to state environmental values with enough force that others will take them seriously. But it is also an attempt to articulate a specific and familiar attitude towards nature. Alone in the woods we find ourselves feeling a sense of gratefulness, of "awe," finally almost of intrusion, a feeling which probably has its closest parallel in those responses to other *people* which make us want to attribute *them* rights. But how closely these feelings are actually parallel remains an open question. Here we first need a careful phenomenology. This may be true even of human rights: real respect for others comes only through the concrete experience and finally "awe" of the other. It is the conditions and nature of this feeling which we really need to understand. Reversing the usual deduction entirely, we might even take rights talk itself as a first and rather crude attempt at just such a phenomenology – but surely we can do better.

Let me conclude by returning to the level of practical problems in environmental ethics. Why, for instance, should we value wilderness? What sort of justification can we give for keeping exploitable land and resources in their natural state? Not surprisingly, it is necessary to begin with a reorientation. Notice that this question is already posed in abstraction from any specific situation. This may itself give rise to absurdities. If we answer that wilderness indeed has intrinsic value, then presumably we are required to go to any lengths to support as much of it as possible, and wherever possible, at least consistent with other intrinsic values. But too many other things of equal or greater importance in the *situation* will not be captured by a hierarchical scheme of intrinsic values. Of course, there are other ways out, perhaps invoking intrinsic principles of such generality that they can be used to justify anything. The response I am urging, however, is the abandonment of these very ways of posing the question. The important questions for pragmatism are the ones posed by specific situations, and while the answers across different situations will probably bear a strong family resemblance, they will not always be the same.

Why should we protect the new Alaskan national parks, for example? Now the answers are much easier: because the new parks are both exceptionally wild and exceptionally fragile; because the non-preservationist pressures in at least this case are exceptionally unworthy, tied largely to the exploitation of energy resources to which there are any number of more intelligent alternatives; perhaps also because their protection is still possible. These arguments do indeed seem to dodge the original question. They do not say why wilderness as such should be protected. On the other hand, one certainly does not have to be an anthropocentrist to doubt whether it *should* be protected "as such." This is why the *exceptional* nature of the Alaskan wilderness makes that particular case so powerful. These "practical" arguments are precisely the kinds offered by the Sierra Club, the Nature Conservancy, and most of the other environmentally oriented organizations. Are these arguments offered merely for lack of better (philosophical?) ones? Or might those organizations actually have a more reasonable position after all?

"What about those people, though, who simply could not care less about wilderness? What about the many cases in which such values simply cannot be assumed? Tame rivers are much nicer than wild ones if one owns a motorboat; exploitation in Alaska might lower our fuel bills and make America more self-sufficient in some vital resources; and so on." Let me respond in several ways. First, even these cases may not be real cases of "could not care less." Nearly everyone recognizes *some* value in nature; think of how often natural scenes turn up on wall calendars and church bulletins. Even motor-boaters like to see woods. Wilderness values may just seem to them less significant than other values at stake in the particular situation. Common ground remains. If we begin by treating others as absolutists, we run the risk of turning them into just what we fear.

But this is only a caricature, and we can instead approach them from a standpoint of complex mutuality. Then, though, if some shared values can indeed be agreed upon, the real issue shifts to the question of alternatives, and this is a recognizably factual issue on both sides, and also negotiable. Motorboats don't have to go everywhere.

The pragmatic approach defended here forswears the search for knockdown arguments that will convince absolutely everyone that natural values are important. We cannot defeat the occasional extremist who sees no value at all in nature. But if this is a defect, it is certainly not unique to pragmatism. No other approach has knockdown arguments to offer either; otherwise, environmental ethics would not be a *problem*. The real difference is that pragmatists are not looking for knockdown arguments; we propose to concern ourselves with defending environmental values in other ways. It is striking, actually, that the search for a proof of the intrinsic value of nature is almost always *post hoc*. Even if someone were finally to discover a knockdown proof, it would not be the reason that most of us who are in search of such a proof do in fact value nature, since our present accounts of natural values differ so markedly. *We* learned the values of nature through experience and effort, through mistakes and mishaps, through poetry and stargazing, and, if we were lucky, a few inspired friends. What guarantees that there is a shortcut? It is wiser to accept the fact that many of our contemporaries, even our most thoughtful contemporaries, hold deeply different, probably irreconcilable, visions of the ideal world.[48] Pragmatism, indeed, celebrates a wide-open and diverse culture; it is the prerequisite of all the central Deweyan virtues: intelligence, freedom, autonomy, growth. What we have yet to accept is its inconclusiveness and open-endedness, its demand that we struggle for our own values without being closed to the values and the hopes of others. The search for intrinsic values substitutes a kind of shadowboxing for what must always be a good fight.

ACKNOWLEDGMENTS

"Beyond Intrinsic Value: Pragmatism in Environmental Ethics" first appeared in *Environmental Ethics* Vol. 7, No. 4 (Winter 1985). I am indebted to Holmes Rushton, III, and to an anonymous reviewer for *Environmental Ethics* for extensive comments on earlier versions of this essay. It has also benefited greatly from a colloquium discussion at the Vassar College Department of Philosophy and from several careful readings by Jennifer Church.

NOTES

1 The confusion of subjectivism with "subject-centrism" is dissected, though not in these terms, by Richard and Val Routley in "Against the Inevitability of Human Chauvinism," in K. E. Goodpaster and K. M. Sayre, eds, *Ethics and the Problems of the 21st Century* (Notre Dame: University of Notre Dame Press, 1979), pp. 42–47.

2 For an extensive list, see David Ehrenfeld, *The Arrogance of Humanism* (Oxford: Oxford University, Press, 1978), Chap. 5; or Holmes Rolston, III, "Valuing Wildlands," *Environmental Ethics* 7 (1985): 24–30.

3 I am equating intrinsic values with ends in themselves, instrumental values with means to ends. For present purposes I think that subtle distinctions between these concepts can be ignored.

4 See W. K. Frankena, "Ethics and the Environment," in Goodpaster and Sayre, *Ethics*, pp. 5–6 and pp. 18–9; and J. Baird Callicott, "Non-anthropocentric Value Theory and Environmental Ethics," *American Philosophical Quarterly* 21 (1984), pp. 299–309.

5 Routley and Routley, "Against the Inevitability," pp. 36–62.

6 Tom Regan, "The Nature and Possibility of an Environmental Ethic," *Environmental Ethics* 3 (1981): 30–34. Frankena, C. I. Lewis, and others use *inherent value* to refer to objects or actions the contemplation of which leads to intrinsically valuable experience. Regan, however, clearly means by *inherent* what Frankena and Lewis mean by *intrinsic*. "If an object is inherently good," he tells us, "its value must inhere in the object itself" (p. 30). Its value does not depend upon experience at all.

7 Evelyn Pluhar, "The Justification of an Environmental Ethic," *Environmental Ethics* 5 (1983): 55–58.

8 G. E. Moore, *Philosophical Studies* (London: Paul, Trench, Trubner, 1922), p. 260.

9 G. E. Moore, *Principia Ethica* (Cambridge: Cambridge University Press, 1903), p. 187.

10 G. E. Moore, "Is Goodness a Quality?" in *Philosophical Papers* (London: Allen and Unwin, 1959), p. 95.

11 Frankena, "Ethics and the Environment," p. 17. Pluhar makes some sharp comments on this claim in "The Justification of an Environmental Ethic," p. 54.

12 Ibid., p. 15. My emphasis.

13 K. E. Goodpaster, "From Egoism to Environmentalism," in Goodpaster and Sayre, *Ethics*, p. 25 and p. 34, his emphasis. Strictly speaking the claim here is only about ethics in the Humean tradition, but he soon allows that the Kantian tradition has still stronger monistic tendencies.

14 Ibid., p. 32.

15 Po-Keung Ip, "Taoism and the Foundations of Environmental Ethics," *Environmental Ethics* 5 (1983): 335–44, and Jay McDaniel, "Physical Matter as Creative and Sentient," *Environmental Ethics* 5 (1983): 291–318.

16 Christopher Stone, *Should Trees Have Standing?* (Los Altos: William Kaufmann, 1974), pp. 52–53.

17 Ehrenfeld, *Arrogance of Humanism*, p. 208.
18 Bryan Norton, "Environmental Ethics and Weak Anthropocentrism," *Environmental Ethics* 6 (1984): 131, 136, 138.
19 J. V. Krutilla and A. C. Fisher, *The Economics of Natural Environments* (Baltimore: Johns Hopkins, 1975).
20 Mark Sagoff, "On Preserving the Natural Environment," *Yale Law Journal* 84 (1974): 205–267; reprinted in Richard Wasserstrom, *Today's Moral Problems* (New York: Macmillan, 1979).
21 Thomas E. Hill, Jr., "Ideals of Human Excellence and Preserving Natural Environments," *Environmental Ethics* 5 (1983): 211–24.
22 See Hill, "Ideals," p. 233, or p. 220: "It may be that, given the sort of beings we are, we would never learn humility *before persons* without developing the general capacity to cherish . . . many [other] things for their own sakes" (my emphasis). Sagoff speaks of our obligation to nature as finally an obligation "to our national values, to our history, and, therefore, to ourselves" (Wasserstrom, *Today's Moral Problems*, p. 620).
23 Regan, "Nature and Possibility," pp. 24–30.
24 Hill, "Ideals," p. 215.
25 Pluhar, "Justification," p. 58.
26 Holmes Rolston, III, "Values Gone Wild," *Inquiry* 26 (1983): 181–207.
27 Moore, *Principia Ethica*, p. 24.
28 Monroe Beardsley, "Intrinsic Value," *Philosophy and Phenomenological Research* 26 (1965): 6. The critique offered here is indebted to Beardsley's fine article.
29 Ibid., p. 13.
30 Ibid., p. 7.
31 C. I. Lewis, *An Analysis of Knowledge and Valuation* (LaSalle, Ill.: Open Court, 1946), pp. 374–75.
32 Rolston was generous enough to comment extensively on an earlier draft of this paper, and I am quoting from his comments. Obviously he should not be held to these exact words, though I think his position here is a natural completion of what he has said in print. See Rolston, "Values Gone Wild" and Holmes Rolston, III, "Are Values in Nature Objective or Subjective?" *Environmental Ethics* 4 (1982): 125–52; reprinted in Robert Eliot and Arran Gare, eds, *Environmental Philosophy* (University Park: Pennsylvania State Press, 1983), pp. 135–165.
33 Rolston, "Are Values in Nature Objective or Subjective?" in Eliot and Gare, *Environmental Philosophy*, p. 158.
34 John Dewey, *Theory of Valuation* (Chicago: International Encyclopedia of Unified Science, 1939), 2:41.
35 Beardsley, "Intrinsic Value," p. 16.
36 See Anthony Weston, "Toward the Reconstruction of Subjectivism: Love as a Paradigm of Values," *Journal of Value Inquiry* 18 (1984): 181–194.
37 Ibid. and R. B. Brandt, *Theory of the Good and the Right* (Oxford: Clarendon Press, 1979), part I.

38 Norton ends up arguing that having ideals need not presuppose the intrinsic value of the things or states of affairs idealized: see Norton, "Weak Anthropocentrism," p. 137.

39 Pluhar, "Justification," p. 60.

40 Ibid., p. 58. This curious inference also mars J. Baird Callicott's otherwise fine survey: see Callicott, "Non-anthropocentric Value Theory," p. 305.

41 See Robin Attfield, "Western Traditions and Environmental Ethics," in Eliot and Gare, *Environmental Philosophy*, pp. 201–230.

42 See Anthony Weston, "Subjectivism and the Question of Social Criticism," *Metaphilosophy* 16 (1985): 57–65.

43 William Wordsworth, "Lines Composed a Few Miles above Tintern Abbey," lines 93–98 and 134–37.

44 H. D. Thoreau, *Walden* (New York: Signet, 1960), p. 128.

45 Rolston, "Values Gone Wild," pp. 181–183.

46 Ibid., pp. 188, 191.

47 For instance, Louis Lombardi, "Inherent Worth, Respect, and Rights." *Environmental Ethics* 5 (1983): 260.

48 A particularly striking example is Steven S. Schwarzchild, "The Unnatural Jew", *Environmental Ethics* 6 (1984): 347–362.

15

SEARCHING FOR INTRINSIC VALUE

Pragmatism and despair in environmental ethics

Eric Katz

I INTRODUCTION

What role does the concept of intrinsic value play in the development of an environmental ethic? Must the principles of an environmental ethic be *grounded* on the existence and recognition of self-sufficient, abstract, and independent value in natural entities? Or is the concern over the existence and explanation of intrinsic value in nature a mistake, a misdirection, a dead end in the field of environmental ethics? In a recent article in this journal, "Beyond Intrinsic Value: Pragmatism in Environmental Ethics,"[1] Anthony Weston argues that the allure of "intrinsic value" is fundamentally misguided, that rather than provide a ground for environmental ethics it dooms the development of a plausible moral argument for environmental protection. Weston claims that even environmental philosophers who base their arguments on the need for intrinsic value recognize the near impossibility of justifying such a theory of moral value. But we can do better than this (he claims). All we need to do is turn to the pragmatic conception of value and apply it to the development of an environmental ethic. This will remove the dualisms of traditional value theory – means/ends, intrinsic/instrumental – that are conceptual obstacles to the real practical understanding of both ethics in general and environmental ethics in particular. We will thus discover that the principles of an environmental ethic are part of the constellation (or web) of interrelated pluralistic values that pragmatists tell us is the heart of ethical theory.

Weston's challenge to the *structure* of the debate in environmental ethics is a powerful one; indeed, I share many of his concerns. None the less, his suggestion for a solution to the "sorry state" of environmental ethics is far from acceptable. A pragmatic value theory and ethics – even if justifiable in itself – would produce an environmental ethic that is irredeemably anthropocentric and subjective. A workable environmental ethic will share many fundamental concepts with pragmatism – e.g., the emphasis on the concrete situation – but it cannot ultimately rest on the values of pragmatism, for these values are inextricably bound up with human desires and interests.

II INTRINSIC VALUE AND PRAGMATISM IN ENVIRONMENTAL ETHICS

How important is the concept of intrinsic value in environmental ethics? Is it really the foundation upon which all ethical obligations regarding the natural environment depend? Weston paints a picture of despair and futility as he surveys the attempts of environmental philosophers to justify the value of natural entities: like Tom Regan, they end in the positing of dubious non-natural ontological properties, or like Thomas Hill, Jr., Mark Sagoff or Bryan Norton, they end in a "second best" appeal to human virtues and ideals.[2] No one, it seems, can give a plausible account and justification of the intrinsic value of non-human natural entities.[3] This despairing failure leads Weston to conclude that environmental ethics is "in a sorry state indeed" (p. 291).

But the difficulties encountered by environmental ethics are not *primarily* the result of the attempted justification of intrinsic value. Weston overemphasizes the importance of this concept in the literature. Environmental philosophers undoubtedly talk a great deal about the concept of intrinsic value – I have been guilty of this myself[4] – but their use of the concept is not as Weston suggests (pp. 286–7). Intrinsic value is not sought as the *ground* of an environmental ethic. Its explication is not pursued as the justification of environmental policies – and thus the failure to articulate it does not lead to the failure of environmental ethics.

The primary justifications for an environmental ethic are instrumental: they attempt to reveal the purposes behind environmental protection, and to show why these purposes are beneficial and moral. But not all instrumental values are acceptable to an environmental ethicist; one major goal of an environmental ethic is to show that exclusively human-centered goals are not capable of justifying environmental policies. Within this project of justification, the intrinsic value of natural entities can be used to limit or to map out the range of appropriate instrumental values. The basic goal of developing a non-anthropocentric justification for environmental policies is *aided* by the existence of non-human intrinsic value. Human purposes, desires, and interests will not be the only possible justifications of action. This supportive function of the concept of non-human intrinsic value is a far cry from the central or primary justification of an environmental ethic. Environmental ethics, in short, does not rest on the development of a theory of a non-human natural intrinsic value; it merely uses this intrinsic value to clarify appropriate instrumental values.

I am here making a claim that is primarily about the *methodology* of environmental ethics, not about its substantive content. Such a claim can only be justified by a detailed examination of the literature – a task obviously beyond the scope of this essay. It may, however, be supported by briefer considerations: (1) a close rereading of Holmes Rolston – who is mentioned quite prominently in Weston's essay – and of J. Baird Callicott on the nature of intrinsic value in environmental ethics; and (2) an examination of the dialectical assumptions of the argument for an environmental ethic supposedly based on intrinsic value.

(1) Although Weston correctly criticizes Rolston for using the concept of "intrinsic value" in a place where it is not needed (p. 300),[5] this particular use is not typical of Rolston's work as a whole. Rolston is much more concerned with the various kinds of *instrumental* value that are provided by wild nature – thus he prepares elegant taxonomies of natural value.[6] Even more importantly, on the theoretical level, Rolston has argued for a dissolving of the distinction between instrumental and intrinsic value. The entire notion of intrinsic value loses its meaning from the perspective of ecological holism, because each natural entity performs some function in the ecological system. "Things do not have their separate natures merely in and for themselves, but they face outward and co-fit into broader natures. Value-in-itself is smeared out to become value-in-together-ness."[7] A pragmatist interested in resolving conceptual dualisms could not ask for more!

Similarly, Callicott has tried to break down and to reinterpret the concept of natural intrinsic value. In a recent series of papers[8] he has developed an axiology based on a model of Humean/Darwinian bio-empathy, and he has further indicated how this model agrees with recent discussions in quantum physics. In brief, Callicott challenges the entire distinction between subject and object, and with it the existence of purely objective properties and purely subjective values for natural entities. The existence of ontologically discrete entities is an illusion, as is the existence of independent intrinsic properties and values. Callicott claims to have "transformed" or "truncated" the traditional concept of intrinsic value because it is no longer entirely independent; it requires the existence of a valuing consciousness.[9]

It is not my purpose here to endorse any specific view of the conceptual structure of intrinsic and instrumental value. More specifically, I am not myself trying to dissolve the categories of "intrinsic" and "instrumental" as Rolston, Callicott – and indeed, Weston – have attempted. My point, again, is formal and methodological: it is about the *use* of the concept of "intrinsic value" in environmental ethics. This brief review of the main ideas of two prominent environmental philosophers – Rolston and Callicott – shows that the concept of intrinsic value is not (as Weston supposes) the ultimate ground upon which all ethical obligation concerning the environment depends; it plays a much more complicated and subtle role in the development of an environmental ethic.

It might be objected that my methodological claim rests on a willful disregard of *other* environmental philosophers who do employ a concept of intrinsic value in the kind of basic way that Weston supposes – such as Tom Regan and Paul Taylor.[10] But this objection is without merit. (i) First, it is my contention that Weston has misconstrued the nature of the debate in environmental ethics, that he has misidentified the purpose of the concept of intrinsic value. The existence of philosophers who do employ a concept of intrinsic value does not damage my claim, as long as I can show that there are important environmental philosophers who do not use the concept of intrinsic value in the manner suggested by Weston. I have done this in my brief review of Rolston and Callicott. (ii) A more controversial reason for ignoring environmental philosophers like Regan and Taylor is that their fundamental value assumptions lead to impractical or incomplete versions of environmental ethics. Regan places value in the subjects of a continuous conscious life; Taylor's biocentric view emphasizes the value of living entities. Neither view is broad enough to include the holistic systems and non-living entities that a comprehensive environmentalism wishes to preserve.

(2) Indeed, this brief criticism of Regan and Taylor leads directly to the second consideration supporting my methodological claim

about environmental ethics. Weston, again, is concerned that the search for intrinsic value is the primary focus of an environmental ethic; but his concern is misdirected. The search for natural intrinsic value is not the ultimate ground of an environmental ethics. Nor *could* it be, for any argument for an environmental ethic based on an articulation of natural intrinsic value would be fundamentally mistaken.

The concept of intrinsic value fails as the source of an environmental ethic for two basic reasons: it implies that individual entities – and not whole systems – are the bearers of value; and it tends to focus attention on anthropocentric values such as sentience and rationality. Weston himself mentions this latter point: "The implicit demand to reduce intrinsic value to a single common denominator may incline us . . . towards the anthropocentric-sentientist end of the range of possible environmental ethics" (p. 289). More directly, Richard Sylvan argues against the notion of intrinsic value as developed by "deep ecologists" such as Arne Naess – i.e. the idea of "self-realization." The idea of developing one's highest potential – whether one be a human, animal, or plant – skews the notion of value not only towards *living* entities, but also towards those that are analogues of human beings.[11] We human philosophers, in short, seem to find that the essential nature of value lies in some aspect of human experience. We can understand this kind of value as intrinsic to our own lives, and thus we suppose it is valuable for non-human entities also. But this articulation of a concept of intrinsic value is nothing but a disguised anthropocentrism, and cannot be the basis of a real environmental ethic.[12]

The search for intrinsic value also biases an environmental ethic towards the perspective of individualism. The entire notion is directed towards the independent properties of discrete individuals.[13] But the most defensible forms of an environmental ethic (e.g., those of Rolston and Callicott) are essentially *holistic*. An environmental ethic, because it deals with environments, must focus its moral concern on the interdependent functioning of the entire ecological system, not merely on the (conceptually) isolated individuals who make up the system. The idea of intrinsic value loses its sense in a holistic system. An emphasis on intrinsic value, indeed, would preclude the development of a holistic environmental ethic. As Rolston notes, "the 'for what it is in itself' facet of *intrinsic* becomes problematic in a holistic web. It is too internal and elementary; it forgets relatedness and externality."[14]

These considerations suggest that the creation of a non-individualistic and non-anthropocentric environmental ethic cannot be based on the search for an intrinsic value that will serve as the ground of moral obligation. But this conclusion is nothing new. Environmental philosophers such as Rolston and Callicott have realized this for quite some time – their methodology, the primary methodology of environmental ethics, is not the methodology that Weston criticizes.

This means that Weston's call for a pragmatic shift in value orientation is pointed in the wrong direction – towards an atypical, individualistic, environmental philosopher.[15] But that alone does not invalidate many of his observations. There is a good deal of truth in a pragmatic conception of value as applied to environmental ethics – we may be dealing here with a mere terminological dispute, for once we see that environmental ethics is primarily concerned with the *instrumental* values of environmental protection – and not, as Weston supposes, the intrinsic value of natural individuals – then many pragmatic elements of moral theory come into play.

Evidence that Weston's critique of intrinsic value and his subsequent "pragmatic shift" is nothing more than a dispute over terms can be found in his discussion of Rolston's value of "sourcehood." Rolston has argued that nature should not be treated as a mere resource to be used, but rather as a *source* of what we value. "One is not so much looking to *resources* as to *sources*, seeking relationships in an elemental stream of being with transcending integrities."[16] For Rolston, nature as a source of value is then itself intrinsically valuable – but Weston thinks that this second claim adds nothing to our understanding and valuation of wild nature. What really matters, Weston says, is that we see nature as a *source* of value and valuable experiences (p. 300). Rolston, however, is saying the same thing! Weston is only disputing the form of Rolston's expression; he agrees with the *content*. Because he is overly concerned with the illusory importance of the concept of intrinsic value, Weston finds it necessary to distance himself from any *terminological* use of the concept of "intrinsic value." He is looking at allies and seeing enemies.

Once clear of this terminological morass, Weston's form of pragmatic value theory coheres very well with the dominant form of instrumental-holistic environmental ethics. First, Weston emphasizes that value for the pragmatist is pluralistic and relational: there are many values found in nature and these values interact in various ways with other values, interests, and desires that we possess. But an adequate environmental ethic does not deny this. It claims that there are many kinds of value found in holistic ecological systems – e.g., diversity, stability, beauty – and that all these contribute to arguments for environmental preservation. No *one abstract* comprehensive "intrinsic" value is sought by philosophers wishing to justify environmental protection. A plurality of practical – albeit ecological – values are articulated by philosophers and environmental scientists: the values that contribute to the well-functioning of the ecological system.

Second, pragmatic value is irrevocably tied to specific concrete situations: there is no good in itself; there are only good situations in the real world. Thus Weston denies meaning to the question "Why preserve wilderness?" As a question of practical policy it is too abstract. For the pragmatist the real question is "Why preserve *this* wilderness?" – what is it about this particular natural region which interacts with our pluralistic set of interrelating values (pp. 301–302)? But any workable environmental ethic does the same. Because an articulation of precise ecological values is a necessary requirement for protectionist policy, the environmental philosopher must examine specific environmental systems to determine their degree of worth. Environmental philosophy must be informed by ecological science. This science is not one of abstract principles; it specifically analyzes concrete problems.[17]

In this way, certain key elements of pragmatic value do fit within a certain kind of environmental ethic. Nevertheless, environmental ethics does not require – as Weston supposes – a *shift* towards pragmatism, for the most adequate and justifiable forms of an environmental ethic already make use of many pragmatic elements. What I have argued so far is that Weston's criticism of "intrinsic value" employs a distorted picture of the methodology – the formal argumentative structure – of environmental ethics. Once beyond that distorted picture one can see the basic similarities between an adequate environmental ethic and a pragmatic ethic. But as I argue in the concluding section, environmental ethics cannot be subsumed under pragmatism *per se*. Despite compelling and comforting analogies, pragmatism and environmental ethics must part company over the role of *human* interests in the determination of value.

III   PRAGMATISM AND HUMANITY

Weston explicitly denies that pragmatism is committed to a "crude anthropocentrism," but despite his denial, he ties pragmatic value to "a certain kind of desiring," and he adds, "possibly only humans desire in this way" (p. 285). Further on, he begins his "defense of environmental values" by focusing on the "feelings" awakened by the experience of nature as "essential starting points" (p. 298). And the entire pragmatic project is designed to "articulate . . . the relation of [environmental] values to other parts of our system of desires . . . and to the solution of concrete problems" (p. 299).

The human desire for certain kinds of natural experience, and indeed, certain kinds of human experiences of nature, thus become the *ground* of a pragmatic environmental ethic. But this is clearly shaky ground. The interrelated web of desires, values, and experiences that make up each individual human life is not common to all humans. There is, in short, *no common ground* for the start of rational negotiation. Weston's dogmatic insistence that there are not really cases where people "do not care less" about wilderness, and that "common ground remains" is the hollow echo of an empty position (p. 302). He clearly misses the fact that some people do not care at all about the experience of nature.

And why must we insist that they do? Why must an ethics of the environment – or any ethics, for that matter – be based on a certain kind of favorable experience? The ethical obligation to tell the truth is not based on the subjective experience of truth telling, nor on the avoidance of the experience of lying. One need not experience adultery to know that it is ethically incorrect. Why then must the ethical obligation to protect the natural environment – *if such an obligation exists* – be based on favorable natural experiences? If environmental protection is morally correct, it is so regardless of the experiences produced by interacting with nature. If some people do not respond to nature in a "positive" environmentalist way, that is no excuse for them to violate the obligation to protect the environment. Similarly, the dislike of monogamous marriage does not justify promiscuous adultery; the dislike of truth telling does not justify telling a lie. Ethical obligations do not derive their force from favorable experiences.

The insistence that an environmental ethic is grounded on the experiences felt in interacting with nature leads inevitably to a kind of subjective relativism: those agents who do not feel the "awe" and "respect" and "wonder" of nature will have no good reason – no reason at all?! – to protect it. This is not a question of needing a "careful phenomenology" of human experiences regarding nature, as Weston suggests (p. 301). The fact is that some people do not experience nature in a positive way. They *do* want their motorboats and hibachis and condos to go everywhere.[18] As Weston admits, many people hold "deeply different, probably irreconcilable, visions of the ideal world" (p. 303). But rather than this being a source of strength for pragmatism – rather than it being an expression of Deweyan moral autonomy – it condemns it to the swamp of subjective relativism. Whatever the individual agent experiences as useful to himself is valuable, is morally obligatory – even the destruction of wild natural entities and systems.

Weston anticipates this subjectivist criticism of a pragmatic environmental ethic, for he begins his article with a brief claim distinguishing "subjectivism" from "subject-centrism." Pragmatism, he admits, is *subjective* – it "makes valuing an activity of subjects, possibly only of human subjects" (p. 285). But that does not mean it is subject-centrist: subjects – human beings – are not necessarily "the sole or final objects of valuation" (p. 285). Human subjects can recognize value that is "world-directed;" i.e. values external to human beings. "Subjectivism," Weston concludes, "does not imply . . . subject-*centrism*."

Weston is undoubtedly correct about the distinction between subjectivism and subject-centrism, but unfortunately, that does not mitigate my criticism of pragmatism. It is good old-fashioned subjectivism that is the problem here. Pragmatism places the value of the natural environment squarely on the *experiences* of human beings interacting with nature: the desires and feelings of human subjects. Pragmatic valuers may be "world-directed" instead of self-interested, but all that means is that they value experiences, desires and feelings that arise in external relations, interactions beyond their immediate selves. The pragmatic value of nature – as with all pragmatic value – is irrevocably tied to human experiences. This kind of value cannot be the basis for a stable environmental ethic, for different human beings are going to value different objects, experiences, and feelings in nature. Pragmatism may not rest on a "crude anthropocentrism" of value, but it will result in a *relativism* of value – not everyone will value the "correct" environmentalist experience of the natural world.

The key point here is that human desires, interests, or experiences cannot be the source of moral obligations to protect the environment. Human desires, interests, and experiences are only contingently related to the continued existence of wild nature as such.[19] If environmental policy is based on an "articulation" of human desires and experiences related to a plurality of human values, then it becomes extremely important *who* is articulating the values: *whose* desires and experiences are being used as the source of moral obligations? Environmental policy will depend on the "feelings" of the decision makers at the particular time the policy is established. The ever-changing flux of human feelings concerning the natural environment does not appear to me to be a secure or reliable "common ground" for establishing an environmental ethic.

These criticisms of the use of human experiences in the development of an environmental ethic do not, of course, apply only to pragmatism – they apply with equal force to any environmental ethic based on human interests (e.g., certain kinds of utilitarianism). But what is particularly distressing in the case of pragmatism is that many of the non-anthropocentric elements of the value theory agree with a workable environmental ethic – it is only the emphasis on human interests that destroys pragmatism's effectiveness. Although an environmental ethic must be based on a plurality of instrumental natural values and on a specific analysis of concrete ecological systems, it cannot be grounded on the ever-changing subjective feelings of humans as they experience nature.

There are many reasons for despair when one surveys the field of environmental ethics; there are many reasons to believe that "we are in a sorry state indeed." Given the present state of twentieth-century metaethics, the possibility of any justification for substantive applied ethics seems remote. Justification of environmental values inevitably raises the problem of reconciling facts with values. A workable environmental ethic must incorporate detailed scientific information about the operations of natural systems; yet, little interdisciplinary dialogue between scientists and philosophers has been successful. Finally, a truly environmental ethic must reorient our value systems away from individuals and towards species, systems and communities; it requires a radical transformation of ethical vision.

However, one problem that is not a cause for despair is the pursuit of the concept of intrinsic value for natural entities. This concept is not the ground of all environmental obligation. Intrinsic value plays only a small role in the formation of an environmental ethic – it serves to limit the exclusive reliance on anthropocentric instrumental values. The existence of intrinsic value needs to be acknowledged, to serve as the limit to anthropocentric instrumentalism; but this value need not be totally articulated or justified, for it is not the ground of all obligation. The problem of intrinsic value in nature is a problem that does not require a solution; it is enough to know that some kind of non-anthropocentric value exists, even if the description of this value remains unclear. What is clear, nevertheless, is that we cannot accept the solution offered by anthropocentric pragmatism. Basing our environmental obligations on the human "system of desires" offered us by the pragmatic theory of value would doom an environmental ethic to the contingent feelings of people who "experience nature" in the "correct" manner. *That* method of justifying ethical obligations is a prescription for real despair in the development of an environmental ethic.

ACKNOWLEDGMENT

"Searching for Intrinsic Value" first appeared in *Environmental Ethics* Vol. 9, No. 3 (Fall 1987).

NOTES

1 Anthony Weston, "Beyond Intrinsic Value: Pragmatism in Environmental Ethics," *Environmental Ethics* 7 (1985): 321–339. Reprinted in this volume, pp. 285–306. All page references in the text refer to this volume.

2 Weston refers specifically to one article by each of these philosophers: Thomas Hill, Jr., "Ideals of Human Excellence and Preserving Natural Environments," *Environmental Ethics* 5 (1983): 211–224; Mark Sagoff, "On Preserving the Natural Environment," *Yale Law Journal* 84 (1974): 205–267; and Bryan G. Norton, "Environmental Ethics and Weak Anthropocentrism," *Environmental Ethics* 6 (1984): 131–148.

3 It is of some interest that Weston omits discussing the work of J. Baird Callicott, one of the leading writers on the concept of "intrinsic value" in environmental ethics. Is it because Callicott "dissolves" the problems of intrinsic value in a way that undermines Weston's argument? See the text below for more discussion.

4 See, e.g., "Organism, Community, and the 'Substitution Problem'," *Environmental Ethics* 7 (1985): 241–256.

5 The reference to Rolston is "Values Gone Wild," *Inquiry* 26 (1983): 181–183.

6 See, e.g., Holmes Rolston, III, "Can and Ought We to Follow Nature?" *Environmental Ethics* 1 (1979): 7–30, and "Valuing Wildlands," *Environmental Ethics* 7 (1985): 23–48.

7 Rolston, "Are Values in Nature Objective or Subjective?" *Environmental Ethics* 4 (1982): 147.

8 J. Baird Callicott, "Hume's *Is/Ought* Dichotomy and the Relation of Ecology to Leopold's Land Ethic," *Environmental Ethics* 4 (1982): 163–174; "Non-anthropocentric Value Theory and Environmental Ethics," *American Philosophical Quarterly* 21 (1984): 299–309; "On the

Intrinsic Value of Non-human Species," in Bryan G. Norton, ed., *The Preservation of Species* (Princeton: Princeton University Press, 1986), pp. 138–172; and "Intrinsic Value, Quantum Theory, and Environmental Ethics", *Environmental Ethics* 7 (1985): 257–275.

9 Callicott, "On the Intrinsic Value of Non-human Species," pp. 142–143.

10 See, e.g., Tom Regan, "The Nature and Possibility of an Environmental Ethic," *Environmental Ethics* 3 (1981): 19–34; and Paul W. Taylor, "The Ethics of Respect for Nature," *Environmental Ethics* 3 (1981): 197–218.

11 Richard Sylvan, "A Critique of Deep Ecology," *Radical Philosophy*, no. 40 (Summer 1985): 11.

12 Compare the argument of John Rodman, "The Liberation of Nature?" *Inquiry* 20 (1977): 83–145.

13 In one of the best treatments of intrinsic value in nature, Andrew Brennan discusses the possibility of attributing intrinsic value to systems as well as to individuals. See his "The Moral Standing of Natural Objects," *Environmental Ethics* 6 (1984): 35–56. Weston fails to mention this article in his critique.

14 Rolston, "Are Values in Nature Objective or Subjective?" p. 146.

15 Again let me emphasize that it is a serious mistake for Weston to use the work of Tom Regan as paradigmatic of environmental ethics as a discipline. Regan simply is *not* a "mainstream" environmental philosopher. He has a different agenda: the moral treatment of animals (indeed, the *higher* animals). Perhaps Weston has a distorted view of the methodology of environmental ethics because he focuses on the wrong philosopher. A good critical discussion of Regan's treatment of environmental ethics can be found in J. Baird Callicott's review of Tom Regan, *The Case For Animal Rights*, in *Environmental Ethics* 7 (1985): 365–72.

16 Rolston, "Values Gone Wild," p. 183.

17 There is always a danger, however, in being too concrete. Focusing only on specific situations may lead to a kind of *ad hoc*, contingent, moral thinking. Moral judgment, if it is to be at all defensible, must be based on some commonly accepted general principles – in the case of environmental protection, these might include beauty, diversity, and stability. These principles then must be applied to the specific situation – or at least, they must be found there – before a judgment about a particular situation can be made.

18 See the penetrating discussion by Mark Sagoff, "Do We Need A Land Use Ethic?" *Environmental Ethics* 3 (1981): 293–308.

19 Eric Katz, "Utilitarianism and Preservation," *Environmental Ethics* 1 (1979): 357–364.

16

UNFAIR TO SWAMPS
A reply to Katz

UNFAIR TO FOUNDATIONS?
A reply to Weston

Anthony Weston and Eric Katz

WESTON'S REPLY

Recently I argued in this journal that it may well be self-defeating to seek intrinsic values in nature upon which to "ground" an environmental ethic.[1] The preoccupation with "grounding" in any case misses what environmental ethics really needs (so I claimed) and has perhaps been struggling towards: a much more sensitive inquiry into our actual environmental values and their possibilities and synergies, in short an understanding of values themselves as dynamically interdependent systems such that we could almost speak of an "ecology" of values. Eric Katz, responding to my article,[2] grants the critique, and by speaking of the "mapping" function of the concept of intrinsic value he seems to point in somewhat the same alternative direction, but he rejects pragmatism as unacceptably anthropocentric, and argues that the concept of intrinsic value is not at all as central in non-anthropocentric environmental ethics as I claimed. Is he right?

Not about the "search for intrinsic value" in recent environmental ethics. J. Baird Callicott – to cite a writer who Katz takes to be especially central – begins one of his major recent articles by insisting that "the central and most recalcitrant problem for environmental ethics is the problem of constructing an adequate theory of intrinsic value for non-human natural entities and for nature as a whole,"[3] and writes in a summary statement of another major article that

319

an adequate value theory for nonanthropocentric environmental ethics must provide for the intrinsic value of both individual organisms and a hierarchy of superorganismic entities – populations, species, biocoenoses, biomes, and the biosphere. It should provide differential intrinsic value for wild and domestic organisms and species . . . and it must provide for the intrinsic value of our present ecosystem, its component parts and complement of species.[4]

In the first piece cited, Callicott undertakes to explore quantum theory, to challenge the very subject–object distinction, and to embrace a form of psychological egoism, precisely in order to vindicate such a concept of intrinsic value. The foundations of science, metaphysics, and finally selfhood itself are rethought in this extraordinary philosophical exertion, but the centrality of the concept of intrinsic value is taken for granted. I don't see it as the rather secondary concern Katz suggests.

For *Katz* it is understandably secondary. In his well-known article "Organism, Community, and the 'Substitution Problem'," Katz argues that the intrinsic value of individual entities is a good reason to prefer a "community" model of the natural environment over an "organism" model: only the community model can honor the intrinsic value of the individuals who/which make it up.[5] The concept of intrinsic value thus allows him to sort out the appropriate form of holism for environmental values; this, I think, is the "mapping" function to which he refers. Katz evidently thinks that we do not need to appeal to intrinsic value in order to vindicate the value of the natural community itself. But other holists do not agree: Callicott, in the above citation, explicitly appeals to the intrinsic value of "superorganismic entities." Perhaps Katz has read Callicott in too Katzian a way.

Even so Katz may be right: perhaps the concept of intrinsic value will become progressively less central in environmental ethics. Perhaps something like his version of holism will even help usher it out. Katz, however, is not entitled to define environmental ethics in such a way that non-holists, for whom intrinsic value remains important, are ruled out. It is an overly narrow view of environmental ethics, to say the least, that "ignores" (his word) environmental philosophers like Paul Taylor and Tom Regan because they are not, or not sufficiently, holistic. Most of deep ecology gets ruled out soon thereafter (pp. 310–311). Maybe Katzian holists need not make the concept of intrinsic value so fundamental, but many other philosophers concerned with natural values do, and ruling them out by fiat accomplishes nothing.

Katz reserves his sharpest criticism for the pragmatic alternative I drew out of the critique of the concept of intrinsic value. He repeatedly refers, for example, to the "swamp" of "subjective relativism." Old bugbears come romping from their caves. According to me, apparently, "whatever the individual agent experiences as useful to himself is valuable, is morally obligatory" (p. 315). Apparently, for pragmatism, "environmental policy will depend on the 'feelings' of the decision makers at the time the policy is established" (p. 316: his scare quotes). There is more in this vein.

But these really are just bugbears. Usefulness, in the first place, is indeed *one* source of value, along with two or three dozen others, but usefulness certainly does not exhaust our concerns. The suggested view of "feelings," meanwhile, is an extremely stereotyped and simplified one. Any plausible subjectivism (e.g., Dewey's, or, indeed, Hegel's) insists that our subjectivity is structured and indeed largely constituted by much more permanent and stable sets of values, exemplars, background beliefs, and commitments, etc., all of which a decision maker naturally appeals to, and should appeal to.

Moreover, Katz is unfair to swamps, and this point is more serious than it might seem. Swamps are enormously complex and creative ecosystems – think of the Everglades – and they are places that we can make our way around quite well if we are sufficiently careful, respectful of their powers, and appropriately equipped. Suppose that ethically we are indeed in a sort of swamp: is that so terrible a fate? What sort of ecosystem would Katz prefer?

Swamps do lack guaranteed firm footing, which is Katz's worry about pragmatism's appeal to our actual values in all their complexity and diversity. "Different human beings," he says, "are going to value different objects, experiences, and feelings in nature" (p. 315); and this is not, he thinks, "a secure or reliable enough 'common ground' for establishing an environmental ethics." Fair enough: *if* one thinks of ethics as the sort of project that might yield real truth, independent of what anyone thinks or feels (and if one adds the all-too-human assumption that that truth is relatively simple – for why mightn't the truth itself be swampy?), then of course pragmatism looks defeatist. Pragmatism denies that "ethical truth" in this sense is a coherent notion, though, and it is not clear how many environmental philosophers still subscribe to it.[6] But otherwise some sort of "swamp" is more or less our actual situation, and the fact that pragmatism attempts to deal with it, so far from being an objection, only reflects its *realism*.

People do differ significantly about natural values, and our common ground is patchy (I share certain values with X, somewhat different values with Y, etc.) rather than uniform across society as a whole. As in real swamps, a lot of the goings-on are unfamiliar. But there is also much to be done. We can try to establish more shared ground: natural history and poetry, for example, may "enrich our [understanding] to the point that appropriate care simply *emerges*", as Jim Cheney, extending some work of Carol Gilligan's, has suggested.[7] We can draw on what I have called the "dissonant" parts of our inherited system of values to call into question the dominant ones: I cite some examples in my original paper. We can work out alternative policies that integrate a number of different perspectives, rather than persisting in trying to homogenize them. Bryan Norton, for instance, looking at the actual history of the conservationist– preservationist debate, suggests that the debate has been misconstrued by philosophers who have overlaid the anthropocentrism– non-anthropocentrism distinction upon it: it is more fruitfully conceived, says Norton, as a tension between consumptive and non-consumptive values – what is really at stake is the appropriate domain of "commercial" motives – and between "divergent criteria of ecosystem stability and health." This is indeed the "swamp" – the world of actual political and ecological debates where none of the old lines look so clear, and where vital issues depend upon empirical questions to which the answers are not yet so clear – but it also looks a lot more interesting and promising than the foundationalist debates in ethics towards which Katz in effect reinvites us. Nostalgia for the old world of ethical certainty gets us nowhere (where were natural values *then*?). Let us instead don our swamp gear and wade in.

KATZ'S REPLY

Weston and I clearly agree on the importance of phasing out the notion of "intrinsic value" for natural entities as a foundation of an environmental ethic. We disagree, however, about the alternative set of concepts that will replace natural intrinsic value. He has confidence in the collection of human values associated with the environment, with all their "possibilities and synergies." I do not.

In addition to this central disagreement, we differ on the interpretation, significance, and use of "foundations."

(1) Is Callicott a foundationalist? Only if one concentrates on his earlier (i.e. pre-1985) writings. Callicott has shifted his view, and the "Quantum Theory" paper cited by Weston marks that shift. Callicott is no longer searching for a natural non-human intrinsic value as a ground for an environmental ethic. His "philosophical exertion" regarding the self, the subject–object distinction, and quantum theory is not meant to "vindicate" the idea of intrinsic value but rather to transcend it, to show that the idea is no longer useful or necessary.

(2) Is Katz a foundationalist? As much and as little as Weston is. We both want to enter the "swamp" of competing environmental and non-environmental values, and begin to "build common ground." This task requires foundations – not the immovable foundations of ethical absolutism, but the secure supports for a consistent environmental policy. My warning about the "swamp" of "subjective relativism" is not "nostalgia for the old world of ethical certainty." It is a call for firmer footing than anthropocentric feelings about the pragmatic usefulness of nature. Foundations are often worthwhile; and they need not be absolute. To build a house one starts with a foundation. It is prudent to do so, even in regions where earthquakes cause the ground to move.

ACKNOWLEDGMENT

"Unfair to Swamps" and "Unfair to Foundations?" first appeared in *Environmental Ethics* Vol. 10, No. 3 (Fall 1988).

NOTES

1 Anthony Weston, "Beyond Intrinsic Value: Pragmatism in Environmental Ethics," *Environmental Ethics* 7 (1985): 321–339. Reprinted in this volume.
2 Eric Katz, "Searching for Intrinsic Value: Pragmatism and Despair in Environmental Ethics," *Environmental Ethics* 9 (1987): 231–41. Reprinted in this volume. All page references are to Katz's essay in this volume.
3 J. Baird Callicott, "Intrinsic Value, Quantum Theory, and Environmental Ethics," *Environmental Ethics* 7 (1985): 257–75. Bryan Norton remarks that the distinction between intrinsic and non-intrinsic values "has been given a central role in the accounts of environmental ethicists too numerous to mention," though he does go on to mention some of the most central of them: see his "Conservation and Preservation: A Conceptual Rehabilitation," *Environmental Ethics* 8 (1986): 196, note 1.

4 J. Baird Callicott, "Non-Anthropocentric Value Theory and Environmental Ethics," *American Philosophical Quarterly* 21 (1984): 299–309.
5 Eric Katz, "Organism, Community, and the 'Substitution Problem'," *Environmental Ethics* 7 (1985): 241–256.
6 See my original article in this volume, pp. 290–291, for some comments on Regan's objectivism.
7 Jim Cheney, "Eco-Feminism and Deep Ecology," *Environmental Ethics* 9 (1987): 143.

17

ENVIRONMENTAL PRAGMATISM AS PHILOSOPHY OR METAPHILOSOPHY?

On the Weston–Katz debate

Andrew Light

After reading through this collection of essays one may reasonably wonder why we have included this exchange at the end of the volume. The issues raised here do not exactly accord with all the other sections of the book, and the ground covered seems very preliminary compared to the advanced stage of these issues as presented in other contributions.

But the Weston–Katz debate was one of the first direct forays in the field of environmental philosophy into the questions concerning environmental pragmatism. It is valuable (instrumentally, of course!), because it demonstrates how the arguments around environmental pragmatism emerged as a part of the growth of the field as a whole.

In this brief comment, I want to offer the beginning of a rereading of this debate, based in part on the perspective of environmental pragmatism that emerges out of this volume. I do not mean to mitigate the real differences between the two authors here, but to provide a new context in which their exchange may be read. I will first address the question of the role of intrinsic value in this debate, then reflect on Katz's suggestion that environmental ethics is already sufficiently pragmatist, and finally, overlay a picture of environmental pragmatism which situates these two theorists in the more mature environmental pragmatism at work today.

One issue raised in the exchange and debated at length is the extent to which intrinsic value is or is not a central component of contemporary environmental ethics. For my purposes here the question is not very important.[1] A case for environmental pragmatism of any sort need not depend on it, even though the claim of its importance provided Weston with a good entry into the field to make his argument. In the end, I think Weston implicitly acknowledges this in his reply to Katz by granting that intrinsic value may be slowly falling out of vogue (p. 320).[2] But note, the loss of the importance of intrinsic value in environmental philosophy is not taken as a premise for invalidating the pragmatic turn by either Weston or Katz in the end.

Weston importantly points out in his reply that even if intrinsic value is on its way out for holists, it may not be for individualists, and there is no good reason to simply write off individualist non-anthropocentrists like Regan and Taylor. This is perhaps the most important comment Weston makes in the first part of his reply, and unfortunately Katz does not respond to him. I say unfortunate, because the prejudice of environmental philosophy concerning individualism is something that needs serious attention in the litera-ture. Katz – and others – simply assume that it is a non-issue. Instead of engaging on this point, Katz chooses to spend half of his final reply answering the question of Callicott's commitment to intrinsic value. I will return to the importance of this oversight in a moment.

But first, let me suggest that the dispute here on Callicott's commitment (or not) to intrinsic value between these two theorists is probably unimportant. Weston has given a nice critique of intrinsic value and thus proved the value of a pragmatist approach to environ-mental ethics at least in its ability to seriously question a component of the field embraced by a critical mass of theorists. Perhaps though, to prevent the debate from degenerating into an exegesis on Callicott (or any other single thinker), the real target all along should have been moral *monism* in environmental ethics, rather than intrinsic value.

Callicott, Regan, and many other theorists mentioned by both participants in this exchange, are all moral monists. But as Bryan Norton argues in his contribution to Part 2 of this volume, monism, as a dominant development in the field, is perhaps the biggest barrier to the acceptance of either a pluralist or a pragmatist stance in environmental philosophy. So, whatever Callicott is trying to over-come in his version of quantum mechanics, it is still his particular interpretation of Leopold's land ethic which provides the basis for his position. The land ethic is the monistic theory that produces the single foundation from which all answers to concrete questions must emanate.

It is this attack on monism which has become the central part of some of the pragmatist critiques of mainstream environmental ethics, and those working against this monism can look to Weston's original claims against intrinsic value for support. But while this focus on questioning monism has progressed, there has been no forward move at all against the discrimination in the field which Weston identifies, specifically, against individualists like Regan and Taylor. Moreover, what is most important for environmental pragmatists is the continued prejudice as well against those who would like to revive some weak form of anthropocentrism as a legitimate framework within which to do environmental philosophy.

I do not wish to defend the legacy of anthropocentric philosophy, for it was certainly an important part of getting us in the "sorry state" of environmental conditions in which we find ourselves today. I am making a plea instead for more tolerance in the environmental philosophy community. We must tolerate those who are sincerely attempting to work within the anthropocentric tradition to provide a framework which can seriously address the important changes in thought and action needed to address environmental problems. We have not yet "found all the answers," – so why would anyone refuse to acknowledge valuable work from those who do not count themselves as non-anthropocentric holists? This is particularly disturbing given a moment of reflection on what "finding the answers" entails. Environmental philosophy arose as a response by intellectuals who thought that philosophers are obliged to try to do something to avert overwhelming contemporary environmental problems. Finding the answers is not success in solving some interesting puzzles or winning some intellectual game – it is, rather, success in developing adequate environmental policies. Environmental pragmatists take very seriously the predicate *environmental* in their self-description, and in their sub-discipline: we are devoted to our subject not simply as the object of our work, but because of our duty to the subject. Intellectual intolerance is something that stands in the way of any possible contribution philosophers might make to the solution of these problems. Intolerance thus contributes to these problems.

If I seem to exaggerate such worries, a perusal of introductory environmental ethics textbooks demonstrates this point quite adequately. Callicott's interpretation of Leopold as a non-anthropocentric holist is almost hegemonic, as is the assumption that we environmental ethicists have now arrived at non-anthropocentric holism as the heir apparent of all previous environmental ethics.[3]

There is a clear prejudice in public presentation of the field which is sometimes as bad as the prejudice in main stream philosophy against the legitimacy of environmental ethics.

Given this state of affairs, what do we make of Eric Katz's argument that pragmatist elements of moral theory are already very important in environmental ethics conceived as primarily a search for instrumental values (p. 312). First, he suggests that contemporary environmental philosophy, like pragmatism, is *pluralistic* and *relational* (p. 312). The pluralism Katz is referring to here is the sense in which environmental ethics does not deny that there are "many values found in nature and these values interact in various ways with other values." All these different values, "diversity, stability, beauty," contribute to arguments for environmental preservation.

But to suggest that this is the sense of pluralism that Weston advocates, is, I think, wrong. It is not just that there are different *kinds* of descriptions of value in nature available to us (which surely no one would disagree with), but that there is no single way of *valuing* nature. The distinction here has been more sharply drawn in the debate, again, between monism and pluralism. Consider Gary Varner's response to Callicott's refutation of pluralism:

> It is because an ecosystem has no welfare of its own, that a holistic environmental ethic must be pluralistic. If it is plausible to say that ecosystems (or biotic communities as such) are directly morally considerable – and that is a very big *if* – it must be for a very different reason than is usually given for saying that individual human beings are directly morally considerable (and, perhaps, higher animals or all individual living organisms).[4]

Even without granting Varner's strong conclusion that a holist position (presumably including Katz's) must of necessity be pluralist, the difference between the pluralism of valuation referred to here, and Katz's notion of a pluralism of values should be clear.

We may also notice another level of pluralism, which though related, is somewhat distinct: a metatheoretical pluralism concerning approaches to environmental philosophy in general.[5] Katz does not want to admit the legitimacy of either individualism or anthropocentrism as candidates for a defensible environmental ethics. Weston, arguing the pragmatist side, does want to admit, or re-admit, these candidates. And if there is anything to Varner's claim, that we have some good reason to be skeptical of an argument which would lump all sorts of valuing together (given the demands of things to value suggested by holism), then wouldn't we want to encourage as many different overlapping theories of how to value? Yes, but only if these theories did not cancel each other out. This requires some principles of tolerance in theory construction which I will return to later.

Second, Katz argues that the pragmatist concern, voiced by Weston, to tie the enterprise of environmental ethics to concrete situations, is something that "any workable environmental ethic" does anyway (p. 313). But surely "workable" here caries a good deal of the weight in this position. Not all environmental philosophy is couched in these sorts of concrete terms, and indeed some lively environmental debates have been dealt with by some environmental philosophers in pretty abstract terms.[6] Is "workable" here more philosophically loaded than Katz lets on?

Similarly, Katz concludes that "most adequate and justifiable forms of an environmental ethic already make use of many pragmatic elements" (p. 313). But if what Weston was calling for was more adequacy and justification in environmental ethics – as the heart of his description of what pragmatism brings to the field – then Katz's argument begins to sound vacuously analytic: most adequate and justifiable forms of environmental ethics are adequate and justifiable. To the extent that Katz agrees with Weston's desire to see some sort of metaphilosophical improvement of environmental ethics, and to the extent that Katz implicitly acknowledges that not all environmental ethicists are meeting the demands that Weston would place on them, then Katz may, under some description, agree with some part of just the pragmatism that Weston is proposing. But why doesn't Katz just admit that he agrees with this form of pragmatism? Because Weston also makes arguments for a straightforward philosophical use of pragmatism along side his arguments for a general improvement of the discipline. Katz must therefore reject the designation of "pragmatism" to just the sorts of improvements he too would like to see in the discipline because they seem to be bound with a new 'side' to the environmental ethics debates that he does not wish to join. To the extent that Katz is correct that at least some non-pragmatists are doing good environmental ethics, Weston would indeed be wrong to say that *only* pragmatism can produce an adequate, justifiable and workable environmental ethics. And Katz is wrong in his suggestion that the general improvements Weston wants to make in environmental ethics don't count as pragmatist.

I submit that Katz would have been in a better position to assent to Weston's argument had Weston made the distinction I made earlier in this volume, that there are two sorts of pragmatism: philosophical and metaphilosophical. Metaphilosophical environmental pragmatists treat pragmatism as providing rules and principles within which environmental philosophy should be conducted. These rules and principles promote just the sort of virtues that Katz claims are already at work in the very best environmental philosophy. The crucial aspect of metaphilosophical pragmatism though is the willingness to give up past prejudices against certain forms of theorizing (still embraced in this particular exchange by Katz), and to embrace some sort of pluralism in the assessment and communication of normative issues in environmental concerns. But this is not a dogmatic pluralism, committed to some version of postmodern relativism which admits no *possibility* for moral realism or foundationalism. It is metaphilosophical, and thus not necessarily closed to the idea of formulating a rich foundationalism in environmental ethics. However, metaphilosophical pragmatism is not so wedded to the project of finding a foundation that it would stall the contribution of philosophy to the community of activists and scholars working on environmental problems until the day the foundation is secure. Therefore, this form of environmental pragmatism provides just those principles of tolerance needed to avoid irreconcilable conflicts among overlapping theories of valuation that may be needed, for example, to provide the pluralism Varner claims is necessary to get holism off the ground.[7] Even before that perfect pluralistic balance is achieved, and because of the *environmental* predicate of this form of pragmatism, the metaphilosophical rules will require some give and take among theorists *at least* in how a total picture of valuing the biota, and the collection of individuals, and other humans, is balanced.

On the other hand *philosophical* environmental pragmatism, represented in this volume more through the contributions of those scholars working with the classical American philosophical tradition, is an attempt to generate a new position which engages fully with the already established theories in environmental ethics on their own ground.[8] It is the commitment to this position – a direct application of traditional pragmatism to environmental issues – that Katz wants to avoid, even though he is clearly attracted to the metaphilosophical pragmatist elements in Weston's argument.

I count myself as a metaphilosophical environmental pragmatist, and sometimes a closet philosophical environmental pragmatist, depending on the issue. But importantly, it is my commitment to metaphilosophical pragmatism that gives me the ability to pick and choose the time and place to apply my philosophical pragmatism. So, my objections to the direction of some text books in environmental ethics, is not that they don't acknowledge that the actual heir apparent to environmental philosophy is pragmatism, rather than non-anthropocentric holism, but that such texts propagate the substantive metaphilosophical position that the future of environmental ethics rests in *one* approach. For some questions, I find straightforward pragmatism unduly limiting, and unfortunately too subjective to aid in the formulation of action guiding moral principles. And what about our participants in this debate? How do they come out on this divide?

Weston is certainly a metaphilosophical pragmatist, and this comes across quite nicely in his commitments to ethical "swamps." The picture of how ethics should relate to environmental problems is not tidy. We can be skeptical up front of any sort of monism which while, perhaps, acknowledging the multiplicity of values in nature, none the less attempts to come up with one system of valuing. "We can" instead, as Weston says, "work out alternative policies that integrate a number of different perspectives, rather than persisting in trying to homogenize them" (p. 322). This is the approach we ought to be willing to take in wading into a realm where "vital issues depend upon empirical questions to which the answers are not yet so clear" (p. 322). It is in this context that we can see that Weston's attack on intrinsic value was bound to his metaphilosophical pragmatism more than to his philosophical pragmatism: the pursuit of intrinsic value stands in the way of philosophers wading into the moral swamps that represent the everyday world of environmental policy-making. Commitments to intrinsic value, non-anthropocentrism and moral monism, may, as with the example Weston cites of Norton's analysis of the conservationist–preservationist debate, stand in the way of a charitable understanding of the different sides of important environmental questions.

It must not be forgotten however, that Weston comes to his metaphilosophical views through his substantive philosophical pragmatist perspective. It is that perspective that makes him skeptical of a search for "real truth, independent of what anyone thinks or feels" (p. 321). But there is no reason to believe that a commitment to foundations could not be articulated within a realist approach to the swamps of ethical dilemmas. Foundations, as Katz points out, can be

331

worthwhile, even if they are not absolute, and *at least* the heuristic ideal of foundations is important enough to keep us working in the face of seemingly intractable dilemmas between the different ways of valuing that will be revealed though a recognition of the different sorts of values in nature.[9] I am skeptical that anthropocentrism in principle cannot provide the "firmer footing" that Katz is looking for in his commitment to foundations (p. 323), but a metaphilosophical pragmatist may share with Katz an understanding of foundations that is not reducible to mere nostalgia.

A metaphilosophical pragmatist may also be committed to abstraction in some cases, and a policy of abstraction may serve the interests of environmental advocates in the swamp of public policy-making. Though I can agree with Weston's argument that pragmatism does not engage in moral speculation on general principles but instead theorizes about the concrete situation – thus avoiding claims like wilderness "as such" should be protected (p. 302) – I can none the less imagine situations where I would want to defend just such a claim. In a context where a settled description of wilderness has been agreed upon, and where wilderness areas have been clearly delineated, I can imagine the possibility of at least an important place for claims to the abstract value of wilderness based in a foundational ethical framework. No one could reasonably claim that the mere force of the argument of the importance of wilderness in the abstract, would lead to its preservation (it is an evocative concept, but let's not get too romantic about it). But similarly no one could claim that such an argument would be worthless to make. And certainly, the articulation of the importance of wilderness in the abstract could provide a place where those with divergent interests as to its comparative importance with other social concerns, could find some ground for negotiation.[10]

Katz is certainly not a philosophical pragmatist, but is a fairly consistent metaphilosophical pragmatist. I say "fairly consistent," because even though his devotion to adequacy, justification and concreteness in environmental ethics is metaphilosophically pragmatist, at times he embraces positions which do not accord with this view. Katz's strict adherence to non-anthropocentrism, for example, does not serve what I take to be his metaphilosophical sympathies.[11]

Still, when pushed, I expect Katz would agree to the following. Suppose that it turns out that appeals to anthropocentrism result in the moral consideration of nature that Katz has in mind? What if, for example, the US Congress called a series of hearings on the environment, and a philosopher was able to persuade the government on anthropocentric grounds to implement every one of Katz's policies with respect to the human relationship with nature. Suppose that the reason Congress embraced these views was because they understood the arguments better when expressed in anthropocentric terms. Would not Katz then agree that anthropocentrism had then served the interests of nature in a non-anthropocentric sense, even if it was done for the wrong reasons? Surely he would. Of course it is not at all likely that this would ever happen, but importantly we can imagine it very likely that in relevantly similar cases, but on a much smaller scale, an anthropocentric argument for serving the interests of nature might better serve the non-anthropocentrically conceived interests of nature. Even if anthropocentrism is only helpful at the level of communicating ethical principles, it is in the interests of a "workable" environmental ethics to avoid rejecting it in total.

Another crucial issue on which Katz is silent is Weston's suggestion in both his first paper, and in his reply, that environmental philosophers look to biography and poetry for the articulation of values found in nature. Again, I take it that if put the question directly, Katz would agree that this is probably the case. But a direct affirmation of the point in the exchange would have helped to cement Katz's metaphilosophically pragmatist credentials. The point that Weston is making here is crucial: if we open ourselves to pluralism in valuing nature, then it can't be the case that we can only articulate the values in nature in "philosophy's peculiar, epistemically oriented way" (p. 300). And the openness of metaphilosophical pragmatism to these other ways of expressing value is again evidence of its commitment to a pluralism in the communication of environmental values. If we are not exclusively moral monists, then we will probably be less inclined to take the position that expression of values in nature is exclusively the province of philosophers, and therefore, that the communication of values in nature in non-philosophical language is flawed. Again, we can run the same thought experiment with Katz: if it turns out that poetry is more effective at communicating the multiplicity of values in nature, given his metaphilosophical sympathies, he would probably see this as unproblematic.[12]

And what about other theorists mentioned in this exchange? A quick look at Callicott's current position is appropriate, given the centrality of his work to both Weston and Katz. In contrast to both of these theorists, Callicott today is still neither a metaphilosophical or a philosophical pragmatist. Most recently, we can see that Callicott rejects the theoretical pluralism of philosophical pragmatism, and the metatheoretical pluralism of metaphilosophical pragmatism. This is surprising given Callicott's clear commitment to *cultural* pluralism, as evidenced by his latest book, *Earth's Insights*.[13] In this book, which is a very helpful survey of the environmental philosophies at work in a plethora of intellectual traditions from around the world, Callicott is still committed to a Leopoldian non-anthropocentric, holistic monism.

After providing a very impressive survey of ecological conscious-ness in a variety of different cultural traditions, the penultimate chapter of the book returns to Leopold, which this time provides Callicott with a "post-modern" environmental ethic which will be "firmly ground in ecology and buttressed by the new physics."

Such an ethic is the "one" in what Callicott calls the "one–many problem," or the need to have a single cross-cultural environmental ethic based on ecology and the new physics, while at the same time acknowledging the importance of a multiplicity of "traditional cultural environmental ethics, resonant with such an international, scientifically grounded environmental ethic and helping to articulate it."[14] The one and the many represent respectively, our quality of being one species facing a worldwide crisis, and the historical reality that we are many people from many cultures and different places. These two aspects of the human experience are not at odds for Callicott, but each will be a part of the emerging world ecological consciousness.

The monism at work here eliminates Callicott as a philosophical pragmatist, as does his interpretation of Leopold which, though challenged by Norton since 1988, is nonetheless expressed here as hegemonic.[15] Metaphilosophically though, this move towards cultural pluralism does not get us the rich metatheoretical, pragma-tist pluralism, which I have suggested, is at the heart of both Katz's and Weston's views of what counts as adequate and justifiable envi-ronmental ethics. All these competing world systems in Callicott are read through the lens of his non-anthropocentric holistic version of the land ethic. This theory clearly amounts to what Callicott calls the "Rosetta stone of environmental philosophy," needed to "translate one indigenous environmental ethic into another, if we are to avoid balkanizing environmental philosophy."[16] But using non-anthropocentric holism as a yardstick of environmental philosophy draws more lines towards balkanization than does a metaphilosophical tolerance of a multiplicity of approaches.

I do not claim to be well versed in the traditions discussed in the book, but from the position of metaphilosophical pragmatism I none the less question the validity of Callicott's comparative critiques of some of the views he surveys. His appraisals of these different environmental ethics are often made on the degree to which they do or do not favorably compare with a non-athropocentric, ecologically based ethic. But if we may call into question the singular vision of this goal, then we may call into question some of these appraisals. So, it may just not be the case that "Africa looms as a big blank spot on the world map of indigenous environmental ethics"[17] because of the predominance of anthropocentrism in its various intellectual and religious traditions, *if* a good argument can be made for an anthropocentric environmental ethic. The metaphilosophical pragmatist holds out for such an ethic, or at least seeks other grounds on which to judge the efficacy of different environmental ethics.

This volume has presented examples of the varieties of environmental pragmatism I have outlined here, and focused on different parts of these theories in different ways. Specifically, with respect to a metaphilosophical environmental pragmatism, we can see, for example, an embrace of metatheoretical pluralism (Thompson), the importance of contextualizing problems (Rothenberg), and preparations for contingency (Varner et al. – specifically the contingency that some environmental ethics are indeed appropriate for some issues, and that philosophical pragmatism is not always appropriate for helping to resolve questions over environmental disputes). Many of these elements were at work in the Weston–Katz exchange, and all of them will continue to be of importance to help see environmental ethics – as a relevant participant in the search for *workable* solutions to environmental problems – into the next century.

NOTES

1 This is not to say that the debate over intrinsic value is no longer important. The temptation to appeal to intrinsic value is still very much alive. My point however is that it need not have become the focus of this debate.

2 All page references in the text are from the articles as they appear in this volume.

3 One example can be seen in Zimmerman et al.'s *Environmental Philosophy: From Animal Rights to Radical Ecology* (New York: Prentice Hall, 1994). The environmental ethics section, edited by Callicott, is clearly introduced and structured in order to draw the clear impression

of a steady stream of progress from anthropocentric individualism, to non-athropocentric individualism, to non-anthropocentric holism, which of course, is Callicott's interpretation of Leopold. The section's structure argues that we have left behind these bad old theories and arrived at the right one. Of course, a careful instructor can structure the readings in a course using this textbook so as to avoid this impression. For other reasons which I will not go into here, I use Zimmerman's book and would heartily recommend it to others as well.

4 Varner, "No Holism without Pluralism," *Environmental Ethics* 13:2, Summer 1991, p. 179.

5 The distinction between theoretical and metatheoretical pluralism is provided in the introduction to this volume, p. 4.

6 See my comments on Katz and Elliot on restoration ecology in my contribution to this volume, pp. 174–.

7 Certainly there are limits to this tolerance, some of which I discuss in my previous paper in this collection. The metaphilosophical pragmatist is not equally open to all theories or acts of theorizing.

8 One could argue that all pragmatism is metaphilosophical, outside of a purely applied exegetical form of scholarship. But even so, the distinction I make here is useful as a way of understanding the difference between the varying *projects* represented in the cluster of theories which we call environmental pragmatism. There is nothing in principle that prohibits a philosophical pragmatist from creating a new "side" in the environmental ethics debates (including an anti-theory side), while this would be impermissible for a metaphilosophical environmental pragmatist. My thanks to Anthony Weston for pointing out this worry.

9 Perhaps this dilemma won't seem so intractable once we give up our theoretical blinders. But this question cannot be resolved here.

10 This is precisely the empirical claim that Varner et al. wish to make in their contribution to this volume. This is not to say that, for example, Thompson's claim in this volume (which is very close to Weston's analysis), that abstraction is more harmful than helpful, is wrong. It is only a reminder that it need not be right in every situation.

11 Again, I would cite Katz's work on restoration ecology, which I examine in my contribution to this volume.

12 Katz is not the only philosopher to ignore this very important position when engaged with a pragmatist on the role of philosophy in social and political debates. Thomas McCarthy dismisses out of hand the importance of a similar, more general argument made by Rorty in their exchange a few years ago. See McCarthy's "Ironist Theory as a Vocation: A Response to Rorty's Reply," *Critical Inquiry* 16: 3, Spring 1990, pp. 644–655. It should be noted that both Rorty and Weston may now hold the stronger position that non-philosophical modes of communicating value are *better* than philosophical methods.

13 J. Baird Callicott, *Earth's Insights: A Multicultural Survey of Ecological Ethics from the Mediterranean Basin to the Australian Outback* (Berkeley and Los Angeles: University of California Press, 1994).

14 Ibid., p. 12.

15 This is crucial as Norton interprets Leopold as a pragmatist. See Norton's "The Constancy of Leopold's Land Ethic," reprinted in this volume.

16 *Earth's Insights*, op. cit., p. 186.

17 Ibid., p. 158. One of Callicott's most recent statements on the character of his moral monism can be found in "Moral Monism in Environmental Ethics Defended," *Journal of Philosophical Research*, Vol. XIX, 1994, pp. 51–60. Here Callicott negotiates the criticisms of various pluralists (Wenz, Varner, Brennan, Weston, and Hargrove) by describing his approach to environmental ethics as a form of communitarianism, where, "all our duties – to people, to animals, to nature – are expressible in a common vocabulary of community," and so, "may be weighed and compared in commensurable terms."(53)

I am currently working on a longer paper which will discuss the claims of this piece, among others. For now, I only want to make a few points: (1) Callicott's claim that his version of communitarianism shares many of the advantages of pluralism is somewhat confusing and needs more discussion. I find Callicott's discussion particularly worrisome when I compare it to the full blown 'environmental' communitarianism of Avner de-Shalit. De-Shalit argues that pluralism, at the level of tolerance to competing theories and valuation schemes (what Callicott calls "intrapersonal pluralism," before rejecting it), is an important part of a robust communitarian theory. See Avner de-Shalit, *Why Posterity Matters: Environmental policies and future generations* (London: Routledge, 1995), pp. 61–62. (2) Metaphilosophical pragmatism provides a litmus test against which competing modes of ethical evaluation can be weighed. So the "extreme pluralism," criticized throughout this article by Callicott (the idea that one would adopt Aristotle's theory on one occasion, Kant's on another, etc.) is not what I have in mind as metatheoretical pluralism. My pluralist (and I think Weston's, Varner's and maybe Katz's too) in acknowledging distinct bases for value, would *at least* be consistent in the application of the best (all things considered) moral theory to a particular type of object of valuation, and not change theories with the evaluation of each token. In other words, my metatheoretical pluralist is not an ethical situationalist, and presumably, since he or she is a competent philosopher, would not apply invalid theories willy nilly in a self-defeating manner (as Callicott hints an extreme pluralist would at the end of his paper). And most importantly, my metaphilosophical pragmatist would not simply tolerate "interpersonal pluralism" (embraced by Callicott in this paper), but would advocate multiple overlapping (and consistent) arguments for valuation when advocating policies. (3) Finally, even if Callicott is not opposed, as he says in the last paragraph of this paper, to pluralism at the level of principle or sentiment, his adherence to monistic theory still blocks him from the form of pragmatism that I endorse here, and hence from the metatheoretical pluralism, which when consistently applied, increases the contribution of environmental philosophy to making workable environmental policies. So I still think that even with his communitarianism, his

ANDREW LIGHT

theoretical commitments still muck up his evaluation of the transformative potential of some systems of environmental values.

Thanks to Baird Callicott for providing me with a copy of this insightful paper. Also, thanks to Eric Katz, Anthony Weston and Gary Varner for helpful comments on this last chapter.

INDEX